FROM SYMBOLIC LOGIC... TO MATHEMATICAL LOGIC

CHARLES L. SILVER
Southeastern Louisiana University

Wm. C. Brown Publishers
Dubuque, Iowa • Melbourne, Australia • Oxford, England

Book Team

Editor *Earl McPeek*
Developmental Editor *Jane Parrigin*
Production Editor *Audrey Reiter*
Designer *Barb Hodgson*

 Wm. C. Brown Publishers
A Division of Wm. C. Brown Communications, Inc.

Vice President and General Manager *Beverly Kolz*
Vice President, Director of Sales and Marketing *Virginia S. Moffat*
Marketing Manager *Julie Keck*
Advertising Manager *Janelle Keeffer*
Director of Production *Colleen A. Yonda*
Publishing Services Manager *Karen J. Slaght*

 Wm. C. Brown Communications, Inc.

President and Chief Executive Officer *G. Franklin Lewis*
Corporate Vice President, President of WCB Manufacturing *Roger Meyer*
Vice President and Chief Financial Officer *Robert Chesterman*

Copyright © 1994 by Wm. C. Brown Communications, Inc. All rights reserved

A Times Mirror Company

Library of Congress Catalog Card Number: 93-71126

ISBN 0-697-14175-6

No part of this publication may be reproduced, stored in a retrieval system, or transmitted, in any form or by any means, electronic, mechanical, photocopying, recording, or otherwise, without the prior written permission of the publisher.

Printed in the United States of America by Wm. C. Brown Communications, Inc., 2460 Kerper Boulevard, Dubuque, IA 52001

10 9 8 7 6 5 4 3 2 1

Table Of Contents

Symbol Table ix
Preface xi
Acknowledgements xv
Introduction xvii

Chapter 0   Mathematical Preliminaries 1

  0.1   Facts About Sets 1
  0.2   Vacuous Truth 2
  0.3   Union, Intersection, Difference,
         and Power Set 6
  0.4   Sequence, Cartesian Product, Relation 10
  0.5   Function, Operation 12
  0.6   Complete Induction 16
  0.7   Principle of Complete Induction 19
  0.8   Another Proof that L has Balanced Parentheses 19
  0.9   Least Counterexample Principle 22
  0.10 Mathematical Induction 24

Chapter 1   The Syntax of Sentential Logic 27

  1.1   The Language of Sentential Logic 28
  1.2   Grammar 28
         Unique Readability 31
  1.3   Rules of Derivation for L_{SL} 32
  1.4   Derivational Strategy (using A, MP, CD) 35
  1.5   The CONTRA Rule 38
  1.6   CONTRA Strategy 40
  1.7   Derivations of Sentences Containing '∨', '∧',
         and '↔' 45
  1.8   Strategy for Obtaining Derivations Using All
         Rules (A, CD, MP, INT, and CONTRA) 47
         Theorems So Far 49
  1.9   Summary of Basic Inference Rules 50

Chapter 2   Short-Cut Rules For Sentential Logic 53

  2.1   Short Derivations 53
  2.2   Derived Rules 58
  2.3   Deriving A Rule 59
  2.4   SHOWing A Disjunction vs. Using A Disjunction 65
  2.5   Theorems of Sentential Logic 71
  2.6   Derivations From Assumptions 73
  2.7   Summary of Inference Rules 76
  2.8   (Optional) An Axiomatic Version of Sentential
         Logic 78

Chapter 3 The Semantics of Sentential Logic — 84

- 3.1 Boolean Interpretation — 84
- 3.2 Truth for a Boolean Interpretation — 85
- 3.3 Formal Definition of Truth for a Boolean Interpretation — 87
- 3.4 Boolean Model of a Set — 88
- 3.5 Proving Facts About Models — 91
- 3.6 Compactness — 105

Chapter 4 Connecting The Syntax and Semantics of Sentential Logic — 108

- 4.1 Syntax vs. Semantics — 108
- 4.2 Technical Derivations — 109
- 4.3 Syntactic (Meta)Theorems — 110
- 4.4 (Strong) Soundness — 113
    - Why Bother With Soundness? — 114
- 4.5 Consistency (Proof-Theoretic) — 117
- 4.6 Strong Completeness — 120
    - Lindenbaum's Lemma — 122
    - Overall Reasoning For Strong Completeness — 126
    - The Interpretation Lemma — 127
- 4.7 Compactness (Revisited) — 129

Chapter 5 The Syntax of First-Order Predicate Logic — 133

- 5.1 The First-Order Language L_{PL} — 134
    - Informal Conventions — 137
- 5.2 Rules of Derivation For L_{PL} — 138
- 5.3 Informal "Meanings" & The Inference Rules — 140
- 5.4 Derivations of Theorems in L_{PL} — 143
- 5.5 Informal Justification of the UG Rule — 144
- 5.6 UG Strategy — 145
- 5.7 The Existential Switch Argument — 147
- 5.8 Existential Switch Strategy — 149
- 5.9 Derivations From Assumptions — 158

Chapter 6 Semantics of First-Order Predicate Logic — 160

- 6.1 Naming Functions — 160
- 6.2 Interpretations — 163
- 6.3 Truth For A Naming Interpretation — 165
- 6.4 Bracket Notation — 166
    - Using Bracket Notation — 169
- 6.5 Truth For an Interpretation — 173
- 6.6 Satisfiability — 176

6.7   Validity — 177
6.8   Invalidity — 178
6.9   Particularizations — 180
6.10 Logical Consequence — 186

Chapter 7   Connections Between Syntax & Semantics For First-Order Logic — 187

7.1   Syntactic Metatheorems — 187
        Eliminating EG — 188
        Eliminating ES — 189
7.2   Semantic Metatheorems [*very technical*] — 190
7.3   Relations Between Syntax and Semantics — 206
        Strong Soundness — 206
        Consistency — 208
        Strong Completeness — 209
        Compactness — 214

Chapter 8   First-Order Theories — 216

8.1   Introduction To First-Order Theories — 216
8.2   The Theory of Equality $\mathcal{EQ}$ — 220
8.3   Mathematical Completeness — 225
8.4   Is The Realm of Arithmetic Complete? — 227
8.5   The Theory of Orderings $\mathcal{O}$ — 228
8.6   Operation (or Function) Symbols — 230
8.7   Modified UG Rule — 231
8.8   Characterization of "First-Order Theory (with Equality)" — 233
        Truth Clause for '=' in L_T+ — 234
8.9   Group Theory $\mathcal{G}$ — 236
8.10 Theory of Addition $\mathcal{A}$ — 241
8.11 The Theories of Arithmetic — 243
        Theory $\mathcal{P}$ — 244
        Theory $\mathcal{K}$ — 245
        Theory $\mathcal{Q}$ — 246
8.12 Mechanical, Decidable, & Axiomatizable — 247
8.13 Extension, Conservative Extension — 252

Chapter 9   First-Order Number Theory — 255

9.1   Theory $\mathcal{N}$ ($\mathcal{EQ} + \mathcal{P}$) — 255
9.2   The MATH-IND Rule — 257
9.3   MATH-IND vs. Mathematical Induction — 258
9.4   Theorems of $\mathcal{N}$ — 259
9.5   Definitions of Further Concepts — 268
9.6   The Representability of Numeric Relations in $\mathcal{N}$ — 273

Chapter 10   Models for First-Order Theories [*Advanced*]   278

10.1 Structures — 279
    Structure vs. Interpretation — 281
10.2 First-Order Theory (model-theoretic version) — 282
10.3 Isomorphism & Elementary Equivalence — 286
10.4 Substructures & Elementary Substructures — 294
10.5 Cardinality of First-Order Languages — 300
10.6 Löwenheim-Skolem (-Tarski & Vaught) Theorems — 301
10.7 First-Order Capture & Escape Theorems — 306
    Being a Natural Number is <u>Not</u> First-Order Capturable — 308
10.8 Categoricity & Completeness — 311
10.9 The Ultraproduct Construction [* very advanced and sketchy *] — 315
    Łoš's Theorem — 318
10.10 Chapter Postscript — 319

Chapter 11   Gödel's Theorems [* starts intuitively and becomes more technical *]   321

11.1 Platonism and Semantic Proofs — 321
    English is Inconsistent — 323
    N is Inconsistent (assuming Falsity can be defined in N) — 324
    Truth is Not Definable in N — 324
    Derivability <u>Is</u> Definable in N — 325
    N Must Be Incomplete (since Truth is Not definable, but Derivability is) — 325
    A Sentence of N "Says": 'I am not derivable' — 326
    N Is Incomplete (because a true sentence <u>says</u> it is not derivable) — 327
11.2 Syntactic Proof of Incompleteness — 327
    Banishing Truth From The Argument — 327
    If N Is Consistent, $N \not\vdash \neg \exists y Der(y, Self(\underline{k}))$ — 328
    ω-Consistency — 329
    If N Is ω-Consistent, $N \not\vdash \exists y Der(y, Self(\underline{k}))$ — 330
    Gödel-Rosser Theorem — 330
11.3 Summary of Proof of Incompleteness — 331
11.4 "Mechanical", Representable, and Undecidable — 332
    Representability Test — 332
    Church's Thesis — 334
    Representability Theorem — 335
    Undecidability of First-Order Logic — 335
    Representability for Functions — 336
    Characteristic Functions & Representability — 337

11.5 The Fixed Point Theorem, Formal Definability,
and Truth — 338
Undecidability of Arithmetic — 340
Incompleteness of N (following from FTP) — 341
Definability in N — 341
Tarski's Truth Theorem — 342
11.6 Moore's Presentation of Gödel's Theorem — 343
The Diagonal Set is Not Definable in N
(assuming N is Decidable) — 346
Therefore, N is Incomplete — 346
11.7 Gödel's Second Theorem — 346
Cons Does <u>Not</u> Express the Consistency of N
in <u>All</u> Models of N — 349
11.8 Primitive Recursive and (General) Recursive
Functions — 350
Definitions of Primitive and (General)
Recursive Functions — 350
Rules for Obtaining New Recursive Functions
From Old Ones — 351
Proofs of Recursive Functions — 352
All Recursive Functions Are "Mechanical"
(the converse of Church's Thesis) — 355
Impossibility of Proving "Mechanical" =
"Recursive" — 357
All Recursive Functions are Representable — 358
11.9 Gödel Numbering — 360
11.10 Epilog — 363

Postscript: Chaitin's Theorem — 364

References — 367
Index — 369

SYMBOL TABLE
(where the symbol first occurs)

$\{1,2\}$, 1
$\in$, 1
$\emptyset$, 1
$\subseteq$, 2
$/$, 2
■, 2
$\cup$, 6
$\cap$, 6
A', 7
$\mathbb{N}$, 7
$A - B$, 8
$\bigcup_{i \in I} A_i$, 9
$\mathbb{Z}$, 9
$\mathcal{P}(A)$, 10
$(a_1, a_2, \ldots, a_n)$, 10
$<a_1, a_2, \ldots, a_k>$, 10
A^k, 10
$A \times A \times \ldots \times A$, 10
$A \times B$, 10
$[x]$, 11
$\mathbb{R}$, 12
f, 12
$f(x)$, 12
$\mathbb{Z}^+$, 12
$f: A \to B$, 12
$f|A$, 13
$R|A$, 13
1-1, 13
T, 14
F, 14
O^k, 14
$f^{-1}(B)$, 15
$f \circ g$, 15
$\text{Card}(A)$, 16
$\aleph_0$, 16
$\varphi, \psi, \chi, \ldots$, 16
$\to$, 16
$\vee$, 16
$S(n)$, 19
LCP, 22

L_{SL}, 28
$P, Q, R, \ldots$, 28
$\neg$, 28
$\wedge$, 28
$\leftrightarrow$, 28
Tn, 35
SHOW, 35
Tn, 35
$\checkmark$, 36
$\psi \begin{array}{c} \theta_1 \\ \chi_1 \end{array}$, 56
A1-A3, 78
TTn, 78
$\mathcal{B}$, 84
$\mathcal{B}(\varphi) = T$, 85
$\mathcal{B}(\Gamma) = T$, 88
$\models \varphi$, 89
$<\{P\}, P>$, 109
$<\Gamma', \varphi>$, 110
$\Gamma \vdash \varphi$, 110
$\vdash \varphi$, 110
L_{PL}, 133
$x, y, z, \ldots$, 134
$a, b, c, \ldots$, 134
$\forall$, 134
$\exists$, 134
$F_j^k, G_j^k, H_j^k, \ldots$, 134
R^k, 134
$\delta_1 \delta_2 \ldots \delta_k$, 134
$R^k \delta_1 \delta_2 \ldots \delta_k$, 134
$\forall \alpha \psi$, 135
$\exists \alpha \psi$, 135
$\varphi \alpha / \beta$, 136
$a|b$, 147
$\cdot$, 147
n, 160
$\mathcal{D}$, 160
$n(\beta)$, 160
$n(a \to 1)$, 161

$n(a \to 1)(b \to 0)$, 161
$n(a \to 1)(b \to 0)(a)$, 161
'a' then 'b', 161
$\mathcal{I}$, 163
$<\mathcal{I}, n>$, 164
$<\mathcal{I}, n>(\varphi) = T$, 165
$<\mathcal{I}, n(\beta \to e)>$, 165
$F[0,1]$, 166
$F[e]b$, 166
$F[e, e']$, 166
$\psi[e]$, 166
$<\mathcal{I}, n>(\Gamma) = T$, 176
$\models \varphi$, 177
$\text{AND}x$, 181
$\text{AND}\alpha[\ldots \alpha \ldots]$, 181
$[\ldots \alpha \ldots]\alpha/\wedge$, 181
$\text{OR}\alpha[\ldots \alpha \ldots]$, 181
$[\ldots \alpha \ldots]\alpha/\vee$, 181
$\Gamma \models \varphi$, 186
$\{\alpha/\beta\}$, 202
L_{PL}^*, 209
L_T, 219
$=$, 219
$\mathcal{T}$, 219
L_{T^+}, 220
$\mathcal{EQ}$, 220
L_{EQ}, 220
φx, 220
φxx, 220
φxy, 220
Id, 222
Δ_{EQ}, 222
Δ_T, 222
$\Delta_T \vdash \varphi$, 222
$\mathcal{T} \vdash \varphi$, 222
'$\beta = \delta$' $= T$, 224
$\mathcal{T} \models \varphi$, 225
$\mathcal{O}$, 228
L_O, 228
$>, <$, 228
$A_i^k, B_i^k, C_i^k, \ldots$, 230
A_k^0, 231

$+$, 231
$Thms_T$, 234
$O^k(\tau_1,\ldots,\tau_k)$, 234
Cns_T, 235
$\mathcal{G}$, 236
$\circ$, $*$, ι, 236
$\mathcal{A}$, 241
$\mathcal{A}r$, 243
$\mathcal{P}$, 244
L_P, 244
S, 244
$\circ$, 244
$\underline{0}$, 244
$\tau_1 \neq \tau_2$, 244
$\mathcal{K}$, 245
Q, 245
$\mathcal{N}$, 247
Δ_{Ar}, 248
$\mathcal{T}_{PL}$, 252
$\mathcal{T}_{PL}*$, 252
$\mathcal{T} \subseteq \mathcal{T}'$, 252
$\underline{n}$, 264
$S(S(\ldots S(\underline{0})\ldots))$, 264
$\underline{n+1}$, 264
$S(\underline{n})$, 264
L_{N^+}, 268
$\leq, \geq$, 268
$t\mid r$, 270
$\underline{m+n}$, 274
$\underline{m}+\underline{n}$, 274
$\mathcal{S}$, 279
$\mathbf{S}$, 279
$\mathbf{R}^k$, 280

$\mathbf{O}^k$, 280
$<, +, \cdot$, 280
$\mathbf{0}$, 280
$\mathbf{R}_j^k$, 281
$\mathbf{O}_j^k$, 281
R_j^k, 281
O_j^k, 281
$\mathcal{S}(\Gamma) = \mathbf{T}$, 281
$\mathfrak{M}$, 282
$Cns(\Gamma)$, 283
$Cns(\Delta_T)$, 283
$\mathcal{T}h(\mathcal{S}_1)$, 283
$\mathcal{T}h(\mathcal{C})$, 284
$MOD(\Gamma)$, 284
Σ, 285
$MOD(\Sigma)$, 285
MOD_L, 285
$\mathcal{T}h(MOD_L)$, 285
$e_1, e_2, \ldots \in \mathbf{S}$, 285
$\mathcal{S}_1 \doteq \mathcal{S}_2$, 289
$\mathcal{S}_1 \equiv \mathcal{S}_2$, 289
$\mathcal{S}_1 \subseteq \mathcal{S}_2$, 294
$\mathcal{S}_1 \leq \mathcal{S}_2$, 295
$\mathcal{S}_1 = \mathcal{S}_2\mid L_1$, 299
$\mathcal{F}$, 323
$\mathcal{F}\underline{k}$, 323
$Der(\underline{i},\underline{n})$, 325
$\exists y Der(y,\underline{n})$, 325
$Self(n)$, 326
$Self(\underline{n})$, 326
$\mathcal{G}$, 331
$Form\text{-}Nums$, 332
$Form(x)$, 333
$Der(i,n)$, 333

$String_k$, 333
$Conjunct$, 336
$\varphi_f(x,y)$, 336
$\varphi_{Self}(\underline{n},\underline{m})$, 336
C_R, 337
$<1,2> \in <$, 337
$C_<$, 337
$\ulcorner \psi \urcorner$, 339
$Theorem(x)$, 340
P, 341
R, 341
$\mathfrak{N}$, 341
$Truth(x)$, 342
$Truth$, 342
$\varphi_n x$, 344
$\varphi_D x$, 344
$\mathbf{D}$, 345
$Cons$, 347
$Neg(n)$, 347
$Neg(\underline{n},\underline{m})$, 347
$Cons^*$, 349
$Z(n)$, 350
$\Pi_i^k(x_1,\ldots,x_k)$, 350
$\Pi_1^1(x) = x$, 351
$\mu y(g(x,y))$, 351
Sum, 352
$Times$, 353
$Pre(x)$, 353
$Sbt(x,y)$, 354
$Nz(x)$, 354
$C_>$, 355
GN, 360
$GN(\varphi_i)$, 361
$P_{\underline{n}}^{GN(\varphi_n)}$, 361

PREFACE

This preface is for those who already know some logic and wish to find out some details about the direction and the emphasis of the book, as well as some of its idiosyncracies. (The *Introduction* presents coverage alternatives for mathematics, philosophy, and computer science students.)

The book attempts to build up mathematical reasoning, from scratch, so to speak. It begins with rules of inference for sentential logic, which the student is asked to accept and to apply, in the derivation of theorems. No formal semantics is presented at first. The correct use of a rule is determined by purely syntactic means. As more derivations are obtained, more and more about the meanings behind the rules is explained. (There's a hint that these are "good" rules, but only a hint.) Eventually, a full-fledged semantics for sentential logic (and, later, for first-order logic) is presented (leading to proofs of soundness and completeness), but the formal semantics is presented much later, in a chapter of its own, and treated as an entirely separate topic.

As more complicated theorems are encountered, intuitions about which rules to use in what contexts are carefully cultivated. The system of rules is supplemented by various 'SHOW strategies',[1] which are intended to accomplish three main goals: (1) to help students become skillful in using the rules to derive theorems; (2) to help them understand the rules themselves on a firm intuitive basis; and (3) to pave the way for the presentation of the formal semantics, which then leads to the metaproofs connecting syntax and semantics.

There is a great deal of emphasis on getting students to make the rules "theirs", to learn them as one learns to swim or to drive a car. If the student does assimilate the rules as hoped (and has understood the syntax-semantics distinction) then it is likelier he or she will <u>care</u> about the metaproofs establishing that the rules are "good ones". Moreover, the principles behind the rules are used in the reasoning of the metaproofs themselves. So, another reason for wanting the students to make the rules theirs is so that they will acquire good proof techniques that can be used in any mathematical context.

---

[1] SHOW lines were first presented in Kalish & Montague [1964], where their use was part of the formal system of derivation itself. In the present book, SHOW lines are extraneous to the derivational system and are used only to help develop useful strategies.

To facilitate the transition from rules inside of a derivational system to principles of proof outside any such system, the rules themselves were selected for their intuitiveness and for easy generalization to wider mathematical contexts. Also, in the presentation of the rules, artificial techniques such as flags, arrows, boxes, and indentation schemes were eschewed, in order that the rules would appear "natural" and so that their use in derivations would seem to be only a special case of their wider applicability. Further, a concerted effort was made to link "derivational reasoning" to mathematical reasoning.

The leap from derivations to proofs is made easier by the introduction of "semi-formal proofs", which look a good deal like derivations. First, semi-formal proofs incorporate TO PROVE lines, which are used like SHOW lines in derivations, to help the student reason from the beginning to the end of the argument. Further, the semi-formal proofs (as opposed to standard mathematical proofs) proceed in a line by line fashion, with the reasons for each line written to the right, as in a derivation. Typical ("informal") mathematical proofs are also displayed to show the student the similarities and differences between them, and to help them better understand both.

In its emphasis on rules of proof and proof strategies, the first part of the book can be seen as a kind of exposition on "what is a proof" and as a primer for doing them. Other books, especially in mathematics and computer science, also attempt to teach students to recognize and to construct proofs, though usually within the context of some other subject matter, such as number theory or computer programs. This book attempts to teach students how to reason mathematically by starting out from the more subject-neutral perspective of "pure logic".

There are some further idiosyncratic details that should be mentioned. One novelty of the first-order semantics in this book is the choice to have "logical names", which are symbols that are distinct from both variables and mathematical constants. Mathematical constants are special non-logical symbols that vary from one first-order theory to another. The logical names, by contrast, just denote arbitrary elements in a given domain of discourse, and do not pertain to any particular theory.

This book deviates significantly from most symbolic logic texts in that it does not symbolize arguments from everyday English. It was tempting to cover symbolization and translation (the reverse of symbolization), but it seemed to divert attention from the direct line of reasoning from the formalization of basic rules of inference to the exciting and intriguing results obtained by Gödel. Symbolization and translation seemed to constitute too significant a detour from this path (though it could be argued that translating Gödel's sentence G as "saying" that G itself is unprovable would be easier to get across after covering symbolization and translation).

Hewing a path from the formalization of first-order logic to Gödel's theorems reflects an interest in the mathematical foundations of mathematics (or logic, to the extent it embraces mathematics) and in the philosophical basis of our logical thought. I think students should have some exposure to Gödel's important results before they graduate from college. One major exception to this direct line of coverage is the decision to include a relatively thorough chapter on model theory.

There are many reasons for covering model theory reasonably thoroughly. Other than desiring to balance the coverage equally between the syntax and the semantics of logic, I succumbed to the temptation of including more material on model theory in order to emphasize an alternative approach to logic. Proof theory makes it look as though all that matters are rules allowing us to manipulate strings within a given language — which may well be the proper way of viewing logic, as rule-governed string manipulations. Model theory, though, presents logic as though it consists in exploring the facts about "realities" that are first presented to our imagination. Language, looked at model-theoretically, can be viewed as only a medium for uncovering and expressing features of those realities. I wanted to reflect both orientations in this book.

Gödel's Theorem is explained somewhat idiosyncratically. First, an attempt is made to understand what Gödel may have been thinking at the time he arrived at his result. This was done to make the reasoning more focussed and easier to follow. Second, Gödel's Theorem is presented informally at first, and is then presented more and more formally as the chapter progresses. His result is also obtained from several different directions. Thus, the reader can decide which proof to read, and how much technical detail is desired.

A brief Epilog indicates that there is a sense in which this book represents an epistemic journey, from certain basic principles of reasoning (plus some set theory) to proofs of the limits of a kind of formalism. A further step in this journey is provided by a fascinating result by Gregory Chaitin, which I stumbled upon while finishing this manuscript. An informal account of Chaitin's Theorem was appended as a *Postcript*. Chaitin's Theorem appears to generalize Gödel's result, and may lead to new insights about the nature of mathematics and the limits of our thought.

ACKNOWLEDGEMENTS

    I would like to acknowledge several persons who have in some way helped me with this book. First, I want to thank Tom Higginbotham for encouraging me to write a logic text when it was just a vague idea in my mind. I also want to extend thanks to Tom Schaffter for his help over the years. A good deal of the book's pedagogical emphasis is due to Tom's influence, as are some technical notions. Dick Epstein pointed out various errors and vague points, and made numerous useful suggestions that resulted in more lucid explanations. Peter Eggenberger was extremely helpful in a variety of ways. He went over the text with care and made suggestions for technical improvements and for improvements of approach. I also wish to thank Michael Losonsky for valuable comments on *Gödel's Theorems*, which helped to clarify some murky areas. Thanks go to Greg Chaitin for help with Chaitin's Theorem. My thanks also go to a reviewer from the University of Iowa for his encouraging remarks and for his gentle criticisms that sometimes led to pervasive changes in presentation. So, to Tom, Tom, Dick, Peter, Michael, Greg, and "Iowa", thanks again.
    I also wish to thank Susan Koppenol, who typed much of the manuscript from my handwritten notes, and who also input numerous editorial changes and made certain decisions herself that improved the look of the book.
    Thanks go to Cay Horstmann and his technical staff for furnishing and supporting many versions of ChiWriter, the excellent technical word processor that I used to write and "typeset" the entire text.
    I also want to extend my appreciation to the staff at Wm. C. Brown who have worked with me on this project. They include Earl McPeek, Dwala Canon, Jan Scotchmer, Jane Parrigin, and Audrey Reiter. Audrey and Jane were particularly helpful in speeding up the completion of this project. Thanks.

    Lastly, I wish to say "thank you" to my wife Susan.

INTRODUCTION

In this section, I want to explain the main idea of the book, its level of difficulty, how the book is organized, and some ideas for emphasis in each of the following disciplines: mathematics, philosophy, and computer science. At the end of this Introduction, there is an explanation of how quotation marks are used in the text. The Preface provides a more thorough account of technical matters for those already familiar with the subject.

MAIN IDEA

This book begins as a Symbolic Logic text, and then becomes a Mathematical Logic text. Symbolic Logic deals with patterns of formal reasoning, and Mathematical Logic examines those patterns from a mathematical standpoint.

LEVEL OF DIFFICULTY

The book was written for the student who has some familiarity with a few topics in abstract mathematics, but it is not expected that he or she has completely mastered those topics. For example, a student who has done average or better work in discrete structures would qualify (typically, a computer science or mathematics major), as would someone who has done satisfactorily in a prior course in logic (typically, a philosophy major). The mathematical concepts used in the book can be found in Chapter 0, which can be perused first and referred to later, when necessary. The book starts out much easier than other mathematical logic texts, though the level of difficulty continuously increases as the book progresses. Toward the end of the book, the material is more difficult, and some topics are marked as "advanced". Thus, it is expected (hoped) that the student will acquire a greater understanding of the material and will build up his or her "mathematical sophistication" in step with the increasing difficulty of the book. Most of the exercises are not hard, though some of them are, and the student is expected to be able to do almost all of them.

INTRODUCTION

ORGANIZATION

<u>Symbolic Logic</u>:
    Chapters 1, 2, 3, 5, and 6 cover the main topics of Sentential Logic and First-Order Predicate Logic without identity. Thus, they form a natural unit of study for a Symbolic Logic course. Sections 8.1 and 8.2 introduce identity by way of a "Theory of Equality", and thus could also be presented if there is time, as well as can other theories in this chapter. Some of Chapter 9, which presents arithmetic as a first-order theory, could also be added to a fast-paced Symbolic Logic course. One concern is that section 3.5, Proving Facts About Models, is quite abstract and difficult for beginning students, and should perhaps be skipped in a course on Symbolic Logic. One further concern is that some teachers may not wish to delve as heavily into the semantics of Sentential Logic and First-Order Logic as the book provides for. So, this material may be covered in an abridged fashion, rather than in detail. With these qualifications in mind, here are three possible schedules for a course in Symbolic Logic:

Symbolic Logic

Level 1: 1, 2, 3(abridged, skip 3.5), 5, and 6(abridged).
Level 2: 1, 2, 3 (skip 3.5), 5, and 6.
Level 3: 1, 2, 3 (skip 3.5?), 5, 6, 8.1, 8.2, and 9.

<u>Mathematical Logic</u>:
    Some of the material on Mathematical Logic is advanced and may either be skipped entirely or covered only for the "ideas" contained in these sections. Another "path" that could be taken through the material is to begin to look at the material more and more informally as it gets more difficult. We signify this by writing 'ideas' in parentheses after the chapter number. Of course, the <u>extent</u> to which only the intuitive features of a section are covered will depend upon the instructor and the students (or the student alone if this is a self-study). We offer these possible schedules:

INTRODUCTION

Mathematical Logic

Level 1: 3, 4, 7(skip 7.2), 8, 9, 10(ideas), 11(ideas)
Level 2: 3, 4, 7(skip 7.2), 8, 9, 10(ideas), 11
Level 3: 3, 4, 7, 8, 9, 10, 11

Teachers of Symbolic Logic and Mathematical Logic may, however, wish to emphasize different topics from different perspectives, depending on their particular discipline. Here are some ideas about how Symbolic Logic and Mathematical Logic may be covered in the fields of mathematics, philosophy, and computer science.

Mathematics: Here it is assumed that a main goal is to help the student to be able to follow proofs and to be able to construct them. With this focus in mind, a great deal of emphasis should be placed on derivational techniques, particularly those in Chapter 2, together with their relationships to the proof techniques of Chapter 3. Emphasis should also be place on the derivational rules for the quantifiers, particularly on the difference between UI and UG, and the use of ES in mathematical arguments (in Chapter 5). In *First-Order Theories* (Ch. 8) and *First-Order Number Theory* (Ch 9), the student can gradually shift from formal derivations to informal proofs, depending on the instructor's wishes. At any rate, once Chapter 9 has been mastered, the student should be very well-versed in following and doing proofs. In courses in abstract algebra and number theory, some of the material in Chapters 8 and 9 will appear again, and hopefully will be much easier for the student to master.

Philosophy: Here it is assumed that there are two main goals: (1) to master first-order logic (with identity?), and (2) to acquire an appreciation of metatheoretic results, particularly Gödel's Theorems. Philosophy students should be encouraged to build up a repertoire of derivational and proof techniques, especially if they wish to specialize in logic, the philosophy of mathematics, the philosophy of science, or certain areas in the philosophy of language. They should also pay special attention to the difference between a syntactic approach, such as that followed in Chapter 2, and a semantic aproach to the same material (in Chapter 3). The

INTRODUCTION

distinction between the two kinds of concepts is sharpened in Chapter 4 where so-called "metaproofs" establish important connections between the two <u>kinds</u> of concepts. Chapter 7 establishes the metatheory for First-Order Logic (though probably 7.2 should be skipped, or only skimmed). Gödel's Theorems, presented in Chapter 11, are a must for philosophy students, and this chapter should be studied with some care.

Computer Science: Here it is assumed that one goal is to acquaint the student with formal derivations that can be mechanically tested for correctness (esp. Chapters 1 and 5). (This can later be linked to abstract language specifications in a course on automata theory. As further derivational procedures, the instructor may wish to mention Resolution and Unification.) Another goal is to help computer science students become more competent in applying mathematical methods to formal theories. Material in Chapters 8 and 9 should be helpful for this. The semantics of a first-order language, presented in Chapter 6, should be emphasized for its relationship to data structures (there is more on structures in Chapter 10, but some of this material is more advanced), and the entire idea of there being a semantics-syntax distinction is important for distinguishing the semantics of a programming language from verification techniques used to prove programs correct for that language. In short, this material can be connected to subjects in courses on data structures, programming languages, computer architectures, automata theory, and (especially) AI. Gödel's results (in Chapter 11) can also be quite useful in helping students to grasp the limits of algorithmic specification.

I want to explain very briefly a topic I have agonized over throughout the writing of this book: quotation marks. I use single quotes to mention phrases. For example, the word 'mention' occurred in the previous sentence. I use double quotes in two main ways. One way is to quote something. To quote myself, for example, I said "mention". So, I use double quotes as actual quotes, and single quotes for citing an expression. I also use double quotes to call attention to something, but not to say exactly what that something is. (These are called 'scare-quotes'.) For example, I often write "says" to indicate that 'says' is being used in an unusual way. You may have noticed if you are attentive to

INTRODUCTION

these matters that I should have written 'I often write '"says"' to indicate that 'says' is being used in an unusual way.' instead of what I wrote. This illustrates that in some places I drop single quotes when it seems to me that using them makes matters cloudier, more complicated, and more distracting. I have also decided to use the British-English convention of placing punctuation marks <u>outside</u> quotation marks rather than inside, because that convention is more consistent with the needs of this textbook.

CHAPTER 0

MATHEMATICAL PRELIMINARIES

This chapter contains a synopsis of the mathematical prerequisites for the entire book. It is assumed that you are already familiar with some of this material, but not necessarily all of it. Peruse this chapter now, skipping over what you already know and noting topics you may wish to return to afterwards. Later, if you have any difficulty with material in chapters 1-11, you may wish to study these earlier topics more carefully. You may also wish at that time to consult another text, such as Halmos's <u>Naive Set Theory</u>, for more complete coverage of these topics.

"Complete Induction" is a good example of a technique that you may not wish to read thoroughly the first time through. So, glance through the section on Complete Induction and make a mental note of what it contains and where it is located, *especially if you are unfamiliar with the concept*. Later, when a proof uses Complete Induction, you can return to it and study it more carefully at that time.

0.1 FACTS ABOUT SETS

Here are some facts about sets (or classes) that are helpful. In what follows the capital letters 'A', 'B', 'C', 'X', 'Y' with or without subscripts or primes are used to denote sets. A set is uniquely determined by its elements. For example, $\{1,2\} = \{2,1\} = \{2,1,2\}$, etc. The last three descriptions of sets describe the same set, namely the one containing the element 1 and the element 2. We said 'containing' in the previous example. $\{1,2\}$ **contains** 2 or 2 **belongs to** $\{1,2\}$ or 2 **is an element** (or **a member**) **of** $\{1,2\}$ or 2 is **in** $\{1,2\}$. These phrases are used interchangeably. We write '$2 \in \{1,2\}$' to express this in symbols.

There is a set with no elements, $\{\ \}$, which we call the **empty set** or the **null set**, and we write it as: $\emptyset$. We know there is only one empty set by the above principle that a set is uniquely determined by its elements. If we supposed there were two (or more) null sets $\emptyset_1$ and $\emptyset_2$, we would see that $\emptyset_1 = \emptyset_2$ because each has the same elements, namely none. Thus, there can be only one empty set.

There is an entirely different relation than the "belong to" relation, called the "inclusion" relation. We write 'A ⊆ B' to mean: A is **included in** B, or A is a **subset** of B, or B **includes** A. This means that whatever elements are in A are in B as well. That is,

A ⊆ B means: any element in A is also in B.

More formally,

A ⊆ B means: for any x, if x ∈ A, then x ∈ B.

For example, {1,2,3} ⊆ {1,2,3,4,5,6}, since all elements of {1,2,3} are elements of {1,2,3,4,5,6}. Another way to say this is that there is no element of {1,2,3} that is not in {1,2,3,4,5,6}. On the other hand {a,b,c} ⊈ {b,d,e,f} since there is an element of {a,b,c} that is not in {b,d,e,f}. Actually, there are two, a and c, that are in {a,b,c} and not in {b,d,e,f}.

When A is a subset of B, B is said to be a **superset** of A. If A ⊆ B and A ≠ B, then **A is a proper subset of B**, and **B is a proper superset of A**, meaning that there is at least one element of B that is not in A. If two or more sets have no elements in common, they are said to be **disjoint**.

0.2   VACUOUS TRUTH

A special case of the above definition of "⊆" deserves special attention. In mathematics certain statements are "vacuously true". They are of the form 'all A is B' and are true simply because there are no A's. As an example, let's look at the empty or null set, ∅. We claim that the empty set is a subset of *any set*. Let us prove it. First, let's write what we wish to prove:

THEOREM: ∅ ⊆ A for any set A.

> PF: Remember our definition of "⊆". '∅ ⊆ A for any A' means that any element in ∅ is also in A. Since there are no elements in ∅, the statement 'any element in ∅ is also in A' is vacuously true.   ∎

(The symbol ∎ is used to show when a proof is complete.)

This appeal to vacuous truth is far from satisfying to many people. So, here is another approach to showing ∅ ⊆ A, for any A. This second approach has two merits: (1) it is more convincing than the vacuous truth argument, and (2) it shows how an appeal to vacuous truth can be eliminated in an argument.

The idea is to show that what we want to prove *can't be false*.

THEOREM: ∅ ⊆ A, for any set A.

> PF: Assume the theorem is false. Then there must be some set B that ∅ is *not* included in. Using our definition of "included in", this means that there must be some element in ∅ that is *not* in B. But, this cannot be so, since ∅ has no elements. Thus, the theorem cannot be false. Thus, it must be true. ∎

Some people have qualms about this argument as well, for it uses the principle that **what cannot be false must be true**. We accept this principle as intuitively correct in this book and will use vacuous truth arguments as well as what-cannot-be-false-must-be-true arguments (WCBFMBT).

We will also use vacuous truth arguments outside of the context of intuitive set theory. Here is an example:

TO PROVE: Every person over 30 feet tall has green ears.

> PF: (Our strategy is to consider an arbitrary person over 30 feet tall and show that that arbitrary person has green ears. Watch what happens =>). Take any person over 30 feet tall. There is **no such person**. Thus, what we wish to show is **true vacuously**. ∎

Again, we recast the above vacuous truth argument into a WCBFMBT (what-cannot-be-false-must-be-true) argument:

TO PROVE: Every person over 30 feet tall has green ears.

> PF: (Our strategy is to assume that what we wish to show is false and derive a contradiction from this assumption. Watch =>). Assume what we wish to prove is false. Therefore, by our assumption, there is at least one person who is over 30 feet tall and does **not** have green ears. Since **there is no person over 30 feet tall**, our assumption is contradicted. This shows that what we

wished to establish cannot be false, and therefore must be true. ∎

On the basis of the above reasoning (either by a vacuous truth argument or a WCBFMBT argument), we claim all sentences are true that have the form 'All (Any, Each, Every) A are (is) B' when it is true that there are no A's.

By calling a sentence of the form 'All A are B' <u>vacuously</u> true, we emphasize the fact that it is true merely because there are no A's. If we neglect to mention that the sentence is <u>vacuously</u> true, the truth of 'All A are B' could be taken in some contexts to imply the existence of A's. In ordinary conversation, for example, the sentence 'All Jones's children are juvenile delinquents' normally implies that Jones has children. In mathematical contexts — such as here in this book — no such assumption is intended about the existence of A's. '$\emptyset \subseteq C$' is true but the empty set, $\emptyset$, has no elements.

Exercises:

1. Prove that the following assertion is true by exhibiting (a) a vacuous truth argument, and (b) a WCBFMBT argument:

(i) Any positive integer that is both even and odd is greater than 10.

(ii) Explain what modifications would have to be made to your arguments in (i) to establish that
any positive integer that is both even and odd is less than or equal to 10.

Given our WCBFMBT principle, we can employ the following strategy in trying to prove that something is true:

(1) We assume what we wish to prove is false,
(2) We prove that the assumption given by (1) leads to a contradiction.

Then, on the basis of (1) and (2), we conclude that what we wished to prove must be true. We reason as follows: a contradiction cannot hold, and since assuming the falsity of what we wished to prove led to a contradiction, the assumption of falsity cannot hold. Thus, by WCBFMBT, we prove what we wished.

Here is a famous example. It is said that an ancient Greek from the island of Crete (i.e., a Cretan) once declared: All Cretans always lie. We shall prove that the negation of that statement must be true, *based on the fact that a Cretan said 'All Cretans always lie'*. First, we state what we wish to prove:

TO PROVE: Some Cretan sometimes tells the truth.

> PF: We assume that what we wish to prove is false. So, we assume that all Cretans always lie. If all Cretans always lie, however, then the speaker, a Cretan, lied when he said 'All Cretans always lie'. Thus, it must be true that some Cretan sometimes tells the truth. So our assumption that all Cretans always lie leads to the contradictory fact that some Cretan sometimes tells the truth. Hence, our assumption must be false and thus it is true that some Cretan sometimes tells the truth. ∎

So, the fact that a Cretan says 'All Cretans always lie' leads to the surprising and somewhat puzzling conclusion that at least one Cretan sometimes tells the truth. Our method of arriving at this result was to use the WCBFMBT principle — What Cannot Be False Must Be True. There is a corresponding principle that we will use far less in our proofs, but which also seems correct: What Cannot Be True Must Be False (WCBTMBF).

We could have used WCBTMBF instead of WCBFMBT to arrive at the conclusion that some Cretan sometimes tells the truth by reasoning as follows:

TO PROVE: Some Cretan sometimes tells the truth.

> PF: We assume the truth of 'All Cretans always lie'. Therefore the speaker, a Cretan, must have lied when he said 'All Cretans always lie'. But, if he lied, then some Cretan sometimes tells the truth. Again, we find that assuming the truth of 'All Cretans always lie' leads to its own falsity. Thus, by WCBTMBF, some Cretan sometimes tells the truth. ∎

Notice that even though we now know that some Cretan tells the truth, we don't know which ones or when. We do know, however, that the very Cretan who told us that all Cretans always lie did not tell the truth when saying that. We now know **that** was a falsehood.

0.3 UNION, INTERSECTION, DIFFERENCE, AND POWER SET

Supposing the existence of an implied universe of objects, V, the **union** ($\cup$) and **intersection** ($\cap$) of two sets are defined this way:

$$A \cup B = \{x \mid x \in A \text{ or } x \in B\}$$
$$A \cap B = \{x \mid x \in A \text{ and } x \in B\}$$

The first one says $A \cup B$ is the set of all elements x (of V) such that x is **either** in A or x is in B. And, the second says that $A \cap B$ is the set of all elements x such that x is in A **and** x is in B. Some facts about unions and intersections are:

$$A \subseteq (A \cup B)$$
$$B \subseteq (A \cup B)$$
$$(A \cap B) \subseteq A$$
$$(A \cap B) \subseteq B$$

These indicate that when you take unions you get potentially "bigger" sets and when you take intersections you get potentially "smaller" sets.

The union with the empty set doesn't "add" anything, but the intersection with the empty set "removes" everything, as the following facts attest:

$$A = (A \cup \emptyset)$$
$$(A \cap \emptyset) = \emptyset$$

Also, if we "add" something that was already there, we do not get any more:

If $A \subseteq B$, then $(A \cup B) = B$.

Often we prove that X = Y for two sets X and Y by proving first that $X \subseteq Y$ and second that $Y \subseteq X$. The reason this works is that when $X \subseteq Y$ and $Y \subseteq X$, X and Y must have the same elements.

* $\qquad$ X = Y iff $X \subseteq Y$ and $Y \subseteq X$.

Note: We write 'iff' for 'if and only if'.

THEOREM: If $A \subseteq B$, then $A \cup B = B$.

We now will prove this theorem as follows: First, we assume the hypothesis (the "if" part): '$A \subseteq B$'. Then, to prove '$(A \cup B) = B$', we need to prove both '$(A \cup B) \subseteq B$', and '$B \subseteq (A \cup B)$' (to satisfy definition *). We write this more pictorially as follows:

TO PROVE: 'If $A \subseteq B$, then $(A \cup B) = B$', we first

   Assume: '$A \subseteq B$'. Then, we need

TO PROVE: $(A \cup B) = B$. By * we are required

TO PROVE:
   (1) $(A \cup B) \subseteq B$, and
   (2) $B \subseteq (A \cup B)$.

PF: Assume $A \subseteq B$. (This means that for any element x, if $x \in A$, then $x \in B$.)
To prove (1) means we must prove that for any element x, if $x \in (A \cup B)$, then $x \in B$. Take any x such that $x \in (A \cup B)$. This means $x \in A$ or $x \in B$ (by definition of $\cup$). But, since $A \subseteq B$, if $x \in A$, then $x \in B$. So, whether $x \in A$ or $x \in B$, x still belongs to B. Thus $x \in B$. This proves (1).
To prove (2) we do not have to do anything since it is a fact we listed above (if we doubt this fact, however, we can prove it as well).
We have now proven (1) $(A \cup B) \subseteq B$ and (2) $B \subseteq (A \cup B)$, given our assumption that $A \subseteq B$. Thus, we have proven: If $A \subseteq B$, then $(A \cup B) = B$. ∎

A principle similar to the one we just proved is:

   If $A \subseteq B$, then $A \cap B = A$.

Prove this as an *exercise*.

The **complement** of A, A', consists of those elements **not in** A. For A' to make sense there needs to be a universe of elements either explicitly stated or implied, called V, such that A' is the set of all things in our universe V that are not in A. For example, suppose our universe $V = \mathbb{N}$. $\mathbb{N}$ is the set of all natural numbers: 0,1,2,3,... . That is,

$\mathbb{N} = \{0,1,2,3,\ldots\}$). Suppose $A = \{0,2,4,6,\ldots\}$. Then, $A' = \{1,3,5,\ldots\}$, namely the set of those elements of $\mathbb{N}$ that are *not* in A. The **difference** of two sets can be defined this way:

$$A - B \text{ means: } A \cap B'.$$

So if $A = \{0,2,4,6,\ldots\}$ and $B = \{0,4,8,\ldots\}$, then $A - B = \{2,6,10,\ldots\}$. $B - A =$ what? Stop and answer this now. If you said $B - A = \{2,6,10,\ldots\}$ or if you said $B - A = \{1,3,5,7,\ldots\}$ or anything other than $\emptyset$, you are wrong and should go over this carefully. ($B - A = B \cap A' = \{0,4,8,\ldots\} \cap \{1,3,5,\ldots\} = \emptyset$).

The following principles may be helpful:

(1) $(B - A) \subseteq B$ (i.e., "removing" elements from B will not result in a set exceeding B)

(2) $B \subseteq A$ iff $(A - B') = B$

(3) $B \subseteq ((B - A) \cup A)$ (i.e., if you "remove" elements of A from B and then "add" them back you get a set that includes B.)

(4) $((B \cup A) - A) \subseteq B$ (however, B includes the set you get when you first "add" the elements of A to B and *then* "remove" them.)

(3 & 4) $((B \cup A) - A) \subseteq B \subseteq ((B - A) \cup A)$

(5) $B \subseteq A$ iff $(B - A) = \emptyset$

We prove a special case of (3) that will be helpful later:

(3') $B \subseteq (B - \{x\}) \cup \{x\}$ for any element x.

PF: Suppose $x \in B$. Then $(B - \{x\}) \cup \{x\} = B$. Hence, $B \subseteq (B - \{x\}) \cup \{x\}$.

Now, suppose $x \notin B$. Then $B - \{x\} = B$. That is, nothing is removed. So, $(B - \{x\}) \cup \{x\} = B \cup \{x\}$. Hence, $B \subseteq (B - \{x\}) \cup \{x\}$ in this case as well.

Thus, whether $x \in B$ or whether $x \notin B$, it is true that $B \subseteq (B - \{x\}) \cup \{x\}$. ∎

Exercise:

Prove $((B \cup \{x\}) - \{x\}) \subseteq B$, which is a special case of (4), analogously to the proof of (3′).

Notice that in the proof of (3′) (and earlier), we have made explicit use of the Law of the Excluded Middle. According to that principle, every mathematical statement is either true or false. In the above proof, we assumed that the statement '$x \in B$' must be true or false (and then we showed that the desired conclusion followed in both cases). We assumed earlier the Law of Non-Contradiction for mathematical statements, in terms of which no statement can be <u>both</u> true and false. The two laws together imply that any mathematical statement has one and only one truth value.

Sometimes primes (′) are also used to indicate a different set, as in A, A′, A′′, and so on. The context will always make it clear whether A′ is the complement of A, or just another set.

We define $\bigcup_{i \in I} A_i$ to be the set of all elements belonging to at least one of the sets A_i where $i \in I$. Here is an example: $\bigcup_{k \in \mathbb{N}} A_k$ is well-defined as soon as we say what elements each A_k has, for $k \in \mathbb{N}$. If k is an even number, let $A_k = \{k/2\}$. If k is an odd number, let $A_k = \{-(k+1)/2\}$. That is, if k is odd, add one to k, divide that result by 2, and put a minus sign in front of that result. The set containing this final number is A_k (when k is odd). In the following chart, each set corresponds to the particular A_k above it:

A_0	A_1	A_2	A_3	A_4	A_5	...
{0}	{-1}	{1}	{-2}	{2}	{-3}	...

Notice that $\bigcup_{k \in \mathbb{N}} A_k = \mathbb{Z}$, the set of integers.

The **power set** of A, $\mathcal{P}(A)$, is the set of all subsets of A. That is,

$$\mathcal{P}(A) = \{B \mid B \subseteq A\}$$

For example, $\mathcal{P}(\mathbb{N}) = \{\emptyset, \{1\}, \{2\}, \ldots, \{1,2\}, \{1,3\}, \ldots, \ldots, \mathbb{N}\}$, (i.e., $\mathcal{P}(\mathbb{N})$ is the set of all subsets of the natural numbers).

0.4 SEQUENCE, CARTESIAN PRODUCT, RELATION

A **sequence of k elements** will be written either as $(a_1, a_2, \ldots, a_k)$, or $\langle a_1, a_2, \ldots, a_k \rangle$. This means a_1 is the first element in the sequence, a_2 is the second, and so on. We also call this a **k-tuple**. A **k-tuple on A** is a sequence of k elements $(a_1, a_2, \ldots, a_k)$ such that for $1 < i \leq k$ $a_i \in A$. For some fixed value of $k > 1$, we call the set of all k-tuples on A, the **Cartesian product** or **cross product** of A with itself k times, and write it: A^k, which is short for $A \times A \times \ldots \times A$ for k factors of A. More symbolically, $A^k = \{(a_1, a_2, \ldots, a_k) \mid a_i \in A, \text{ where } 1 \leq i \leq k\}$. Two sequences are equal if and only if they have the same number of terms (i.e., both are k-tuples for some $k > 1$), and each term in one sequence is equal to the corresponding term in the other. Sequences with exactly two terms are called **ordered pairs**. The **Cartesian** (or **cross**) **product** of two sets A and B, written $A \times B$, is the set of all ordered pairs (x,y) such that $x \in A$ and $y \in B$. A set of ordered pairs is a **binary relation**. In other words, a binary relation R is a subset of $A \times B$, where A and B are two sets. When R is a subset of $A \times A$, then R is a (binary) relation **on** A. Let R be the "father of" relation. Then the ordered pair (x,y) is in R if and only if x is father of y. The set from which we get both x and y is the set of people in the world. That is, if x and y are people, then $(x,y) \in R$ if and only if x is father of y. So, for example, (Philip of Macedon, Alexander the Great) $\in R$, and (William Godwin, Mary Wollstonecraft Shelley) $\in R$.

An **equivalence relation** is a binary relation that is **reflexive, symmetric,** and **transitive**. For example, consider the binary relation R on the set $\{a,b,c\}$ such that $R = \{(a,a), (a,b), (b,b), (c,c)\}$. R is **reflexive** if for any element $x \in A$, $(x,x) \in R$. In this example R <u>is</u>, in fact, reflexive since $(a,a) \in R$, $(b,b) \in R$, and $(c,c) \in R$, and

{a,b,c} contains no additional elements besides a, b, and c.
R is **symmetric** if for any elements x,y ∈ A, if (x,y) ∈ R,
then (y,x) ∈ R. R in this example is <u>not</u> symmetric because
(a,b) ∈ R but (b,a) ∉ R. (Note: it doesn't matter for
symmetry that, for example, (c,a) ∉ R, since (a,c) ∉ R
either. Thus, {(a,a), (b,b), (c,c)} <u>would</u> be symmetric.) R
is **transitive** if for x,y,z ∈ A, whenever (x,y) and (y,z) ∈ R,
(x,z) ∈ R as well. At first glance it may look as though R,
as defined above, is not transitive, but it is. For example,
consider (a,b) and (b,b). By definition, if R is transitive
then since both (a,b) ∈ R and (b,b) ∈ R, we should expect
(a,b) ∈ R as well. But, we already know (a,b) ∈ R. So there
is <u>no case</u> where x,y,z ∈ A and both (x,y) ∈ R and (y,z) ∈ R,
while (x,z) ∉ R. Hence, R in our example <u>is</u> transitive.

Since we began by defining an equivalence relation, and
found that R above is <u>not</u> symmetric. Let's define R' to
contain everything R has, plus whatever is needed to make R'
an equivalence relation. We saw above that because
(b,a) ∉ R, R was not symmetric. So let's define the relation
R' on the same set of elements {a,b,c}, such that
R' = {(a,a), (a,b), (b,a), (b,b), (c,c)}. The question is:
Have we now destroyed transitivity by including (b,a) in R'?
The answer is no. So R' is an equivalence relation.

Let R be an equivalence relation on a set A. For each
element, x ∈ A, there is a set [x] consisting of all the
elements of A that bear the relation R to x. [x] is called
the **equivalence class of x** (with respect to R). More
formally: [x] = {y ∈ A| (x,y) ∈ R }. For example, consider
the equivalence relation R' defined a few lines above. The
equivalence classes are as follows: [a] = {a,b}, [b] = {a,b},
and [c] = {c}. Notice that [a] = [b], and that [c] has no
elements in common with [a] (or [b], since they are
identical). It is always the case that two equivalence
classes on the same relation are either the same or disjoint.
A further property is that the union of all these equivalence
classes equals the original set, A. Thus, it is said that
the collection of all equivalence classes (for a given
equivalence relation R on a set A) is a **partition of A**. The
partition of {a,b,c} created by R' is {{a,b}, {c}}. Again,
the sets in the partition are disjoint, and their union is
the original set A. A further property is that the empty
set, ∅, is not permitted to be a member of a partition.

The <u>smallest</u> equivalence relation on a set A is the
Identity Relation (or the Equality Relation) on that set.
For example, for set A = {a,b,c}, the Identity Relation
I = {(a,a), (b,b), (c,c)}.

More generally, just as a set of ordered pairs is a binary relation, a set of ordered k-tuples is a **k-ary relation**. A k-ary relation, R^k, on **A** is a subset of A^k, the cross product of A used as a factor k times. For example, define the ternary relation B (meaning "between") among points on a line as follows:

$(x,y,z) \in B$ iff y is between x and z on the real line.

If we wish to be precise about this, we can say that $B \subseteq \mathbb{R} \times \mathbb{R} \times \mathbb{R}$, where $\mathbb{R}$ is the set of real numbers and $\mathbb{R} \times \mathbb{R} \times \mathbb{R}$ (or $\mathbb{R}^3$) is the set of triples such that each element is a real and $(x,y,z) \in B$ if and only if *either* x is greater than y and z is less than y *or* x is less than y and z is greater than y. More succinctly, $B = \{(x,y,z) \mid x,y,z \in \mathbb{R}$ & either $x < y < z$ or $z < y < x\}$.

0.5 FUNCTION, OPERATION

A **function** is a binary relation R such that if xRy (i.e., $(x,y) \in R$) and xRy' (i.e., $(x,y') \in R$), then $y = y'$. Another way to put this is that if a function, f, contains (x,y) and (x,y'), which are two ordered pairs with the same first element, then their second elements must be the same as well — which really says that for a given left element in an ordered pair (belonging to a function), the right element is unique. We want to say that the second element is "f of" the first element, and we cannot do this if there is not a unique second element for each first element. So, for example, if $(x,y) \in f$, where f is a function, y is "f of" x. That is, $y = f(x)$. For example, take the relation $R \subseteq \mathbb{Z}^+ \times \mathbb{Z}^+$, where $\mathbb{Z}^+$ is the set of positive integers and $(x,y) \in R$ iff y is x^2. Then $R = \{(1,1), (2,4), (3,9),...\}$. We can see that this relation is a function because every positive integer has only one square.

A function is also called a **mapping** or a **map** because it is said to map elements **from** one set **to** elements of (possibly) another set. A function f that maps elements from set A to elements of set B is written as $f: A \rightarrow B$.

The **domain** of a function is the set of first elements in the ordered pairs belonging to the function, and the **range** of that function is the set of second elements in the ordered pairs belonging to the function. For example, if $f = \{(a,b), (b,c), (c,d)\}$, then $\{a,b,c\}$ is the domain of f, and $\{b,c,d\}$ is the range of f. The domain of a function f from A to B is always the set A, but the range of f may be only a proper subset of B. For example, suppose we define $h:\mathbb{N} \rightarrow \mathbb{N}$ such that for any $k \in \mathbb{N}$, $h(k) = 1$ if k is even and $h(k) = 0$ if k is odd. Then $h = \{(0,1), (1,0), (2,1), (3,0), \ldots\}$. The range of h has only two elements, 0 and 1, but the set to which h is mapped has infinitely many elements. When the set to which a function is mapped is the same as the range of that function, the function is said to be **onto**. If function $f:A \rightarrow B$ is onto, then the range of f is all of B. For example, take $f:\mathbb{N} \rightarrow \mathbb{N}$ such that for any $k \in \mathbb{N}$, $f(k) = k+1$. f is **not** onto in this case because the range of f is $\{1, 2, 3, \ldots\}$, which is missing the element 0.

As we have just seen, the range of a function $f:A \rightarrow B$ may be a proper subset of B, though its domain is always all of A. Sometimes we emphasize that its domain is all of A by saying that f is a **total function**. Then we define a **partial function** g from A to B (i.e., $g:A \rightarrow B$) as a (total) function from some subset of A to B. For example, let us define the partial function $g:\{a,b\} \rightarrow \{a,b\}$ such that $g(a) = b$, and that's all. g is a partial function from $\{a,b\}$ to $\{a,b\}$, but g is <u>not</u> a (total) function from $\{a,b\}$ to $\{a,b\}$. However, g <u>is</u> a (total) function from $\{a\}$ to $\{a,b\}$. We can also define a function f' whose domain is "restricted" to a proper subset of the domain of function f. If the domain of f is A, and $A' \subseteq A$, we can define $f'(x)$ to equal $f(x)$ for any $x \in A'$, the smaller domain. That is, for the subset of A for which f' is defined, f' agrees with f. In this case, f' is called **the restriction of f to A'**. More formally, if $A' \subseteq A$, **the restriction of f to A'**, written $f|A'$, is: $f \cap (A' \times B)$. And, of course, since a function is just a special kind of binary relation, this definition holds more generally for a binary relation R, such that $R \subseteq A \times B$. That is (where $A' \subseteq A$), **the restriction of R to A'**, written $R|A'$, is: $R \cap (A' \times B)$.

A **1-1** (pronounced: one-to-one) **function** is one that maps distinct elements of the domain to distinct elements of the range. This function is **not** 1-1: $\{(a,c), (b,c), (c,d)\}$. The reason is that a and b are "distinct" — not equal — yet they both are mapped by the function to the element c. More succinctly, a function $f:A \rightarrow B$ is 1-1 iff whenever $a_1, a_2 \in A$, if $a_1 \neq a_2$ then $f(a_1) \neq f(a_2)$.

A **1-1 correspondence** is a function, that is both 1-1 and onto. Let us define a function f that is a 1-1 correspondence between the k of 'A_k' and the element belonging to A_k in the example explaining $\bigcup_{k \in \mathbb{N}} A_k$ a few pages earlier. We want $f:\mathbb{N} \rightarrow \mathbb{Z}$ such that if k is even,

$f(k) = k/2$, and if k is odd, then $f(k) = -(k+1)/2$. The pairing of elements is as follows:

```
ℕ:   0   1   2   3   4   5   6   7   ...
ℤ:   0  -1  +1  -2  +2  -3  +3  -4   ...
```

Observe that, no two distinct elements of ℕ are mapped to the same element in ℤ and no element in ℤ is missed by the mapping f. So, f is indeed a 1-1 correspondence.

In this book, we are concerned with the truth or falsity of sentences from a given formal language, according to some evaluation of those sentences. Formally speaking, the "evaluation" of the sentences is just a function from the set of sentences to the set of truth values {T,F} ('T' for '**True**' and 'F' for '**False**'). For example, suppose our evaluation function E maps all the sentences in the following set to the value **True**: {'John loves Mary', 'Mary loves Bill', 'Bill loves Consuela'} then, E is a function such that E('John loves Mary') = **T**, whereas (we will suppose) E('Consuela loves Bill') = **F**.

A function is a **unary operation**, meaning that it operates on a single object — usually called the **argument** — and returns a **value**. Looking at a function as a kind of mechanical device accepting inputs and producing outputs, we can say that a function accepts one input at a time (the argument of the function) and produces one output (the value of the function for that argument). A k-ary operation, in general, accepts k inputs and produces a single output. The "black box" representing a k-ary operation is given k objects and produces one in return. Often the term 'function' is used to mean the same thing as 'operation' regardless of the number of inputs. So, we can also speak of a k-ary function, as well as a k-ary operation. When we use 'function' for a k-ary operation, where k is <u>not</u> 1, we will explicitly note this usage.

A function (as a unary operation) is a special kind of binary relation. More formally, a function $f: A \to B$, is really a subset of $A \times B$. If h is a function from the natural numbers to the natural numbers, then $h \subseteq \mathbb{N}^2$. Similarly, a **k-ary operation** O^k on the natural numbers is a subset of $\mathbb{N}^{(k+1)}$, where $\mathbb{N}^{(k+1)} = \mathbb{N} \times \mathbb{N} \times \ldots \times \mathbb{N}$ (k+1 times). More concretely, the 2-ary (binary) operation of "plus" is a subset of $\mathbb{N}^3$, that is, a set of ordered triples of elements from $\mathbb{N}$. If O_+^2 is the plus operation on $\mathbb{N}^3$, then:

$$O_+^2 = \{(0,0,0), \ldots, (1,2,3), (2,1,3), \ldots (7,8,15), \ldots\}.$$

To be a k-ary operation O^k, the following condition must be satisfied: if $(x_1, x_2, \ldots, x_k, y) \in O^k$ and $(x_1, x_2, \ldots, x_k, y') \in O^k$, then $y = y'$. In terms of our example of the "plus" operation O_+^2, whenever $(x_1, x_2, y) \in O_+^2$ and $(x_1, x_2, y') \in O_+^2$, it must be the case that $y = y'$. This rules out, for example, both $(1,2,3) \in O_+^2$ and $(1,2,y') \in O_+^2$, where $3 \neq y'$. All this means is that "plussing" two natural numbers cannot yield more than one result; otherwise "plus" would not be a well-defined operation.

The **image of A' under** f is the range of $f|A'$ and is often written $f(A')$, which is ambiguous if A' is also an <u>element</u> of A (as well as being a subset of A). The **image of an element** x of A under f is just the object to which x is mapped by f (i.e., $f(x)$). Where f is a mapping from A to B, and $B' \subseteq B$, the **inverse image of B' under** f, written $f^{-1}(B')$ is the set of all elements of A that are mapped by f to elements of B'. In symbols: $f^{-1}(B') = \{x \in A | f(x) \in B'\}$. So, $f^{-1}(B')$ is not necessarily a function for two reasons: (1) more than one element of A may be mapped by f to an element of B', and (2) there may be elements of B' to which <u>no</u> elements of A are mapped by f. If $f^{-1}(B')$ <u>is</u> a function, it is called the **inverse function** from B' to A (or to some subset of A).

Suppose we have three (not necessarily distinct) sets A, B, and C, and a function f such that $f: B \rightarrow C$ and a function g such that $g: A \rightarrow B$. The **composition of** f **and** g, written $f \circ g$, is the function defined on any element $x \in A$ as follows: $(f \circ g)(x) = f(g(x))$. Thus, $f \circ g$ is a mapping from set A to set C. For example, if $g(x) = x^2$, and g is a mapping from the integers to the natural numbers, then each integer gets mapped to its square (both n and -n being mapped to n^2). If f is a mapping from the natural numbers to the reals such that $f(x) = \sqrt{x}$ then f returns the positive square root of each natural number. Thus in this case, $f \circ g$ is a mapping from the set of integers to the set of positive reals and 0. Note that $(g \circ f)$ is undefined, because $f(y)$ returns the positive square root of y, which may not be an integer. Hence, $g(f(y))$ would not yield a value.

A k-ary operation O^k on A is a function from A^k to A. If $k = 0$, then a single element of A is picked out. Thus, a **0-ary operation** on A is a unique element from A.

Two sets are **equinumerous** if they have the same number of elements. When the two sets in question are finite, there is no intuitive difficulty in determining whether they have the same number of elements or whether one set has more elements than the other. When the two sets are both infinite, we define them as **equinumerous** if and only if there is a 1-1 correspondence between the two sets (i.e., a 1-1, onto function from one of the sets to the other one). This definition then works for finite sets as well. When two sets are equinumerous, they are said to have the same **cardinality**. If there is a 1-1 function from set A to set B, but no function that is <u>both</u> 1-1 <u>and</u> onto from A to B, then the cardinality of B is greater than the cardinality of A, meaning, intuitively, that B has <u>more</u> elements than A. This is written Card(B) > Card(A). For example, Cantor's famous Diagonal Argument shows that there are more real numbers than there are natural numbers. We write it this way: Card($\mathbb{R}$) > Card($\mathbb{N}$). The expression used most frequently to stand for the cardinality of the natural numbers is the Hebrew letter $\aleph$ followed by the subscript 'o' to produce: $\aleph_0$. ($\aleph_0$ is pronounced "aleph nought".) So, Cantor's result can also be written: Card($\mathbb{R}$) > $\aleph_0$. A set is **denumerable** if it has cardinality $\aleph_0$, meaning that it can be put into a 1-1 correspondence with the natural numbers (or that it is equinumerous with the natural numbers). For example, the set of rational numbers, $\mathbb{Q}$, is denumerable (because there is a 1-1 correspondence between $\mathbb{N}$ and $\mathbb{Q}$). Hence, Card($\mathbb{Q}$) = $\aleph_0$. A set is **non-denumerable** if its cardinality is <u>greater</u> than $\aleph_0$. So, by Cantor's proof, the reals are <u>non</u>-denumerable, since Card($\mathbb{R}$) > $\aleph_0$. A set is called **countable** if it is <u>either</u> denumerable <u>or</u> finite. Thus, the set of natural numbers is countable (since it is denumerable), and all finite sets are countable as well.

0.6 COMPLETE INDUCTION

Rather than defining complete induction in the abstract, let us look at a specific use of it and then highlight the principles used.

In the study of logic we are often going to be concerned with **sentences** of a given language. Here are rules for forming sentences in a language we will call 'L' ('φ' and 'ψ' are Greek letters, pronounced 'phi' and 'psi' respectively):

(1) The symbols 'P' and 'Q' are sentences of L;

(2) If φ and ψ are any sentences of L, then $(\varphi \to \psi)$ and $(\varphi \lor \psi)$ are sentences of L;

(3) Nothing else is a sentence of L (except by clauses (1) and (2)).

Here are examples of sentences of L: P, Q, $(P \to P)$, $(Q \to P)$, $(P \lor (Q \to Q))$, $(Q \to ((P \lor Q) \lor Q))$. (See if you can tell why each is a sentence of L.) Here are **non**-sentences: X, a, #, S, P $\lor$ Q, $(P \to T)$, $(P \lor (Q \to P \lor P))$, $(P \lor (Q \leftarrow P))$. (See if you can tell why each is a **non**-sentence of L.)

What we wish to prove by complete induction is that every sentence of L has the same number of left parentheses ('(') as right parentheses (')').

Here, informally, is our reasoning: For a string of symbols χ (pronounced 'chi') to be a sentence of L, it has to be constructed in accordance with one of the two clauses, (1) or (2). Suppose χ is a sentence by clause (1). Then χ is the symbol 'P' or 'Q', both of which have 0 left and 0 right parentheses. So, we know that if χ is a sentence by clause (1), χ has the same number of left as right parentheses (namely 0).

Now, suppose χ is a sentence by clause (2). Here we see clause (2) adds balanced parentheses to previous sentences. Therefore, χ cannot have unbalanced parentheses due to clause (2).

This concludes our informal proof. We see by (3) that nothing else is a sentence, except by (1) and (2). Since sentences by (1) have <u>no</u> parentheses and sentences by (2) add matching parentheses to previous sentences, there is no sentence-forming clause that can create sentences with unbalanced parentheses. We conclude that every sentence χ must have the same number of left parentheses as right parentheses. ∎

We believe that you, the reader of this text, should be officially warned about the material to follow. In an effort to instill certain principles pertaining to more formal proofs, we prove the same result about balanced parentheses again and again. The reason for this is to provide examples of these formal proofs for later reference. It is **not** thought that the more formal proofs that follow are "better" proofs of the result about balanced parentheses. For this result the above "informal proof" is probably best. The more formal proofs that follow are for later reference, so that you can master all the gory details of Complete Induction. Therefore, considering the purpose of these more formal proofs, you may want to skim them now and return to them later.

Now, for a more formal proof:

PF: We suppose n is the total number of occurrences of the symbols '$\rightarrow$' and '$\vee$' in the sentence χ, where n $\geq$ 0 (there may be **no** occurrences of '$\rightarrow$' or '$\vee$' in a sentence. 'P', for example, has none.)

Base Case: n = 0. If there are no occurrences of '$\rightarrow$' or '$\vee$', then χ is either 'P' or χ is 'Q' (these are the only two ways χ can have 0 occurrences of '$\rightarrow$' or '$\vee$'). In both of these two cases χ has the same number of left as right parentheses (namely, 0).

Induction Case: n > 0 (so χ must be obtained by clause (2)). We **assume** that for any sentence that has **fewer** occurrences of '$\rightarrow$' and '$\vee$' than n that our result holds — namely, that left and right parentheses are balanced. Again: we **assume** this and call this assumption the **Inductive Hypothesis**. Then, we wish to show, based on this assumption, that χ itself has balanced left and right parentheses.

Since the number of '$\rightarrow$' and '$\vee$' in χ is greater than 0 (i.e., n > 0), χ must be a sentence by clause (2). That is, χ is a sentence of the form $(\varphi \rightarrow \psi)$ or $(\varphi \vee \psi)$. Suppose χ is of the form $(\varphi \rightarrow \psi)$. By our Inductive Hypothesis **each** of the component sentences φ and ψ has the same number of left and right parentheses **because each of them has fewer occurrences of** '$\rightarrow$' **and** '$\vee$' **than** χ. (χ has 1 more than the total of φ and ψ put together, because '$\rightarrow$' is added to get χ). So, if parentheses are balanced in each of φ and ψ (by our Inductive Hypothesis), then the sentence $(\varphi \rightarrow \psi)$ has one more left one and one more right one. So $(\varphi \rightarrow \psi)$ has balanced left and right parentheses as well. $(\varphi \rightarrow \psi)$ is χ in this case, so χ has balanced left and right parentheses. The case where χ is of the form $(\varphi \vee \psi)$ is exactly the same, and is left as an *exercise*.

■

This proof was labored over in order to furnish a detailed example of Complete Induction. Often the proofs we provide will be less drawn out and more like the informal one presented first. But sometimes we will need to be rigorous just to make sure nothing has been missed.

0.7 PRINCIPLE OF COMPLETE INDUCTION

In what follows, **S**(n) is a statement about n and n is any natural number. That is, n ∈ {0,1,2,...}.

If
  (1) We prove: **S**(0) is true, and
  (2) We prove: **If S**(k) is true for **all** k < n, **then S**(n) is true, for n > 0.

Then
  We have proved: For **all n S**(n) is true.

In our example, **S**(n) was the statement: Any sentence of L having n total occurrences of the symbols '→' and 'v' has the same number of left parentheses as right parentheses.

We proved:

  (1) **S**(0), namely: Any sentence...having 0 total occurrences...has balanced left and right parentheses.

  (2) If **S**(k) for all k < n, then **S**(n) is true as well, namely: if...all sentences having k occurrences of '→' and 'v' have balanced left and right parentheses, where 0 ≤ k and k < n, then a given sentence χ with n occurrences of '→' and 'v' has balanced left and right parentheses.

THEREFORE:

For any sentence of L, say φ, having any number of occurrences of '→' and 'v', φ has the same number of left as right parentheses.

0.8 ANOTHER PROOF THAT L HAS BALANCED PARENTHESES

There is another very natural method of proving that every sentence of L has balanced (the same number of left as right) parentheses that is **equivalent** to the method of Complete Induction. As before, first we will look at the proof. Then we will look at the principles used in the proof.

TO PROVE: Any sentence χ of L has balanced left and right parentheses.

PF: Assume it is false. (We arrive at a contradiction from this assumption, and that shows that what we wish to prove is true by the WCBFMBT principle.) That is, we assume there is **at least one** sentence of L having **un**balanced parentheses. Let χ_1 be a sentence with the smallest number of occurrences of '$\rightarrow$' and '∨' that has **un**balanced left and right parentheses.

:BEGIN INTERRUPTION $\Longrightarrow$

Let us interrupt this proof for a moment to explain the above sentence. (After this paragraph has been thoroughly understood, the proof should be read again and **this** paragraph skipped.) If some sentences have **un**balanced parentheses (i.e., **if** what we wish to prove is false), then, we can put the sentences with **un**balanced parentheses in a list. In the front of the list we put all those sentences with the **least** number of occurrences of '$\rightarrow$' and '∨'. For example, suppose that **all** sentences with $0,1,2,3,\ldots$, up to 7,648 occurrences of '$\rightarrow$' and '∨' **do** have balanced parentheses, but **some** sentences with 7,649 occurrences do not. Then, we put at the front of our list (in some order, we do not care which) those sentences with **un**balanced parentheses having 7,649 occurrences of '$\rightarrow$' and '∨'. We then take the first sentence having 7,649 occurrences of '$\rightarrow$' and '∨' with **un**balanced parentheses and call it χ_1. That is, χ_1 has **un**balanced left and right parentheses, but **all** sentences on the list have **balanced** parentheses, including those with **fewer** occurrences of '$\rightarrow$' and '∨'. Now that a more vivid picture of χ_1 has been presented, return to the beginning of the proof above this interruption and then skip this paragraph.

:END INTERRUPTION $\Longleftarrow$

Case 1: Suppose χ_1 is a sentence by clause (1). That is, χ_1 is either 'P' or 'Q'. This cannot be, since neither 'P' nor 'Q' has **any** parentheses, much less **un**balanced ones. So, it must be that χ_1 is a sentence by case (2).

Case 2: χ_1 is a sentence of the form $(\varphi \rightarrow \psi)$ or $(\varphi \vee \psi)$ by clause (2). Suppose χ_1 is of the form $(\varphi \vee \psi)$. Just to be more specific, we will suppose there are **more left parentheses** in χ_1, than right ones. Since χ_1 is $(\varphi \vee \psi)$, let us remove the outermost left parenthesis and right parenthesis and look at φ and ψ. Since we removed **balanced** parentheses from χ_1 when there were **more left parentheses** in χ_1 to begin with, there must be more left parentheses in either φ or ψ (or both). Suppose there are more left parentheses in φ than right parentheses. That is, suppose that φ is at least partly responsible for χ_1 having **un**balanced parentheses (where, in this case, χ_1 is $(\varphi \vee \psi)$). But this supposition leads to a contradiction. The number of occurrences of '$\rightarrow$' and '$\vee$' in φ is **at least one less** than in $(\varphi \vee \psi)$ since $(\varphi \vee \psi)$ has an additional occurrence of the symbol '$\vee$' between φ and ψ. But, remember, we began by taking $(\varphi \vee \psi)$ (as χ_1) to be a sentence with the **smallest** number of unbalanced occurrences of '$\rightarrow$' and '$\vee$'. **NO** sentence with fewer occurrences of these symbols can have **un**balanced parentheses. So, φ **cannot possibly have unbalanced** parentheses (much less, more left ones than right ones). And the same holds for ψ. Since both φ and ψ have fewer total occurrences of '$\rightarrow$' and '$\vee$', $(\varphi \vee \psi)$ must have balanced parentheses as well. (The same conclusion holds for the case when χ_1 is of the form $(\varphi \rightarrow \psi)$.)

So, if χ_1 has unbalanced parentheses, it is **not** a sentence by clause (1) (For in this case there are 0 parentheses), and it is **not** a sentence by clause (2) either (since then some sentence with fewer occurrences of '$\rightarrow$' and '$\vee$' than χ_1 must have unbalanced parentheses, which is impossible). Clause (3) says that nothing else is a sentence except by clauses (1) and (2). Therefore, χ_1 **cannot possibly** have unbalanced parentheses. Therefore, all sentences of L have balanced left and right parentheses. ∎

Whew! **That** was a long and drawn out explanation. We remind you that our future proofs employing Complete Induction will not contain this much detail. Please return to these examples to remind yourself of how to proceed if a later proof seems difficult to grasp or to construct. A proof by Complete Induction can also be viewed as a proof using the Least Counterexample Principle.

0.9 LEAST COUNTEREXAMPLE PRINCIPLE

Let $S(k)$ be a statement about the number k and let us assume that for some natural number the statement is false. This means there must be a **least** natural number for which it fails, say n. That is, we assume $S(n)$ does **not** hold. Then

(1) We **prove** n cannot be 0. (That is, we prove $S(n)$ **holds** when n = 0.)

(2) We **prove** n cannot be greater than 0 either. We prove this by showing that a contradiction follows. Specifically, we prove if NOT-$S(n)$, then for some k < n, NOT-$S(k)$. This contradicts the fact that n is the **least** natural number for which it fails.

From (1) and (2) we conclude **there is no number n for which $S(n)$ fails**, therefore, $S(n)$ is true for all n.

Let us take one more (hopefully the final) look at the proof by the least counterexample principle (**LCP**) that the number of left parentheses and right parentheses are balanced in every sentence of L.

Our "n" in that proof was the least number of occurrences of '$\rightarrow$' and 'v' in sentences of L with **un**balanced parentheses. χ_1 had n occurrences of those symbols, but all sentences with fewer occurrences of those symbols had **balanced** parentheses.

So, 'NOT-$S(n)$' expresses the fact that **there is** a sentence of L having n occurrences of '$\rightarrow$' and 'v' that does **not** have balanced parentheses. For k < n, $S(k)$ is true. That is, **all** sentences of L having k occurrences of '$\rightarrow$' and 'v' have **balanced** parentheses.

To show how our proof falls under the LCP category, we write:

Assumption: Some sentence χ_1 of L has n occurrences of '$\rightarrow$' and 'v', and χ_1 does **not** have balanced parentheses. Yet, **all** sentences of L with 0,1,2,...,k occurrences of '$\rightarrow$' and 'v', for k < n, **do have balanced parentheses**.

(1) We proved n cannot be 0. (When there are 0 occurrences of '→' and '∨', there are no parentheses, and thus they balance.)

(2) We proved when n > 0, if χ_1's parentheses do **not** balance, then there must be **another** sentence, φ, with k occurrences of '→' and '∨', where k < n, such that φ's parentheses do **not** balance as well.[1]

Therefore, our proof of this result can be seen as an application of LCP. Again, we mention that future proofs of this type will not normally contain so many details. If these details are desired, they can always be provided. (If they cannot be provided, the "proof" is **not** an instance of LCP.)

In our two formal proofs that sentences of L have balanced parentheses, we let n range over the natural numbers, where n was the total number of occurrences of the connectives '→' and '∨'. Let us call this measure of complexity of a sentence φ, **the height of** φ.

Thus, 'P' and 'Q' are of height 0, whereas '(P → Q)', '(P → (Q → P))', '((P ∨ Q) → (Q → P))' are, respectively, of height 1, 2, and 3.

Many proofs that we do later proceed by Complete Induction on the height of a sentence of a language that is somewhat more complex than L. L_{SL}, for example, has more connectives than L, and L_{PL} has the same connectives as L_{SL} plus two quantifiers. The **height** of a sentence φ of L_{PL} is the number of occurrences of connectives and quantifiers in φ.

Let us define **the order of** φ, for sentence φ of L as follows:

(i) If φ is 'P' or 'Q', φ is of order 1;

---

[1] Note the similarity between (2) above and the corresponding case for the principle of Complete Induction. In that case we proved that for k < n if all sentences of k occurrences of '→' and '∨' have balanced parentheses, then any sentence of n occurrences of those symbols must have balanced parentheses as well.

(ii) If φ is of the form $(\psi \rightarrow \chi)$ or $(\psi \vee \chi)$ and the maximum order of ψ and χ is n, then the order of φ is n + 1.

It should be clear that the order of a sentence φ will not necessarily be the same as its height. The **order** of 'P' and 'Q' is 1 (the **height** is 0). The **orders** of 'P $\rightarrow$ Q', 'P $\rightarrow$ (Q $\rightarrow$ P)', and '((P $\vee$ Q) $\rightarrow$ (Q $\rightarrow$ P))' are, respectively, 2, 3, and 3 (the **heights** are, respectively 1, 2, and 3). The main difference between the two measures — other than the fact that one starts at 0 and the other at 1 — is that the height of $(\psi \rightarrow \chi)$ (or $(\psi \vee \chi)$) is the **sum** of the height of ψ **and** the height of χ, **plus** 1, whereas the order of $(\psi \rightarrow \chi)$ (or $(\psi \vee \chi)$) is the **maximum** order of ψ and χ **plus** 1.

Exercise:

To better understand complete induction and LCP, prove that any sentence of **order** n has balanced parentheses using each method. Note that in this case n ranges not over the natural numbers, but over the positive integers, since the lowest possible order is 1. (The proofs are not very different; just be sure to indicate at the appropriate places the **order** of the sentences being considered, based on the definition of the order of a sentence.)

0.10 MATHEMATICAL INDUCTION

A slightly different principle of induction that is more frequently used than Complete Induction is **mathematical induction**. It is more normally defined for $\mathbb{Z}^+$, the set of positive integers than for $\mathbb{N}$, the set of natural numbers. So, we begin by stating **the principle of mathematical induction** for $\mathbb{Z}^+$:

If
(1) We prove: **S**(1) is true, and
(2) We prove: For any $k \in \mathbb{Z}^+$ if **S**(k) is true, then **S**(k+1) is true,

Then
We have proved: For all $n \in \mathbb{Z}^+$, **S**(n) is true.

MATH PRELIMINARIES   25

The stock example of mathematical induction is the proof that the sum 1 + 2 + 3 +...+ n is always equal to n(n+1)/2. (This numerical fact was supposedly intuited instantly by the great mathematician Carl Friedrich Gauss when he was eight years old.) We begin the proof by noticing that the series 1,2,3,...,n can have any number of terms greater than or equal to 1. So, we start with our lowest (or base) case: 1. S(1) is the statement that 1 = 1(1+1)/2. We check that 1 is indeed equal to 1(1+1)/2, which it is. At this point we have succeeded in performing action (1) above.

Now, we attempt (2). First, we <u>assume</u> <u>S(k)</u> <u>holds</u> for some arbitrary k. This assumption is our Inductive Hypothesis.

That is, we <u>assume</u>: 1+2+3+...+k = k(k+1)/2.

To finish (2) it is necessary

TO PROVE: 1+2+3+...+k+[k+1] = [k+1]([k+1]+1)/2.

Since the sum of the first k terms equals k(k+1)/2 by our Inductive Hypothesis (i.e., our assumption), this means that we have

TO PROVE: k(k+1)/2 + [k+1] = [k+1]([k+1]+1)/2.

(On the left side of the above equation we substituted 'k(k+1)/2' for '1+2+3+...+k' because their values are equal according to our Inductive Hypothesis.)

At this point, proving that k(k+1)/2 + [k+1] = [k+1]([k+1]+1)/2 is a simple matter of algebra. The only thing to make sure of is that we manipulate each side of the equation independently. (Otherwise, we may inadvertently assume the truth of what we want to prove.) So (manipulating the left side according to correct principles of algebra), we get:

k(k+1)/2 + [k+1] = k(k+1)/2 + 2[k+1]/2
= (k(k+1) + 2(k+1))/2 = (k+2)(k+1)/2.

Now, by correct manipulation of the right side of the equation to be proved, we get:

(k+1)([k+1]+1)/2 = (k+1)(k+2)/2 = (k+2)(k+1)/2.

Since the left side is now exactly the same as the right side, the proof is concluded. We have proved by mathematical induction that for any positive integer n,

$$1 + 2 + 3 + \ldots + n = n(n+1)/2.$$ ∎

CHAPTER 1

THE SYNTAX OF SENTENTIAL LOGIC

TOPICS: *the language of sentential logic, its grammar, sentences, atomic sentences, molecular sentences, negation, double negation, disjunction, conjunction, conditional, antecedent, consequent, biconditional, contradictory, unique readability, main connective, rules of derivation:* A, CD, MP, CONTRA, INT, *contradiction, theorem, derivation, derivational strategy (using* A, MP, CD*),* CONTRA *strategy, contrapositives, derivations using* INT, *strategy using* A, CD, MP, CONTRA *and* INT.

In this chapter we are going to make logical inferences using an artificial language called "the language of sentential logic". The word 'sentential' just means 'whole sentences', so the language of sentential logic is really the language of whole sentences. This language has many symbols in it, and our first task will be to explain which symbols placed next to each other make up sentences, just as in English words and punctuation marks can be placed together to form sentences. For example, one rule of English grammar is that a declarative sentence begins with a capital letter and ends with a period. But, there are many more grammatical rules in English, some of them very complicated, and there are also many exceptions to the rules. By contrast, the rules of grammar for sentential logic are very simple, and there are no exceptions to the rules.
After the rules for forming sentences have been explained, various rules of inference for sentential logic will be explained. For example, in English, if you say, "If it rains today, I will take an umbrella to school with me", and it <u>does</u> rain, then if you stick to your word you will take an umbrella with you. Someone else who hears you say what you'll do if it rains and then sees that it is raining, can correctly <u>infer</u> (provided you stick to your word) that you'll carry an umbrella with you. In sentential logic there is a very similar inference that we will describe after

giving a precise description of the language. First, let us describe the language of sentential logic:

1.1 THE LANGUAGE OF SENTENTIAL LOGIC

This entire chapter and the next one are about the syntax of an artificial language called sentential logic. The **syntax** (of a language) concerns the ways symbols are put together to form defined units of that language. The most important defined unit of sentential logic is "The Sentence". A sentence, as you'll see below, consists of certain recognizable arrangements of symbols. Another syntactic unit of sentential logic is a derivation. A **derivation** consists of consecutively numbered lines on which sentences and numbers have been written in accordance with certain rules of inference. Both a sentence and a derivation can be easily recognized by visual inspection and by following certain simple mechanical rules.

In order to say what a sentence of sentential logic is, we first specify the symbols (or the alphabet) of that language. Then we define certain strings of those symbols as being sentences. Afterwards, we explain the rules of inference that determine how sentences can be used in derivations.

The **Language of Sentential Logic (L_{SL})** consists of the following **sentence letters** (considered to be non-logical symbols):

$$P, Q, R, S, T, P_1, Q_1, R_1, S_1, T_1, P_2, \ldots$$

together with the following logical symbols:

$$(\ )\ \rightarrow,\ \neg,\ \wedge,\ \vee,\ \leftrightarrow.$$

We now provide the grammar for this artificial language, L_{SL}.

1.2 GRAMMAR

For this simple language, the only grammatical unit we care about is: the sentence.

A **sentence of** L_{SL} is one of the following:

(i) A sentence letter,
(ii) If ψ is a sentence, then $\neg\psi$ is a sentence,
(iii) If ψ and χ are both sentences, then the following are all sentences:
        (a) $(\psi \rightarrow \chi)$,
        (b) $(\psi \wedge \chi)$,
        (c) $(\psi \vee \chi)$,
        (d) $(\psi \leftrightarrow \chi)$.

Nothing else is a sentence of L_{SL} except by clauses (i)-(iii). '$\rightarrow$', '$\neg$', '$\wedge$', '$\vee$', '$\leftrightarrow$' are **connectives**.

Here are some examples of sentences of L_{SL}:

$$P, Q_{1000}, \neg P, (Q_{30} \rightarrow P_9), \neg(T \leftrightarrow (\neg Q \rightarrow R)).$$

The first two items in the above list are sentences because they are sentence letters. '$\neg P$' is a sentence because (ii) says that the result of writing '$\neg$' followed by a sentence is a sentence. Since 'Q_{30}' and 'P_9' are sentences, so is '$(Q_{30} \rightarrow P_9)$' by (iii)(a).

The last item on the above list is a sentence, as the following argument shows:

(1) 'Q' is a sentence letter, so '$\neg Q$' is a sentence by (ii).
(2) 'R' is a sentence because it is a sentence letter.
(3) Thus, by (iii)(a) '$(\neg Q \rightarrow R)$' is a sentence.
(4) 'T' is a sentence because it is a sentence letter.
(5) By (iii)(d) '$(T \leftrightarrow (\neg Q \rightarrow R))$' is a sentence. And,
(6) '$\neg(T \leftrightarrow (\neg Q \rightarrow R))$' is a sentence by (ii).

Exercises:

Show that these are sentences of L_{SL}:
1. $((P \wedge \neg Q) \vee S)$
2. $(\neg P \rightarrow \neg\neg Q)$

Show these are <u>not</u> sentences of L_{SL}:
3. $(\neg (\neg P) \rightarrow Q)$
4. $(P \rightarrow Q \rightarrow R)$

Sentences of L_{SL} are either **atomic** or **molecular**. Sentence letters are all atomic sentences; all others are molecular.

Given any sentence φ, the sentence gotten by placing the symbol '¬' in front of φ, i.e., ¬φ, is called: **the negation of φ**. So, ¬¬φ is the **double negation of** φ and the negation of ¬φ. Question: Is φ the negation of ¬φ? Answer: no, ¬φ is the negation of φ, but φ is <u>not</u> the negation of ¬φ. (¬¬φ is the negation of ¬φ). ('¬' corresponds roughly to "not" in English.)[1]

A sentence of the form ($\varphi \vee \psi$) is called a **disjunction**, and φ and ψ are its **disjuncts**. ('$\vee$' means something like "or" in English.)

($\varphi \wedge \psi$) is called a **conjunction**, and φ and ψ are its **conjuncts**. (The meaning of '$\wedge$' is something like "and" in English.)

($\varphi \rightarrow \psi$) is a **conditional**. φ is the **antecedent** and ψ the **consequent** of the conditional. (The rough meaning of '$\rightarrow$' is "If ..., Then _ _ _ ".)

($\varphi \leftrightarrow \psi$) is a **biconditional**. ('$\leftrightarrow$' corresponds roughly to "just in case" or "if and only if" in English, though you do not often hear or see the phrase "if and only if" outside of mathematical contexts.) There is no widely accepted term to describe φ and ψ in the sentence ($\varphi \leftrightarrow \psi$), so we will use phrases like '**the left sentence**' (of the biconditional) and '**the sentence on the right**' (of the biconditional).

A **contradictory** of a sentence φ is gotten by either adding or removing a negation symbol from the front of φ. For example, if φ is the sentence '¬P', then one contradictory of '¬P' is '¬¬P' and another contradictory of '¬P' is 'P'. Then '¬P' and '¬¬P' are **contradictories of each other**, and '¬P' and 'P' are also **contradictories** of each other. We will often write '**CONTRA(φ)**' as short for 'a contradictory of φ'. Notice that sentences that are not negations have only one contradictory, whereas negations have two contradictories.

Exercises:

1. If φ is the sentence '(P $\rightarrow$ Q)' and ψ is the sentence '(Q $\vee$ R)', then what sentences are:
    (a) ($\varphi \rightarrow \psi$),

---

[1] The informal remarks in parentheses are added only to suggest connections with phrases of English and are not intended as definitions of the "meaning" of these symbols.

(b) $(\neg\varphi \land \psi)$, and
   (c) $\neg\neg(\psi \rightarrow \neg\varphi)$?
2. List the contradictories of '$\neg(P \lor Q)$'.
3. List the contradictories of '$(P \lor Q)$'.
4. Is '$(P \lor Q)$' a contradictory of '$\neg\neg(P \lor Q)$'?
5. Is '$\neg\neg(P \lor Q)$' a contradictory of '$(P \lor Q)$'?
6. If φ is the sentence '$\neg P$', what two sentences can CONTRA(φ) be?
7. Is it true that if two sentences are contradictories of each other that one can be written as φ and the other $\neg\varphi$?

Unique Readability

Notice that no sentence of L_{SL} can be read "in more than one way". That is, a sentence is *one* of the following: a sentence letter, a negation, a disjunction, a conjunction, a conditional, or a biconditional. No sentence of L_{SL} could be, say, a disjunction *and* a conjunction because parentheses eliminate any potential for ambiguity.

Suppose we did *not* use parentheses in sentences. Then, we could have the following ambiguous sentence:

$P \lor Q \land R$ Ambiguous

'$P \lor Q \land R$' can be considered to be a conjunction if we read '$P \lor Q \land R$' as '$((P \lor Q) \land R)$'; and it can be considered to be a disjunction if we read '$P \lor Q \land R$' as '$(P \lor (Q \land R))$'.

The sentences of L_{SL}, however, are **uniquely readable**; there is never more than one way to read a given sentence. This is a fact that can be proved formally (by Complete Induction on the number of connectives in a sentence). (See the exercises at the end of this chapter.)

Since sentences of L_{SL} are uniquely readable, there is always a **main connective** for a molecular sentence. The **main connective** of a molecular sentence is the one that was added last in the formation of the sentence. The main connective of '$\neg P$' is '$\neg$'. The main connective of '$((P \lor Q) \land R)$' is '$\land$', while the main connective of '$(P \lor (Q \land R))$' is '$\lor$'.

Exercises:

Which occurrence of '$\rightarrow$' is the main connective in the following sentences:
   1. $((P \rightarrow \neg P) \rightarrow \neg P)$,

2. $((P \rightarrow Q) \rightarrow (\neg Q \rightarrow \neg P))$,
3. $(P \rightarrow ((P \rightarrow Q) \rightarrow (Q \rightarrow R)))$.

NOTE: We will frequently omit the external parentheses of sentences. (For example, we will write 'P ∧ Q' for '(P ∧ Q)'.) But the "official" sentence is still '(P ∧ Q)'. This will not create any ambiguity. For example, '(P ∨ Q) ∧ R' is still clearly distinguishable from 'P ∨ (Q ∧ R)'. We will never omit any internal parentheses.

1.3 RULES OF DERIVATION FOR L_{SL}

A **derivation in** L_{SL} consists of consecutively numbered lines (beginning with the number 1) on each of which (1) there is a sentence of L_{SL} and (2) to the right of each sentence there are some numbers[2] (or, no numbers at all in certain special cases) such that the sentences and the numbers conform to the following rules of inference:

(1) **Assumption (A)**. Any sentence may be placed on a line with the number of that line written to the right. The sentence on the line is hereby called **an assumption** and the number written to the right (which is the same number as the line on which the sentence is placed) is called **its assumption number**.

(2) **Conditional Derivation (CD)**. If a sentence ψ is on a line, then $\varphi \rightarrow \psi$ may be placed on a later line, where φ is any sentence whatsoever. As assumption numbers of $\varphi \rightarrow \psi$, take all the assumption numbers of ψ, but remove the assumption number of φ if φ was an assumption.

(3) **Modus Ponens (MP)**. If sentences φ and $\varphi \rightarrow \psi$ are on two lines, then ψ may be placed on a later line with the assumption numbers of φ and $\varphi \rightarrow \psi$ taken together.

(4) **Contradiction (CONTRA)**. If sentences ψ and $\neg\psi$ are on two lines, then any sentence φ may be placed on a later line taking the assumption numbers of both ψ and $\neg\psi$, but removing the assumption number of CONTRA(φ) if CONTRA(φ) was an assumption.

---

[2]Strictly speaking, these are numerals, not numbers.

(5) **Interchange (INT)**. If a sentence φ is one of the pairs listed under (a), (b), or (c) below, and φ is on a line, the other sentence of the pair may be placed on a later line with the numbers on the right of φ written to the right of the new line.

(a) $(\psi \lor \chi)$ $\qquad$ $(\neg\psi \rightarrow \chi)$
(b) $(\psi \land \chi)$ $\qquad$ $\neg(\psi \rightarrow \neg\chi)$
(c) $(\psi \leftrightarrow \chi)$ $\qquad$ $(\psi \rightarrow \chi) \land (\chi \rightarrow \psi)$

We now offer several examples of derivations:

Example 1:

1. P $\qquad$ 1, A
2. $P \rightarrow Q$ $\qquad$ 2, A
3. Q $\qquad$ 1, 2, MP

(Notice that the letters abbreviating each rule were placed to the right even though they are not required by the rules.)

Where a rule mentions a sentence being "on" a line, it is <u>not</u> meant to include the case where that sentence is a proper part of another sentence. For example, in example 1 'P' <u>is</u> "on" line 1, but 'P' is <u>not</u> "on" line 2.

Example 2: (An extension of example 1)

1. P $\qquad$ 1, A
2. $P \rightarrow Q$ $\qquad$ 2, A
3. Q $\qquad$ 1, 2 MP
4. $(P \rightarrow Q) \rightarrow Q$ $\qquad$ 1, CD
5. $P \rightarrow ((P \rightarrow Q) \rightarrow Q)$ $\qquad$ CD

Notice that it is fairly easy to check that each line of example 2 above was written in accordance with the rules of derivation. Line 1 for example is:

1. P $\qquad$ 1, A.

'P' was placed on line 1 in accordance with rule A, and number 1 (really: numeral '1') was written to the right of sentence 'P'. We also, wrote 'A' as a reminder that the Assumption rule was used. Line 2 uses the Assumption rule also.

Line 3 in the above derivation is an example of the use of MP. 'Q' was put on line 3 in accordance with this rule. To check this, we can see that 'P' was on line 1, and 'P → Q' was on line 2. Both of these lines are earlier than line 3. So, we can put 'Q' on line 3 with the assumption numbers of 'P' and the assumption numbers of 'P → Q' to the right of line 3.

We can think of line 4, which uses the rule CD, as taking place in two steps. First, we took sentence 'Q' from line 3 and made it the consequent of the conditional whose antecedent is 'P → Q'. If you look back at the statement of the rule CD, you will see that ψ is the sentence 'Q' and φ is 'P → Q' in this case. The second part of this rule is that the assumption numbers of this conditional, '(P → Q) → Q', are to be the assumption numbers of 'Q' with the assumption number of '(P → Q)' removed <u>if</u> 'P → Q' <u>was an assumption</u>. Since '(P → Q)' <u>was</u> an assumption on line 2, we remove 2 from the assumption numbers, and only the number 1 is left. So, that's why line 4 looks like this:

  4. (P → Q) → Q   1,CD.

Line 5 is arrived at by CD also. The new conditional on line 5, 'P → ((P → Q) → Q)', has 'P' as its antecedent, which is the assumption on line 1. So, since 'P' was <u>assumed</u>, we remove the number 1 from the assumption numbers of line 5. This leaves us with no assumption numbers whatsoever on line 5.

Informally, the "idea" behind the CD rule is similar to a principle we used in doing proofs in high-school geometry. In geometry, we normally proved hypothetical statements, such as 'If two angles of a triangle are equal, then the triangle is isosceles' by first assuming the hypothesis. In this case, the hypothesis to be assumed is 'two angles of a triangle are equal'. After the hypothesis has been assumed, the task is to prove that the conclusion is true, that the triangle is isosceles. Once the conclusion has been proved true, then we are permitted to write 'QED', signifying that the proof has been completed.

So, the first line of our geometry proof would have begun with the hypothesis, 'Two angles of a triangle are equal' and the last line would have held the conclusion: The triangle is isosceles. If we had wanted to make the underlying logic more explicit, we could have added one more line to the proof, consisting of the whole sentence: If two angles of a triangle are equal, then the triangle is isosceles. But, in geometry class, that additional line was unnecessary.

In a sentential logic derivation, however, the derived sentence must <u>itself</u> appear on the last line. Therefore, the CD rule is used to rebuild the original sentence. We first assume the antecedent of a conditional we wish to derive, and then we try to SHOW the consequent. Once the consequent has been SHOWn, we write an additional line consisting of the entire conditional sentence. The CD rule permits us to do this.

A Reminder: The CD rule permits us to remove <u>only</u> the (single) assumption number of an assumed sentence. Removing more than one number is <u>incorrect</u>, and it is also <u>incorrect</u> to remove the assumption number of an antecedent that is <u>not</u> an assumption.

The sequence of lines from 1 to 5 in the example above is called "a derivation of 'P $\rightarrow$ ((P $\rightarrow$ Q) $\rightarrow$ Q)'", since 'P $\rightarrow$ ((P $\rightarrow$ Q) $\rightarrow$ Q)' is on line 5, and since each line was obtained by following the rules specified above. In addition, since 'P $\rightarrow$ ((P $\rightarrow$ Q) $\rightarrow$ Q)' has no assumption numbers, it is called a **theorem**. More generally, a **derivation of** (a sentence) φ is a derivation with φ on the last line of the derivation. If there is a derivation of φ with no number to the right of φ, then φ is a **theorem** (of sentential logic). So, we now have our first theorem:

T1:   P $\rightarrow$ ((P $\rightarrow$ Q) $\rightarrow$ Q)

Notice that we can *check* that 'P $\rightarrow$ ((P $\rightarrow$ Q) $\rightarrow$ Q)' is a theorem by assuring ourselves that the 5 lines listed above constitute a derivation of 'P $\rightarrow$ ((P $\rightarrow$ Q) $\rightarrow$ Q)' with no numbers to the right. But, *checking* a derivation is easier than actually *creating* one. How do we *construct* a derivation?

1.4 DERIVATIONAL STRATEGY (using A, MP, CD)

To construct a derivation requires a plan or strategy. The strategy followed to derive 'P $\rightarrow$ ((P $\rightarrow$ Q) $\rightarrow$ Q)' was fairly straightforward. Loosely speaking, we just assumed "if"s and then got our "then". Remember that an "if" of a conditional sentence is the <u>antecedent</u> and the "then" is the <u>consequent</u>. The "if" of 'P $\rightarrow$ ((P $\rightarrow$ Q) $\rightarrow$ Q)' is 'P'. So, we assumed it. Then, we wished to show the "then" of that conditional, which is '(P $\rightarrow$ Q) $\rightarrow$ Q)'. Since that sentence is <u>itself</u> a conditional, we assume <u>its</u> "if" (i.e., its antecedent), namely 'P → Q'. <u>Its</u> "then" (i.e., its consequent) is 'Q', which is the final "then" we wish to show. Armed with our two assumptions, 'P' and 'P → Q', we get our final "then" by Modus Ponens (MP). We then use Conditional Derivation (CD) to eliminate the numbers to the right and to build 'Q' first to '(P → Q) → Q' and then to 'P → ((P → Q) → Q)', and we are done.

The strategy, then, to derive a conditional sentence, $\varphi \to \psi$, is to assume the antecedent, φ, and try to derive the consequent, ψ.

We can reflect our strategy by writing **SHOW lines** indicating what we wish to derive. If what we wish to "SHOW" is a conditional, we will normally assume the antecedent of that conditional and try to SHOW the consequent.[3]

```
        SHOW P → ((P → Q) → Q)
  1.  P                                        1,A
                   SHOW (P → Q) → Q
  2.  P → Q                                    2,A
                   SHOW Q
  3.  Q                                        1,2,MP
```

As soon as we have 'Q' on a line, as we do on line 3, we put a check mark ('√') in front of the SHOW line that says 'SHOW Q', and we try to SHOW the sentence in the next SHOW line above '√SHOW Q'. Our derivation now looks like this:

```
        SHOW P → ((P → Q) → Q)
  1.  P                                        1,A
                   SHOW (P → Q) → Q
  2.  P → Q                                    2,A
                   √SHOW Q
  3.  Q                                        1,2,MP
```

Now, when we write '(P → Q) → Q' on line 4 we put a check mark in front of the SHOW line 'SHOW (P → Q) → Q'. The derivation now looks like this:

---

[3]The idea of using SHOW lines originates with Kalish & Montague [1964], as well as some of the techniques for using them.

```
        SHOW P → ((P → Q) → Q)
   1.   P                                              1,A
                                        √SHOW (P → Q) → Q
   2.   P → Q                                          2,A
                                        √SHOW Q
   3.   Q                                              1,2,MP
   4.   (P → Q) → Q                                    1,CD
```

All that is left to do now is put the final "SHOW sentence", 'P → ((P → Q) → Q)' on line 5 and check off the top SHOW line. The final derivation, incorporating our SHOW strategy, looks like this:

```
   √SHOW P → ((P → Q) → Q)
   1.   P                                              1,A
                                        √SHOW (P → Q) → Q
   2.   P → Q                                          2,A
                                        √SHOW Q
   3.   Q                                              1,2,MP
   4.   (P → Q) → Q                                    1,CD
   5.   P → ((P → Q) → Q)                              CD
```

Notice that the last three lines of the derivation are the three SHOW lines from <u>bottom</u> to top.

Exercises:

Provide derivations of T2, T3, and T4, incorporating the SHOW lines as demonstrated above. <u>All</u> theorems in this chapter that do not have derivations provided should be derived as *exercises*, even when they are not explicitly listed as such.

T2:   (P → Q) → ((Q → R) → (P → R))
T3:   (Q → R) → ((P → Q) → (P → R))
T4:   (P → (Q → R)) → ((P → Q) → (P → R))
T5:   P → P

```
        √SHOW P → P
   1.   P                     1,A
                              SHOW P
   2.   P → P                 CD
```

T6:   P → (Q → P)

It is tempting to use CD to remove an assumption number even when the antecedent is <u>not</u> an assumption. That use is **incorrect**. Remember: CD can be used to remove an assumption number <u>only</u> <u>when</u> the antecedent of the conditional obtained is <u>itself</u> <u>an</u> <u>assumption</u>. And then, <u>only</u> that <u>one</u> assumption number can be removed. In the next chapter we will show what can happen when CD is used INCORRECTLY.

1.5 THE CONTRA RULE

The above examples illustrate the use of the rules Assumption, Modus Ponens, and Conditional Derivation. The following derivation of T7 illustrates the use of the CONTRA rule.

When φ is on one line of a derivation and $\neg\varphi$ is on another line of that derivation, we shall say a **contradiction** has been derived or shown (in that derivation).

T7 $(\neg P \rightarrow \neg Q) \rightarrow (Q \rightarrow P)$

$\checkmark$SHOW $(\neg P \rightarrow \neg Q) \rightarrow (Q \rightarrow P)$
1. $\neg P \rightarrow \neg Q$ 1,A
      $\checkmark$SHOW $Q \rightarrow P$
2. Q 2,A
      $\checkmark$SHOW P
3. $\neg P$ 3,A
      $\checkmark$SHOW A CONTRADICTION
4. $\neg Q$ 1,3, MP
5. P 1,2 CONTRA
6. $Q \rightarrow P$ 1, CD
7. $(\neg P \rightarrow \neg Q) \rightarrow (Q \rightarrow P)$ CD

ANALYSIS: Our strategy through lines 1 and 2 above was the same as before: to derive a conditional, assume its antecedent and try to show its consequent. If the consequent is <u>itself</u> a conditional, assume <u>its</u> antecedent and try to show <u>its</u> consequent. But after line 2, we embark on a new strategy. We assume a contradictory of what we wish to show and then try to show a contradiction.

After writing 'SHOW P' below and to the right of line 2, we assume the contradictory of 'P', '¬P', on line 3. The next SHOW line reminds us that we wish to show a contradiction. We then see that '¬P' and '¬P → ¬Q' yield '¬Q' by MP, so we write that result on line 4. We now have our contradiction, which is what we said we wanted in our last SHOW line, so we are essentially finished with this derivation.

But, line 5 is a bit trickier to follow than the earlier lines, so let us go through it carefully. Suppose we have gotten down to line 4 and realize lines 2 and lines 4 are contradictory. By rule CONTRA we put 'P' on line 5, take all the assumption numbers of line 2 and 4 (they are: 1, 2, 3), but we remove the number 3 because 3 is the assumption number of '¬P', which is the contradictory of 'P'. ('lines 2 and lines 4' is a convenient way to refer to the <u>sentences</u> on these lines.)

At this point we wish to make sure that it is absolutely clear that CONTRA was used correctly on line 5 of the above derivation. So, we are going to go over it again: lines 2 and 4 are contradictory, since the sentences on those lines, 'Q' and '¬Q' are contradictories of each other. The CONTRA Rule tells us we can "put any sentence φ on a later line". So, on Line 5, we put 'P'. After substituting 'P' for φ and 'Q' for ψ in the statement of the CONTRA Rule, the rule says that the assumption numbers of the new line are the "assumption numbers of both 'Q' and '¬Q', *minus the assumption number of* '¬P' *if* '¬P' *was an assumption*". The assumption number of 'Q' is 2, and the assumption numbers of '¬Q' are 1 and 3. Taken together, we have the assumption numbers 1, 2, and 3. But, *we can remove the assumption number (singular,* **not** *"numbers") of* '¬P' *if* '¬P' *was an assumption.* '¬P' <u>was</u> assumed on line 3. Thus, its assumption number is 3. Hence, we remove 3 from the list of assumption numbers: 1, 2, 3. Then Line 5 has assumption numbers 1 and 2. Lines 6 and 7, then, are the two remaining SHOW lines *from bottom to top*.

This is the technical analysis of the use of CONTRA. The idea behind CONTRA is quite intuitive. Loosely speaking, the idea is this: we wish to derive something, so we assume one of its contradictories. That assumption leads to a contradiction. Therefore, what we wished to derive must be true.

This principle is used in mathematics proofs all the time. A standard proof that $\sqrt{2}$ is irrational, for example, begins with the assumption that $\sqrt{2}$ is <u>not</u> irrational. Then the proof goes on, and fairly soon a contradiction is reached. The contradiction establishes that the assumption that $\sqrt{2}$ is <u>not</u> irrational was mistaken, and thus, $\sqrt{2}$ *is* irrational.

1.6 CONTRA STRATEGY

Our strategy in using CONTRA will be the same as in mathematics: we assume a contradictory of what we wish to derive, get a contradiction, and conclude (without the assumption) that what we wished to derive must be true.

We then add this strategy to our previous strategy, which was to assume the antecedent of a conditional that we wish to derive. Our new strategy is: when you cannot see how to derive what you wish, assume one of its contradictories and try to get a contradiction.

At this point, you should try your hand at using this CONTRA strategy by deriving the following theorems:

T8: $(P \rightarrow Q) \rightarrow (\neg Q \rightarrow \neg P)$
T9: $(P \rightarrow \neg Q) \rightarrow (Q \rightarrow \neg P)$
T10: $(\neg P \rightarrow Q) \rightarrow (\neg Q \rightarrow P)$

The consequents of the four theorems of this form are the **contrapositives** of their antecedents. To get the **contrapositive** of $(\varphi \rightarrow \psi)$, first switch φ and ψ, and if φ or ψ has no negation symbol in front, put one in; if φ or ψ has at least one negation symbol in front, remove one of them. Thus, the contrapositive of $(\varphi \rightarrow \psi)$ is the sentence of the form $(CONTRA(\psi) \rightarrow CONTRA(\varphi))$ with the least number of negation symbols.

Here are a few more theorems, some of whose derivations are provided (the rest should be done as *exercises*):

T11: $\neg\neg P \rightarrow P$

```
    √SHOW ¬¬P → P
    1.   ¬¬P                              1,A
              √SHOW P                     2,A
    2.   ¬P
              √SHOW A CONTRADICTION
    3.   P                                1, CONTRA
    4.   ¬¬P → P                          CD
```

ANALYSIS: In the above derivation of T11, since '¬¬P' and '¬P' are contradictories, any sentence can be put on line 3 by CONTRA. Since '¬P' is a contradictory of 'P', the sentence on line 3, the assumption number of '¬P' can be removed. Thus we have only 1 as assumption number of line 3.

T12:   P → ¬¬P

The following two theorems exemplify the idea that from a contradiction you can get anything. In this case getting 'P' on one line and '¬P' on another constitutes our contradiction. Our "anything" sentence is 'Q'.

T13:   P → (¬P → Q)

       √SHOW P → (¬P → Q)
1. P   1,A
                  √SHOW ¬P → Q
   2. ¬P   2,A
                  √SHOW Q
   3. ¬Q   3,A
                  √SHOW A CONTRADICTION
   4. Q   1,2 CONTRA
   5. ¬P → Q   1,CD
   6. P → (¬P → Q)   CD

The idea after writing 'SHOW Q' was to assume the contradictory of 'Q', which is '¬Q', and try to get a contradiction. Then, using CONTRA, we can show Q. As it turns out, though, line 3 did not really help us. So, lines 1,2,4,5, and 6 (skipping 3) constitute a "more elegant" derivation.

T14:   ¬P → (P → Q)
T15:   (¬P → P) → P
T16:   (P → ¬P) → ¬P
T17:   ¬(P → Q) → P
T18:   ¬(P → Q) → ¬Q

The general idea behind writing SHOW lines among the lines of a derivation is to provide direction for the derivation. More specifically what we are doing when we write a SHOW line is expressing our confidence that _if_ we were able to obtain that line without making any new assumptions, we would be able to finish the derivation. Thus, 'SHOW φ' abbreviates 'SHOW φ from the assumptions made so far'. Assuming 'SHOW φ' appears immediately below line k of a derivation, 'SHOW φ' is short for 'SHOW φ from assumptions made in lines through k only'.
    Let us look again at the derivation of T13. At line 1 in that derivation the latest SHOW line — the only SHOW line in the derivation so far — is the statement to show T13

itself, namely: SHOW P → (¬P → Q). When we reach line 3, the derivation looks like this:

SHOW P → (¬P → Q)
1.   P   1,A
                                           SHOW ¬P → Q
2.   ¬P
                                           SHOW Q
3.   ¬Q

At this stage in the above derivation (line 3) the latest SHOW line is 'SHOW Q' and the latest SHOW sentence is 'Q'. The next latest SHOW sentence is '¬P → Q', and the last SHOW sentence is the theorem itself.

It is important in constructing a derivation to always try to show the latest SHOW line and then show the next latest and then the next next latest, and so on. Consider the use of SHOW lines in the following derivation of T19.

T19:   ((P → Q) → P) → P   (Peirce's Law)

√SHOW ((P → Q) → P) → P
1.   (P → Q) → P   1,A
                                √SHOW P
2.   ¬P   2,A
                                √SHOW A CONTRADICTION
                                √SHOW P → Q
3.   P   3,A
                                √SHOW Q
4.   Q   2,3,CONTRA
5.   P → Q   2,CD(3,4)
6.   P   1,2,MP(1,5)
7.   P   1,CONTRA(2,2,6)
8.   ((P → Q) → P) → P   CD(1,7)

ANALYSIS: We write 'SHOW ((P → Q) → P) → P' at the very top, meaning that <u>if</u> we could show '((P → Q) → P) → P' from <u>no</u> assumptions, we would be finished. Let us move to the next SHOW line.

The second SHOW line is 'SHOW P'. Since we have assumed only '(P → Q) → P', we know that <u>if</u> we can obtain 'P', we can discharge the assumption '(P → Q) → P' by CD and get the entire theorem.

When we write our third SHOW line, 'SHOW A CONTRADICTION', we are at the stage in the derivation where '(P → Q) → P' has been assumed and so has '¬P'. We are thinking along these lines: "<u>if</u> we get a contradiction from these two assumptions alone, we can write 'P', the sentence of the SHOW line above this one, and discharge the extra assumption, '¬P', by CONTRA. Then we get the entire theorem by CD."

We look at what sentences we have so far after writing 'SHOW A CONTRADICTION'. We have only '(P → Q) → P' and '¬P'. But we see that <u>if</u> we could get '(P → Q)' without assuming anything else, we could get that SHOW line (i.e., we could SHOW A CONTRADICTION) by getting 'P'. What's more, we realize we <u>can</u> get 'P → Q' from '¬P' alone, because we remember T14, '¬P → (P → Q)'. This insight motivates our writing 'SHOW P → Q'.

After writing 'SHOW P → Q', we naturally assume 'P'. We then write our last SHOW line, 'SHOW Q', knowing that <u>if</u> we get 'Q' we can get 'P → Q' by CD, which will give us our contradiction, which will give us 'P', which will give us the entire theorem, as reasoned above.

We look at our last SHOW line, 'SHOW Q', and we look at our assumptions so far and realize we <u>can</u> get 'Q', by CONTRA, from the fact that we have assumed contradictories, namely 'P' and '¬P'.

So, we get 'Q', check off the last SHOW line and work our way up the SHOW lines, checking them off and writing each SHOW line. Specifically, we write line 5, which is the SHOW line above 'SHOW Q'. (Notice that we have added line numbers in parentheses to show which previous lines were used.) Then we check off 'SHOW P → Q'. We then look to the first unchecked SHOW line above this one, which is 'SHOW A CONTRADICTION'. Since our motivation in getting 'P → Q' was to use it together with '(P → Q) → P' by MP to get 'P', which contradicts '¬P', we get our contradictory sentence on line 6.

We then check off 'SHOW A CONTRADICTION', since lines 6 and 2 give it to us, and we then look at the unchecked SHOW line above this one. It says 'SHOW P', which our contradiction enables us to get, by CONTRA, so we write 'P' on line 7 and check off 'SHOW P'.

At this point, we use CD to discharge assumption 1, we check off the final SHOW line and we are done.

Let us emphasize a point about our SHOW strategy in the previous derivation of Peirce's Law. Suppose after having written line 3 of the derivation we notice that lines 2 and 3 contradict each other and that an unchecked SHOW line above line 3 is: SHOW A CONTRADICTION. We may be tempted to check off that SHOW line and skip over the latest SHOW line immediately above line 3, 'SHOW P → Q'. That would be a mistake. Remember: 'SHOW A CONTRADICTION' is really an abbreviation of 'SHOW A CONTRADICTION from assumptions made in lines 1 and 2 only', since that SHOW line appears immediately after line 2. Thus, it is incorrect to check off 'SHOW A CONTRADICTION' by using the additional assumption of line 3.

Look at what happens when we check off 'SHOW A CONTRADICTION' in the derivation of Peirce's Law after adding the assumption on line 3:

```
       SHOW ((P → Q) → P) → P
   1.   (P → Q) → P                              1,A
                            X      √SHOW P                    X
   2.   ¬P                                       2,A
                            X      √SHOW A CONTRADICTION      X
                                   SHOW P → Q
   3.   P                                        3,A
   4.   P                                        3,CONTRA
   5.   ((P → Q) → P) → P                        3,CD
```

Once we use assumption 3 to SHOW A CONTRADICTION and erroneously check off that line, we are stuck with that assumption. We can make the same mistake again, writing 'P' on line 4 and checking off 'SHOW P', but line 4 gives us no more information than we already had on line 3, where 'P' was *assumed*. Thus, after we perform CD on line 5, we are still stuck with assumption 3. The moral is to show the latest SHOW sentence first. Otherwise, it may not be possible to check off the SHOW lines above the latest one in a step-by-step way and finish the derivation.

In the derivation of T19 above, we put line numbers in parentheses to indicate the previous lines that were used to get the current line. For example, line 6 is:

   6.   P   1,2,MP(1,5).

The assumption numbers 1,2, indicate that the assumptions of line 1 and 2 are needed to get 'P'. The numbers, 1,5, in parentheses refer to the previous **lines** that were used in performing Modus Ponens. Thus, the numbers in parentheses are **line numbers**, whereas the other numbers are assumption numbers. Line numbers (in parentheses after the rule) clarify which previous lines have been used if there is any uncertainty.

1.7 DERIVATIONS OF SENTENCES CONTAINING '∨', '∧', and '↔'

Now we wish to obtain derivations of sentences with the symbols '∨', '∧', and '↔'. So far, every sentence in a derivation contained only the connectives '¬' and '→'.

The simplest derivation involving the '∨' symbol requires the INT(∨) Rule:

T20:   (¬P → Q) → (P ∨ Q)

    √SHOW (¬P → Q) → (P ∨ Q)
    1. ¬P → Q   1,A
                                    √SHOW (P ∨ Q)
    2. P ∨ Q   1,INT(∨)
    3. (¬P → Q) → (P ∨ Q)   CD

There are analogous derivations using '∧' and '↔'.

T21:   ¬(P → ¬Q) → (P ∧ Q)
T22:   ((P → Q) ∧ (Q → P)) → (P ↔ Q)

    √SHOW ((P → Q) ∧ (Q → P)) → (P ↔ Q)
    1. (P → Q) ∧ (Q → P)   1,A
                                    √SHOW (P ↔ Q)
    2. P ↔ Q   1,INT
    3. (P → Q) ∧ (Q → P) → (P ↔ Q)   CD

Some other theorems containing the '∧', '∨', and '↔' are:

T23:   P → (Q → (P ∧ Q))

    √SHOW P → (Q → (P ∧ Q))
    1. P   1,A
                                    √SHOW Q → (P ∧ Q)
    2. Q   2,A
                                    √SHOW P ∧ Q
                                    √SHOW ¬(P → ¬Q)
    3. P → ¬Q   3,A
                                    √SHOW A CONTRADICTION
    4. ¬Q   1,3,MP
    5. ¬(P → ¬Q)   1,2,CONTRA

6.	P ∧ Q	1,2,INT(∧)
7.	Q → (P ∧ Q)	1,CD
8.	P → (Q → (P ∧ Q))	CD

ANALYSIS: We begin the derivation of T23 by noticing that it is a conditional. To derive a conditional, we assume the antecedent and try to derive the consequent. The antecedent 'P' is thus assumed on line 1. Now, we try to SHOW 'Q → (P ∧ Q)', the consequent of T23. To remind us of this we first write:

$$\text{SHOW} \quad Q \to (P \land Q)$$

under line 1. (Once we derive this line we put a check mark in front of the word 'SHOW'. At the moment 'SHOW' has no check in front of it). To SHOW 'Q → (P ∧ Q)' we assume 'Q', since 'Q' is the antecedent of 'Q → (P ∧ Q)'. Next, we need to derive 'P ∧ Q', so we write:

$$\text{SHOW} \quad P \land Q$$

We know that to show 'P ∧ Q' we need to use INT(∧), so we try to show the interchange of 'P ∧ Q', which is '¬(P → ¬Q)'. <u>This</u> we try to derive by assuming <u>its</u> contradictory on line 3. Then we write:

$$\text{SHOW A CONTRADICTION}$$

We get our contradiction on line 4 (together with line 2), and we check off our last SHOW line. This gives us our next-to-last SHOW line on line 5. We put a check by this SHOW line and then look at the next SHOW line above. We get this one easily on line 6 by INT(∧), which was our plan. So check the SHOW of:

$$\text{SHOW} \quad P \land Q$$

too. Next, we look above to our final SHOW line, the one at the top:

$$\text{SHOW} \quad Q \to (P \land Q)$$

We get this result on line 7 by using CD. Now, all we must do is use CD again to get the final SHOW line and the theorem.

Our strategy, as before, is to add superfluous SHOW lines to reflect the strategy followed. Then, we check each SHOW line from *bottom to top* as we arrive at each preliminary result. Thus, we are essentially finished the derivation when we can get the bottom SHOW line because we can then normally finish the derivation simply by checking off the rest of the SHOW lines while adding each SHOW sentence to the derivation.

1.8 STRATEGY FOR OBTAINING DERIVATIONS USING ALL RULES (A, CD, MP, INT, AND CONTRA)

1. When trying to show a conditional, $\varphi \rightarrow \psi$, write this:

$$\begin{array}{lll} & \text{SHOW } \varphi \rightarrow \psi & \\ n. \quad \varphi & & n, A \\ & \text{SHOW } \psi & \end{array}$$

Then try to derive ψ using MP and CD (and sometimes using CONTRA <u>without</u> having assumed a contradictory).

2. If you have written:

$$\text{SHOW } \psi$$

and you cannot proceed further using MP, CD, or CONTRA. Then continue this way:

$$\begin{array}{lll} & \text{SHOW } \psi & \\ k. \quad \text{CONTRA}(\psi) & & k, A \\ & \text{SHOW A CONTRADICTION} & \end{array}$$

3. If the negation of a conditional, $\neg(\varphi \rightarrow \psi)$, appears on a line, try to show the conditional, $\varphi \rightarrow \psi$, by employing strategy 1. — especially if the latest SHOW line is 'SHOW A CONTRADICTION'.

4. If the latest SHOW line is 'SHOW A CONTRADICTION', look to see whether there is some sentence ψ such that (1) <u>if</u> you can show ψ you will have your contradiction, <u>and</u> (2) you think (hope) you can show ψ, then continue this way:

$$\begin{array}{l} \text{SHOW A CONTRADICTION} \\ \text{SHOW } \psi \end{array}$$

5. (A generalization of 4.) If the latest SHOW line is 'SHOW φ' and you think (1) you can show φ <u>if</u> you can show ψ, <u>and</u> (2) you think you may be able to show ψ, then continue this way:

$$\text{SHOW } \varphi$$
$$\text{SHOW } \psi$$

6. To derive either $\varphi \vee \psi$, $\varphi \wedge \psi$, or $\varphi \leftrightarrow \psi$, try to SHOW the sentence that corresponds to it by INT, and then use INT to get the original sentence. For example:

$$\text{SHOW } \varphi \vee \psi$$
$$\text{SHOW } \neg\varphi \rightarrow \psi$$

Use the strategies described above to derive the following theorems:

T24: $P \rightarrow (Q \rightarrow (Q \wedge P))$
T25: $(P \wedge Q) \rightarrow P$

	√SHOW $(P \wedge Q) \rightarrow P$	
1.	$P \wedge Q$	1, A
2.	$\neg(P \rightarrow \neg Q)$	1, INT($\wedge$)
	√SHOW P	
3.	$\neg P$	3, A
	√SHOW A CONTRADICTION	
	√SHOW $P \rightarrow \neg Q$	
4.	P	4, A
	√SHOW $\neg Q$	
5.	$\neg Q$	3,4 CONTRA
6.	$P \rightarrow \neg Q$	3, CD
7.	P	1, CONTRA (2,6)
8.	$P \wedge Q \rightarrow P$	CD (1,7)

T26: $(P \wedge Q) \rightarrow Q$
T27: $P \rightarrow (P \vee Q)$
T28: $Q \rightarrow (P \vee Q)$
T29: $(P \leftrightarrow Q) \rightarrow ((P \rightarrow Q) \wedge (Q \rightarrow P))$

Theorems So Far

T1: $P \to ((P \to Q) \to Q)$
T2: $(P \to Q) \to ((Q \to R) \to (P \to R))$
T3: $(Q \to R) \to ((P \to Q) \to (P \to R))$
T4: $(P \to (Q \to R)) \to ((P \to Q) \to (P \to R))$
T5: $P \to P$
T6: $P \to (Q \to P)$
T7: $(\neg P \to \neg Q) \to (Q \to P)$
T8: $(P \to Q) \to (\neg Q \to \neg P)$
T9: $(P \to \neg Q) \to (Q \to \neg P)$
T10: $(\neg P \to Q) \to (\neg Q \to P)$
T11: $\neg\neg P \to P$
T12: $P \to \neg\neg P$
T13: $P \to (\neg P \to Q)$
T14: $\neg P \to (P \to Q)$
T15: $(\neg P \to P) \to P$
T16: $(P \to \neg P) \to \neg P$
T17: $\neg(P \to Q) \to P$
T18: $\neg(P \to Q) \to \neg Q$
T19: $((P \to Q) \to P) \to P$
T20: $(\neg P \to Q) \to (P \vee Q)$
T21: $\neg(P \to \neg Q) \to (P \wedge Q)$
T22: $((P \to Q) \wedge (Q \to P)) \to (P \leftrightarrow Q)$
T23: $P \to (Q \to (P \wedge Q))$
T24: $P \to (Q \to (Q \wedge P))$
T25: $(P \wedge Q) \to P$
T26: $(P \wedge Q) \to Q$
T27: $P \to (P \vee Q)$
T28: $Q \to (P \vee Q)$
T29: $(P \leftrightarrow Q) \to ((P \to Q) \wedge (Q \to P))$

1.9 SUMMARY OF BASIC INFERENCE RULES

A: n. φ n, A

MP:
$$\varphi \rightarrow \psi$$
$$\varphi$$
$$\therefore \psi$$

CD:
$$\frac{\psi}{\therefore \varphi \rightarrow \psi}$$

     n. φ n, A
$$\psi$$
$$\therefore \varphi \rightarrow \psi \quad [-n]$$

CONTRA:
$$\psi$$
$$\neg \psi$$
$$\therefore \varphi$$

     n. CONTRA(φ)   n, A
$$\psi$$
$$\neg \psi$$
$$\therefore \varphi \quad [-n]$$

INT:
$$\frac{\varphi \vee \psi}{\neg \varphi \rightarrow \psi}$$
$$\frac{\varphi \wedge \psi}{\neg(\varphi \rightarrow \neg \psi)}$$
$$\frac{\varphi \leftrightarrow \psi}{(\varphi \rightarrow \psi) \wedge (\psi \rightarrow \varphi)}$$

- The double lines indicate that the inference can go in both directions.

- In each case the sentence below the line (or above and below the double lines) has all the assumption numbers of the sentence(s) above the line.

- '[-n]' means to remove the number n from the assumption numbers.

Chapter Exercises:

1. A set of sentences is "inconsistent" (we cover this more formally in Chapter 4) if some sentence and its negation can be derived from that set. Being "derived from that set" means that the derivation ends with the assumption numbers of sentences in that set remaining, rather than no assumption numbers. Here's an example of two derivations:

A.
  1. P → Q   1,A
  2. P   2,A
  3. Q   1,2,MP

B.
  1. Q → ¬Q   1,A
      SHOW ¬Q
  2. Q   2,A
  3. ¬Q   1,CONTRA

The above two derivations show that the set {P → Q, P, Q → ¬Q} is inconsistent because both 'Q' and '¬Q' can be derived from sentences of that set. In derivation **A** above, the assumption numbers remaining are of sentences 'P → Q' and 'P', which are both in the given set. And, in derivation **B**, the remaining assumption is 'Q → ¬Q', which is also in that set. Show, by providing derivations, that the following sets are inconsistent:

  A) {P, ¬P}
  B) {(P ∧ Q), (Q ∧ ¬P)}
  C) {¬(P → ¬P), ¬(Q → P)}
  D) {P → (P → ¬P), ¬P → Q, Q → P}
  E) {¬(P ∨ Q), ((P ∧ ¬Q) ∨ (Q ∧ ¬P))}
  F) {¬((P → Q) ∧ (Q → P)), (P ↔ Q)}

2. Prove that any sentence of L_{SL} has exactly the same number of left parentheses ('(') as right parentheses (')'). [See the section on Complete Induction and the Least Counterexample Principle in Chapter 0, *Mathematical Preliminaries* for proofs of this fact for a simpler language, L.]

3.(a) Prove that no initial part φ^* of a sentence φ of L_{SL} that is not identical with φ, is itself a sentence of L_{SL}.

(b) Give an example of an end part $\varphi^\#$ of a sentence φ of L_{SL} that is not identical with φ but *is* a sentence of L_{SL}.

4. Conclude on the basis of (1) and (2) that every sentence φ of L_{SL} is uniquely readable.

CHAPTER 2

SHORT-CUT RULES FOR SENTENTIAL LOGIC

TOPICS: *Short derivations, Theorem Rule, substitution instance, derived rules*: CTB, CONJ, SIMP, BC, ADJ, MT, DN, MTP, SOC, BD, *derivations from assumptions, (optional) an axiomatic system of sentential logic, Ax-Sent.*

This chapter presents a new batch of inference rules that are very intuitive and convenient to use. Since they can all be derived from A, MP, CD, CONTRA, and INT of the previous chapter, no new theorems can now be derived that couldn't be derived before. But, even though the system is not more powerful than before (in the sense of there being more theorems derivable), the new rules will make finding derivations easier. In addition, the principles behind these rules are closely connected to patterns of reasoning used in mathematical proofs in general.

Mastering these principles requires a good deal of practice. And, as before, not all theorems are explicitly assigned as *exercises*, but all of them should be derived anyway, or at least you should be able to convince yourself that you could provide a derivation of any theorem you skip. Becoming skillful at constructing derivations will make mathematical proofs easier to follow and easier to do.

In the previous chapter, we produced what we now call "full" derivations, derivations in which every step was written down. In this chapter, we begin to omit certain steps of the derivations when it is absolutely clear how to fill in the missing steps. We call them "short derivations".

2.1 SHORT DERIVATIONS

A **short derivation** of φ is an initial segment of a derivation of φ containing at least one checked SHOW sentence, but missing others. Let us look at a short derivation of T25 (there is a full derivation of T25 in the previous chapter):

```
        SHOW (P ∧ Q) → P
    1.    P ∧ Q                    1,A
    2.    ¬(P → ¬Q)                 1,INT(∧)
```

		SHOW P	
3.	¬P		3, A
		SHOW A CONTRADICTION	
		√SHOW P → ¬Q	
4.	P		4, A
		√SHOW ¬Q	
5.	¬Q		3, 4, CONTRA
6.	P → ¬Q		3, CD

We can see that line 6 contradicts line 2, so 'SHOW A CONTRADICTION' can be checked off, 'P' can be written on line 7 and 'SHOW P' can be checked off. Then, the final sentence can be written on line 8 and 'SHOW (P ∧ Q) → P' can be checked off. Thus, this short derivation can easily be expanded to a "full" derivation of '(P ∧ Q) → P'.

But, we could have stopped even earlier. We could have stopped the derivation at line 5, after checking off only SHOW ¬Q. The rest of the SHOW sentences are easy to fill in. We will begin writing short derivations like the above one when the missing lines are easy to fill in.

We can reduce a derivation in another way, if desired, by removing some number of SHOW lines. Since the purpose of the SHOW lines is to reflect the structure of a derivation, removing them makes the derivation more difficult to follow. If we remove all SHOW lines from a short derivation, we may wind up with a derivation that is fairly incomprehensible.

Consider the following truncated derivation of (P ∧ Q) → P' without SHOW lines:

1.	P ∧ Q	1, A
2.	¬(P → ¬Q)	1, INT(∧)
3.	¬P	3, A
4.	P	4, A
5.	¬Q	3, 4, CONTRA
6.	(P → ¬Q)	3, CD

Notice that the removal of the SHOW lines renders the derivation difficult to follow. We could have stopped at line 4, which would make the derivation even more mysterious. We will always retain enough SHOW lines so that the structure of the derivation remains clear and the reasoning is easy to follow.

In the previous chapter you may have noticed that sometimes in deriving a theorem of L_{SL}, you had to in essence re-derive an earlier theorem. The derivation of T23, for example, requires the derivation of T17 all over again. Similarly, the derivation of T24 requires the re-derivation of T18. We wish to eliminate this redundancy by introducing a **Theorem Rule**, permitting previous theorems to appear on any line of a derivation of a later theorem.

For example, let us re-derive T23. Suppose we had the Theorem Rule available to us at that time. Then we could put T17 on a line of a short derivation as follows:

```
        SHOW  P → (Q → (P ∧ Q))
   1.   P                                 1,A
   2.   Q                                 2,A
                        √SHOW P ∧ Q
   3.   P → (Q → ¬(P → ¬Q))               T17
   4.   Q → ¬(P → ¬Q)                     1,MP(1,3)
   5.   ¬(P → ¬Q)                         1,2,MP(2,4)
   6.   P ∧ Q                             1,2,INT(∧)(5)
```

Here is a simple, straightforward use of the Theorem Rule:

T30: ¬P ∨ P

```
   1.   ¬¬P → P                           T11
   2.   ¬P ∨ P                            INT(∨)
```

We would like to derive T31, 'P ∨ ¬P', the same way we derived T30. But, we cannot proceed as straightforwardly. Here's how the derivation <u>should</u> look. But, there is a problem, indicated by '??':

T31: P ∨ ¬P

```
   1.   ¬P → ¬P                           ??T5
   2.   P ∨ ¬P                            INT(∨)
```

The problem is that '¬P → ¬P' is <u>not</u> T5. T5 is 'P → P'. What we would <u>like</u> to say is that '¬P → ¬P' is an **instance** of T5. The idea is that a derivation of '¬P → ¬P' would look <u>exactly</u> like the derivation of 'P → P', except that the sentence '¬P' would be used in the derivation instead of 'P'.

CHAPTER 2

Let us look at the derivation of '¬P → ¬P':

1. ¬P   1,A
2. ¬P → ¬P   CD

Since it is the same derivation as the one for 'P → P' except that '¬P' appears in this one, rather than 'P', we say that '¬P' is a **substitution instance** of 'P' in T5.

Formally, φ **is a substitution instance of** ψ iff φ is obtained from ψ by uniformly replacing sentence letters in ψ by sentences.

Note carefully that to get a substitution instance, we substitute sentences for **sentence letters**. We do **not** substitute sentences for **molecular sentences** — sentences that are not sentence letters.

Example 1: 'Q → (P → Q)' is a substitution instance of 'P → (Q → P)'. 'Q' is substituted for 'P' and 'P' is substituted for 'Q'.

Example 2: '¬P → ((Q v R) → ¬ P)' is also a substitution instance of 'P → (Q → P)', where '¬P' is substituted for 'P' and '(Q v R)' is substituted for 'Q'.

Example 3: '¬P → ((Q v R) → P)' is <u>not</u> a substitution instance of 'P → (Q → P)' because the substitution of '¬P' for 'P' is *not* uniform, meaning that it was not done in all cases.

When we put a substitution instance of a previous theorem on a line of a later derivation we will write a horizontal line with the sentence letters of the original sentence ψ above the line, separated by commas, and directly below each sentence letter we will write the replacement sentences (also separated by commas).

For example (taken from Example 2, above),
'¬P → ((Q v R) → ¬P)' = 'P → (Q → P)' $\frac{P, Q}{\neg P, (Q \vee R)}$

Hence if ϕ is a substitution instance of ψ, then

$$\phi = \psi \frac{\theta_1, \theta_2, \ldots, \theta_n}{\chi_1, \chi_2, \ldots, \chi_n}$$

where each sentence letter occurring in ψ is a θ_i and χ_i is the sentence uniformly replacing θ_i in ψ.

Theorem Rule (T): any substitution instance of a previous theorem may now be placed on a line.

Now, we can return to the derivation of T31 and use the appropriate substitution instance of T5:

T31:   P ∨ ¬P

    1.   ¬P → ¬P   T5, $\dfrac{P}{\neg P}$

    2.   P ∨ ¬P   INT(∨)

T32: ¬(P ∧ ¬P)
T33: (P ∧ ¬P) → Q
T34: (P ∨ Q) ↔ (Q ∨ P)

The strategy for deriving T34 is to first show '(P ∨ Q) → (Q ∨ P)', for which T10 is helpful. Once we have that '(P ∨ Q) → (Q ∨ P)' is a theorem, we also have '(Q ∨ P) → (P ∨ Q)', since that sentence is a substitution instance of '(P ∨ Q) → (Q ∨ P)', substituting 'Q' for 'P' and 'P' for 'Q'. Once both '(P ∨ Q) → (Q ∨ P)' and '(Q ∨ P) → (P ∨ Q)' have been derived, they can be "conjoined" by T23, where the two sentences '(P ∨ Q) → (Q ∨ P)' and '(Q ∨ P) → (P ∨ Q)' are substituted, respectively, for 'P' and 'Q' in T23.

In the following derivation we write numerical superscripts above matching parentheses to make it easier to read particularly long sentences:

√SHOW (P ∨ Q) ↔ (Q ∨ P)
      √SHOW ((P ∨ Q) → (Q ∨ P)) ∧ ((Q ∨ P) → (P ∨ Q))
        √SHOW (P ∨ Q) → (Q ∨ P)
        √SHOW (Q ∨ P) → (P ∨ Q)
1. P ∨ Q   1,A
      √SHOW Q ∨ P
        √SHOW ¬Q → P
2. ¬P → Q   1,INT(∨)
3. (¬P → Q) → (¬Q → P)   T10
4. ¬Q → P   1,MP(2,3)
5. Q ∨ P   1,INT(∨)
6. (P ∨ Q) → (Q ∨ P)   CD(5)

7. $(Q \vee P) \rightarrow (P \vee Q)$ T, line 6.(i.e., line 6 has no numbers, so it is a theorem)
$$\frac{P \;,\; Q}{Q \;,\; P}$$

8. $\overset{1}{\Big(}(P \vee Q) \rightarrow (Q \vee P)\overset{1}{\Big)} \rightarrow \overset{2\;3}{\Big(} \overset{1}{\big(}(Q \vee P) \rightarrow (P \vee Q)\overset{3}{\big)}$
    $\overset{4\;5}{\rightarrow \Big(} \big((P \vee Q) \rightarrow (Q \vee P)\overset{5}{\big)} \wedge \overset{6}{\big(}(Q \vee P) \rightarrow (P \vee Q)\overset{6\;4\;2}{\big)\Big)\Big)}$

$$\text{T23} \;\; \frac{P, \qquad\qquad\qquad Q}{(P \vee Q) \rightarrow (Q \vee P), \;\; (Q \vee P) \rightarrow (P \vee Q)}$$

9. $\overset{3}{\big(}(Q \vee P) \rightarrow (P \vee Q)\overset{3}{\big)} \rightarrow \overset{4\;5}{\Big(} \overset{5}{\big(}(P \vee Q) \rightarrow (Q \vee P)\big) \wedge$
    $\overset{6}{\big(}(Q \vee P) \rightarrow (P \vee Q)\overset{6\;4}{\big)\Big)}$ MP(6,8)

10. $\overset{5}{\big(}(P \vee Q) \rightarrow (Q \vee P)\overset{5}{\big)} \wedge \overset{6}{\big(}(Q \vee P) \rightarrow (P \vee Q)\overset{6}{\big)}$
    MP(7,9)

11. $(P \vee Q) \leftrightarrow (Q \vee P)$ INT($\leftrightarrow$)(10)

T35: $(P \wedge Q) \leftrightarrow (Q \wedge P)$
['$P \wedge Q$' is by INT($\wedge$) '$\neg(P \rightarrow \neg Q)$' and '$Q \wedge P$' is by INT($\wedge$) '$\neg(Q \rightarrow \neg P)$'. T9 should come in handy here.]

T36: $P \leftrightarrow \neg\neg P$

2.2 DERIVED RULES

At this point we would like to add some **derived rules** in addition to the Theorem Rule to make derivations less cumbersome. We used T23 to prepare for the biconditional in deriving T34. But, this takes too many steps, once the idea of how to proceed is clear. The following theorem improves matters somewhat:

T37: $(P \rightarrow Q) \rightarrow ((Q \rightarrow P) \rightarrow (P \leftrightarrow Q))$

        1. $P \rightarrow Q$ 1,A
        2. $Q \rightarrow P$ 2,A
        3. $(P \rightarrow Q) \rightarrow ((Q \rightarrow P) \rightarrow ((P \rightarrow Q) \wedge (Q \rightarrow P)))$

$$\text{T23} \;\; \frac{P \;\;,\;\; Q}{P \rightarrow Q, \; Q \rightarrow P}$$

```
    4.  (Q → P) → ((P → Q) ∧ (Q → P))      1,MP(1,3)
    5.  (P → Q) ∧ (Q → P)                  1,2,MP(2,4)
    6.  P ↔ Q                              1,2,INT(∧)(5)
```

T37 says, in effect, that to derive a biconditional, first derive one direction and then the other. By T37, then, you have the biconditional.

To see how useful T37 is, let us suppose it was available to us in the derivation of T36 (Note, it was <u>not</u> available to us then; we are <u>supposing</u> it was in order to see how much more direct the derivation <u>would have been</u>.):

```
    √ SHOW P ↔ ¬¬P                         (The second time)
    1.  P → ¬¬P                            T12
    2.  ¬¬P → P                            T11

    3.  (P → ¬¬P) → ((¬¬P → P) → (P ↔ ¬¬P))
                                      T37    P  ,  Q
                                             ─────────
                                             P  , ¬¬P
    4.  (¬¬P → P) → (P ↔ ¬¬P)              MP(1,3)
    5.  P ↔ ¬¬P                            MP(2,4)
```

Now, in order to derive a biconditional we still would need to, essentially, repeat the above derivation each time, citing T37 and the appropriate substitution instance. Instead of doing this over and over, we will derive a new rule to give us the biconditional once we have both conditionals.

2.3 DERIVING A RULE

Deriving a rule is different from deriving a theorem. When we derive a theorem, we arrive at a sentence of L_{SL} on a line of a derivation with no assumption numbers to the right. In deriving a rule we show that previous inference rules enable us to arrive at a sentence of a given form on a line of a derivation, provided that we have obtained other specified forms on earlier lines. For example, we wish to prove if a sentence of the form $\varphi \rightarrow \psi$ is on one line of a derivation and a sentence of the form $\psi \rightarrow \varphi$ is on another line of a derivation, then our available rules enable us to arrive at the sentence of the form $\varphi \leftrightarrow \psi$ on a later line of a derivation, where the assumption numbers are those of both previous sentences taken together. First, let us suppose

$a_1, a_2, \ldots, a_p$ are the assumption numbers of $\varphi \to \psi$ and
$b_1, b_2, \ldots, b_q$ are the assumption numbers of $\psi \to \varphi$.

$$\begin{array}{lll} k. & \varphi \to \psi & a_1, a_2, \ldots, a_p \\ n. & \psi \to \varphi & b_1, b_2, \ldots, b_q \end{array}$$

TO PROVE: $\varphi \leftrightarrow \psi$

$$\begin{array}{ll} m+1. & (\varphi \to \psi) \to \big((\psi \to \varphi) \to ((\varphi \to \psi) \wedge (\psi \to \varphi))\big) \\ & \qquad\qquad\text{T23} \\ m+2. & (\psi \to \varphi) \to \big(((\varphi \to \psi) \wedge (\psi \to \varphi))\big) \\ & \qquad\qquad a_1, a_2, \ldots, a_p, \text{MP}(k, m+1) \\ m+3. & (\varphi \to \psi) \wedge (\psi \to \varphi) \qquad a_1, a_2, \ldots, a_p, b_1, b_2, \ldots, b_q \\ & \qquad\qquad \text{MP}(n, m+2) \\ m+4. & \varphi \leftrightarrow \psi \qquad\qquad a_1, a_2, \ldots, a_p, b_1, b_2, \ldots, b_q \\ & \qquad\qquad \text{INT}(\leftrightarrow)(m+3) \end{array}$$

The above proof is an example of what we shall call a **derivation form**. A derivation form is very similar to a derivation. The major difference (though not the only difference) is that a derivation form has the <u>forms</u> of sentences on lines, not actual sentences of L_{SL}. These "forms" are variables ranging over sentences that dictate the kinds of sentences that can appear. Another difference is that a derivation form does not necessarily begin at line 1, as a derivation does. And, just as we use variables ranging over sentences in a derivation form, we use variables over numbers as well. The above derivation form begins at line k, where 'k' ranges over all positive integers as does 'n'. (It does not matter whether line n or line k comes first.) The variables 'a_1', 'a_2', ..., 'a_p' and 'b_1', 'b_2', ..., 'b_q' range over positive integers as well, standing in for assumption numbers.

If we wished to convert the above derivation form into part of an actual derivation, we would have to substitute sentences of L_{SL} for 'φ' and 'ψ' (which are not sentences of L_{SL} but variables ranging over sentences) and we would need to have the counterparts of steps k through m+3 appearing in the actual derivation. This derivation form justifies our use of the rule **Conditionals To Biconditional**, which lets us skip steps m through m+2, as indicated above. The derivation of T23 requires eight more lines and uses only our basic rules. So, to turn the form above into one using only basic rules, we would delete line m and insert in its place lines r through r+7 for the eight lines in the derivation of T23.

Ultimately, then, rule CTB lets us skip ten lines (r through r+7 and lines m+1 and m+2). Conversely, if we wished to turn any derivation using CTB into one using only basic ones, we would restore the eight lines for the derivation of T23 plus the two MP lines.

Rule **Conditionals To Biconditional (CTB)**: when $\varphi \rightarrow \psi$ is on a line and $\psi \rightarrow \varphi$ is on a line, $\varphi \leftrightarrow \psi$ may be placed on a later line with the assumption numbers of $\varphi \rightarrow \psi$ and $\psi \rightarrow \varphi$ taken together.

Let us show 'P $\leftrightarrow$ ¬¬P' a third time. This time we use CTB:

```
√ SHOW   P ↔ ¬¬P              (For the third time)
1.    P → ¬¬P                  T12
2.    ¬¬P → P                  T11
3.    P ↔ ¬¬P                  CTB(1,2)
```

In a similar vein, we shall add some more derived rules, each of which can be eliminated in favor of the theorem (or theorems) justifying it (which can, in turn, be eliminated in favor of our basic rules).

Rule **Conjunction (CONJ)**: When φ is on a line and ψ is on another line, $\varphi \wedge \psi$ may be placed on a later line with the assumption numbers of the previous two lines (justified by T23, T24).

Rule **Simplification (SIMP)**: When $\varphi \wedge \psi$ is on a line either conjunct, φ or ψ, may be placed on a later line with the same assumption numbers for $\varphi \wedge \psi$ (justified by T25, T26).

Given SIMP, we may as well derive another convenient rule, saying in effect that if you have $\varphi \leftrightarrow \psi$, you have $\varphi \rightarrow \psi$ and you have $\psi \rightarrow \varphi$. First, let us derive the two relevant theorems:

T38:   (P ↔ Q) → (P → Q)
T39:   (P ↔ Q) → (Q → P)

Rule **Biconditional-Conditional (BC)**: If $\varphi \leftrightarrow \psi$ is on a line, you may place either $\varphi \rightarrow \psi$ or $\psi \rightarrow \varphi$ on a later line with the same assumption numbers (justified by T38, T39).

Rule **Adjunction** (**ADJ**): When φ is on a line <u>or</u> ψ is on a line, φ ∨ ψ may be placed on a later line with the same assumption numbers as the sentence (either φ or ψ) on the earlier line (justified by T27, T28).

Rule **Modus Tollens** (**MT**): When a conditional φ → ψ is on one line of a derivation and a <u>contradictory</u> of the consequent, CONTRA(ψ), is on another line, then a contradictory of the antecedent, CONTRA(φ), may be placed on a later line with the assumption numbers of φ → ψ and CONTRA(ψ) taken together (<u>four</u> theorems, T7 - T10 justify the four distinct forms of MT).

Rule **Double Negation** (**DN**): When φ is on a line, ¬¬φ may be placed on a later number with the assumption numbers of φ, <u>and vice versa</u> (justified by T36).

Exercises:

1 - 6. Provide a derivation form for each of: CONJ, SIMP, DC, ADJ, MT, and DN.

Here are several examples of theorems requiring SIMP and MT. The first one, '((P ∨ ¬Q) ∧ Q) → P', says, in effect: "If you have P or not-Q <u>and</u> you have Q, then you must have P". The reasoning behind the derivation of this sentence using SIMP and MT reflects the content of that sentence very clearly. To derive it <u>without</u> these rules, however, obscures its content and is more laborious. First, we will derive it <u>without</u> these two new derived rules:

```
          SHOW  ((P ∨ ¬Q) ∧ Q) → P
    1.    (P ∨ ¬Q) ∧ Q                         1,A
              √ SHOW  P
    2.    ¬P                                   2,A
    3.    ((P ∨ ¬Q) ∧ Q) → (P ∨ ¬Q)            T25  P   , Q
                                                    P ∨ ¬Q, Q
    4.    P ∨ ¬Q                               1,MP(1,3)
    5.    ¬P → ¬Q                              1,INT(∨)(4)
    6.    ¬Q                                   1,2,MP(2,5)
    7.    ((P ∨ ¬Q) ∧ Q) → Q                   T26  P   , Q
                                                    P ∨ ¬Q, Q
    8.    Q                                    1,MP(1,7)
    9.    P                                    1,CONTRA(6,8)
```

(We continue to omit the final CD step, as before.) Now, we will derive '((P ∨ ¬Q) ∧ Q) → P' using SIMP and MT, which reflect this reasoning: "If you either have P or you have not-Q, and you have Q (meaning: you don't have not-Q), then you must have P":

T40: ((P ∨ ¬Q) ∧ Q) → P
    1. (P ∨ ¬Q) ∧ Q   1,A
    2. P ∨ ¬Q   1,SIMP
    3. Q   1,SIMP
    4. ¬P → ¬Q   1,INT(∨)(2)
    5. P   1,MT(3,4)

Line 2 "says": "You have P or you have not-Q". Line 3 is a contradictory of "not-Q". Therefore: (by line 5) "You must have P".

T41: ((P ∨ Q) ∧ ¬Q) → P
T42: ((¬P ∨ ¬Q) ∧ Q) → ¬P
T43: ((¬P ∨ Q) ∧ ¬Q) → ¬P

    The reasoning of T40-T43 can be embodied in an additional derived rule:

    Rule **Modus Tollendo Ponens** (**MTP**): If the disjunction φ ∨ ψ appears on a line of a derivation and a *contradictory* of *one* of the disjuncts appears on another line, then the *other* disjunct may be placed on a later line with the assumption numbers of the original disjunction plus the assumption numbers of the contradictory of that disjunct.

    Actually, MTP goes further than T40-T43 permit, because MTP can be used to get ψ from φ ∨ ψ and CONTRA(φ), (whereas T40-T43 support obtaining φ from φ ∨ ψ and CONTRA(ψ)). In order to justify these additional applications of MTP, we need to use T34 as well, which "says" in essence that the order of the disjuncts does not matter (disjunction is commutative).

Exercises:

  <u>Without</u> using MTP, formulate and derive two of the four additional theorems that require T34, as explained above. [The analog of T41 would be '((P ∨ Q) ∧ ¬P) → Q', for example.]

Let us derive one of the forms of the MTP rule in order to indicate the steps that are left out. We will derive this form of MTP:

If $\neg\varphi \vee \psi$ is on one line of a derivation and φ is on another line, then ψ may be placed on a later line with the assumption numbers of the two earlier lines taken together.

As before, we create a derivation form on which the two earlier lines containing $\neg\varphi \vee \psi$ and φ appear. Then we show all the missing steps between these two lines and the later line on which ψ appears:

```
k.     ¬φ ∨ ψ              a₁,a₂,...,aₚ
n.     φ                   b₁,b₂,...,b_q
n+1.   ¬¬φ → ψ             a₁,a₂,...,aₚ  INT(→)(k)
n+2.   ¬φ                  n+2,A
n+3.   ¬¬φ                 b₁,b₂,...,b_q  CONTRA(n,n+2)
n+4.   ψ                   a₁,a₂,...,aₚ,b₁,b₂,...,b_q  MP(n+1, n+3)
```

In the case above, MTP saves us three steps, n+1 through n+3.

Exercises:

Derive these two other forms of MTP using <u>only</u> A, MP, CD, CONTRA, and INT:
1. If $\varphi \vee \psi$ and $\neg\varphi$, then ψ.
2. If $\neg\varphi \vee \psi$ and $\neg\psi$, then $\neg\varphi$.

[In both of the above cases, the assumption numbers of the conclusion are to be the assumption numbers of the two earlier sentences in accordance with the statement of MTP above.]

We have not yet provided examples of the derived rules CONJ and ADJ. The following theorem illustrates a straightforward application of CONJ in the left-to-right direction:

T44: $((P \wedge Q) \to R) \leftrightarrow (P \to (Q \to R))$

```
                SHOW  ((P ∧ Q) → R) → (P → (Q → R))
         1.   (P ∧ Q) → R              1, A
                                SHOW  P → (Q → R)
         2.   P                        2, A

                                SHOW  Q → R
         3.   Q                        3, A
                                √SHOW R
         4.   P ∧ Q                    2, 3, CONJ
         5.   R
```

(The right-to-left direction illustrates SIMP. Do this as an *exercise*.)

The following theorem (Leibniz's *Praeclarum Theorema*) illustrates <u>both</u> CONJ <u>and</u> SIMP:

T45:   (P → R) ∧ (Q → S) → ((P ∧ Q) → (R ∧ S)

This next theorem (the *idempotence* law of 'v') illustrates ADJ in the left-to-right direction and MTP in the right-to-left direction:

T46:   P ↔ (P v P)
T47:   (P → Q) → ((P v Q) → Q)

2.4 SHOWING A DISJUNCTION vs. USING A DISJUNCTION

Strategy: A useful strategy to follow in trying to show a disjunction, $\varphi \vee \psi$, is to try to show its interchange, $\neg\varphi \to \psi$. The steps to follow are:

```
                        SHOW φ v ψ
                        SHOW ¬φ → ψ
         n.   ¬φ           n, A
                        SHOW ψ
```

(Then, after the line 'SHOW ψ', you may wish to begin a CONTRA derivation, assuming $\neg\psi$ on line n + 1.) The idea behind this strategy is to allow you to make another assumption, $\neg\varphi$, and to require you to SHOW only the smaller sentence ψ (as opposed to ($\varphi \vee \psi$)). To derive the following two theorems the above strategy may prove helpful. For example, in deriving T49 below, it is difficult to know where to start. The conditional '¬(P → Q) → (Q → R)' at least establishes a starting point:

T48: (P → Q) → ((P ∨ R) → (Q ∨ R))
T49: (P → Q) ∨ (Q → R)

The strategy for <u>deriving a disjunction</u> does nothing to help you <u>use a disjunction</u> in a derivation. That is, suppose you already <u>have</u> φ ∨ ψ on a line. How can you *use* it to help you show whatever it is you are trying to show?

One helpful rule to use with φ ∨ ψ is MTP. If φ ∨ ψ is on one line and CONTRA(φ) is on a second line, you may put ψ on a later line (<u>or</u> if CONTRA(ψ) is on a second line, you may put φ on a later line). Another helpful strategy is called: **a Separation of Cases Argument.** But, before we can discuss that form of argument, we need another theorem.

T50: ((P → R) ∧ (Q → S)) → ((P ∨ Q) → (R ∨ S))

The idea of T50 is that **if** <u>both</u> conditionals 'P → R' and 'Q → S' hold, **then** if one of the two antecedents holds then one of the two consequents must hold also. That is, if 'P ∨ Q' holds, then 'R ∨ S' must hold as well.

Rule **Separation Of Cases (SOC)**: If <u>both</u> φ → χ and ψ → θ are on lines of a derivation, then (φ ∨ ψ) → (χ ∨ θ) may be placed on a later line with the assumption numbers of the two previous lines taken together. (See 1a on next page for a diagram of the rule.)

We also accept three further forms of the **Separation Of Cases (SOC)** rule. If (φ → χ) and (ψ → θ) are on lines of a derivation <u>and</u> (φ ∨ ψ) <u>is on another line</u>, then by the SOC rule explained above we may place (φ ∨ ψ) → (χ ∨ θ) on a later line. (This is form 1a.) But, since (φ ∨ ψ) also appears on an earlier line, we can place (χ ∨ θ) itself on a still later line by MP. This is, then, **a second form of SOC**. (See 1b, next page.)

There are also two special cases of SOC. Suppose instead of having (φ → χ) and (ψ → θ) on two previous lines, we have (φ → χ) and (ψ → χ). That is, suppose χ = θ. Intuitively, (φ → χ) means that if φ holds then χ holds <u>and</u> (ψ → χ) means if ψ holds then χ holds. So if <u>either</u> φ <u>or</u> ψ holds, then χ holds. So, by **a special case of SOC 1a** we get (φ ∨ ψ) → χ. (Another way to see this is that by SOC 1a we get (φ ∨ ψ) → (χ ∨ χ). But, (χ ∨ χ) holds just in case χ itself holds.) This is SOC 2a. SOC 2b comes from SOC 1b, where again $\chi = \theta$.

The four forms of SOC are:

1a. $\varphi \rightarrow \chi$
    $\psi \rightarrow \theta$
    $\therefore (\varphi \vee \psi) \rightarrow (\chi \vee \theta)$

1b. $\varphi \rightarrow \chi$
    $\psi \rightarrow \theta$
    $\varphi \vee \psi$
    $\therefore \chi \vee \theta$

2a. $\varphi \rightarrow \chi$
    $\psi \rightarrow \chi$
    $\therefore (\varphi \vee \psi) \rightarrow \chi$

2b. $\varphi \rightarrow \chi$
    $\psi \rightarrow \chi$
    $\varphi \vee \psi$
    $\therefore \chi$

(The assumption numbers for each form of SOC are the assumption numbers of all the sentences above the line taken together.)

The dominant form of SOC is 1a (which is justified by T50). The other forms either have an additional assumption or constitute the special case when $\chi = \theta$.

Here is an interesting use of a separation of cases argument:

T51: $((P \rightarrow Q) \wedge (\neg P \rightarrow Q)) \rightarrow Q$

1.	$(P \rightarrow Q) \wedge (\neg P \rightarrow Q)$	1,A
2.	$P \rightarrow Q$	1,SIMP
3.	$\neg P \rightarrow Q$	1,SIMP
4.	$(P \vee \neg P) \rightarrow Q$	1,SOC(2,3)
5.	$P \vee \neg P$	T31
6.	Q	1,MP(4,5)

Given our SOC Rule, one strategy to follow in deriving a conditional whose antecedent is a disjunction is to derive the consequent from *both* disjuncts and use SOC to get the desired result. For example, we wish to show the left-to-right direction of the following theorem:

$$((P \wedge Q) \vee (\neg P \wedge \neg Q)) \leftrightarrow (P \leftrightarrow Q)$$

by a SOC argument. The argument will be that '$(P \wedge Q) \rightarrow (P \leftrightarrow Q)$' holds, and '$(\neg P \wedge \neg Q) \rightarrow (P \leftrightarrow Q)$'

holds as well. Hence, by SOC, the left-to-right direction is secured.

In order to use SOC as explained in the above paragraph, we need to first derive the two theorems alluded to. They are both interesting in and of themselves. (T52 is gotten essentially from T6 and T53 from T14).

T52:   (P ∧ Q) → (P ↔ Q)
T53:   (¬P ∧ ¬Q) → (P ↔ Q)

Now, the left-to-right direction of T54 is a straightforward application of SOC. The right-to-left direction can be done using SOC also, with the help of T31.

T54:   ((P ∧ Q) ∨ (¬P ∧ ¬Q)) ↔ (P ↔ Q)
T55:   (P ∨ (Q ∨ R)) ↔ ((P ∨ Q) ∨ R)
T56:   (P → (Q ∨ R)) ↔ ((P → Q) ∨ (P → R))
T57:   ¬(P ∧ Q) ↔ (¬P ∨ ¬Q)

   √SHOW   ¬(P ∧ Q) → (¬P ∨ ¬Q)
      1.   ¬(P ∧ Q)   1,A
          √SHOW (¬P ∨ ¬Q)
          √SHOW   ¬¬P → ¬Q
      2.   ¬¬P   2,A
          √SHOW   ¬Q
      3.   Q   3,A
          √SHOW A CONTRADICTION
      4.   P   2,DN
      5.   P ∧ Q   2,3
      6.   ¬Q   1,2, CONTRA
      7.   ¬¬P → ¬Q
      8.   ¬P ∨ ¬Q
      9.   →

     10.   ¬P ∨ ¬Q   √SHOW   (¬P ∨ ¬Q) → ¬(P ∧ Q)
                                                      √SHOW   ¬(P ∧ Q)
     11.   P ∧ Q
     12.   P
     13.   ¬Q
     14.   Q
     15.   ¬(P ∧ Q)
     16.   →
     17.   ↔

T58:   ¬(P ∨ Q) ↔ (¬P ∧ ¬Q)

The following seven theorems are quite helpful. Each succeeding pair shows how one of the connectives '→', 'v', and '∧' relates to the two others when supplemented by '¬'. For example, the first pair has '→' on the left and one of the two others on the right, together with the requisite negation symbols. The second pair is "about" 'v' and the third is about '∧'.

T59 (P → Q) ↔ (¬P v Q)
T60 (P → Q) ↔ ¬(P ∧ ¬Q)
T61 (P v Q) ↔ (¬P → Q)
T62 (P v Q) ↔ ¬(¬P ∧ ¬Q)
T63 (P ∧ Q) ↔ ¬(P → ¬Q)
T64 (P ∧ Q) ↔ ¬(¬P v ¬Q)
T65 ¬(P → Q) ↔ (P ∧ ¬Q)

T65 "says" that if you don't have 'if P then Q' then you have 'P and not Q' and vice versa. The "vice versa" part is: if you have 'P and not Q' then you can't have 'if P then Q'.

The left-to-right direction of T65 can come in handy in a derivation, for it gives us a conjunction, which can be broken down into its separate conjuncts by SIMP.

The following four theorems express important facts about '↔', the biconditional, and provide part of the basis for an additional derived rule, **Biconditional Detachment**:

T66 (P ↔ Q) ↔ (Q ↔ P)
T67 ¬(P ↔ Q) ↔ (¬P ↔ Q)
T68 ¬(P ↔ Q) ↔ (P ↔ ¬Q)
T69 (¬P ↔ Q) ↔ (P ↔ ¬Q)

A **detachment rule** for a given binary connective allows us to detach the sentence on one side of the connective (or its contradictory), given the sentence on the other side of the connective (or its contradictory), provided we also have the full sentence containing the connective. We have two kinds of detachment rules for '→':

$\varphi \to \psi$
φ
─────────
∴ ψ MP

$\varphi \to \psi$
CONTRA(ψ)
─────────
∴ CONTRA(φ)   MT

We have two kinds of detachment rules for 'v':

$$\frac{\varphi \lor \psi}{\therefore \varphi} \quad \text{MTP} \qquad\qquad \frac{\varphi \lor \psi}{\therefore \psi} \quad \text{MTP}$$

For '∧' we use SIMP alone to detach either conjunct. We do not need an additional assumption for '∧' as we do for the '→' and the '∨'.

We now wish to introduce four forms of **Biconditional Detachment** in order to place '↔' on the same footing as the other three binary connectives. The four forms of BD are:

$$\frac{\varphi \leftrightarrow \psi}{\therefore \varphi} \qquad \frac{\varphi \leftrightarrow \psi}{\therefore \psi} \qquad \frac{\varphi \leftrightarrow \psi}{\therefore \text{CONTRA}(\psi)} \qquad \frac{\varphi \leftrightarrow \psi}{\therefore \text{CONTRA}(\varphi)}$$

(The assumption numbers for each form of BD are the assumption numbers of all sentences above the line taken together.)

Rule **Biconditional Detachment (BD):** When $\varphi \leftrightarrow \psi$ is on a line and the sentence on one side of '↔' is also on a line, you may put the sentence that is on the other side of '↔' on a later line with the assumption numbers of the two previous lines. Similarly, if the contradictory of the sentence on one side of '↔' is on a line, you may put the contradictory of the sentence that is on the other side of '↔' (assuming, of course, that the biconditional itself is on a line).

We now put in one list all the theorems arrived at so far, plus several additional ones. The conscientious reader should derive these additional ones as *exercises*:

2.5 THEOREMS OF SENTENTIAL LOGIC

T1 $P \rightarrow ((P \rightarrow Q) \rightarrow Q)$
T2 $(P \rightarrow Q) \rightarrow ((Q \rightarrow R) \rightarrow (P \rightarrow R))$
T3 $(Q \rightarrow R) \rightarrow ((P \rightarrow Q) \rightarrow (P \rightarrow R))$
T4 $(P \rightarrow (Q \rightarrow R)) \rightarrow ((P \rightarrow Q) \rightarrow (P \rightarrow R))$
T5 $P \rightarrow P$
T6 $P \rightarrow (Q \rightarrow P)$
T7 $(\neg P \rightarrow \neg Q) \rightarrow (Q \rightarrow P)$
T8 $(P \rightarrow Q) \rightarrow (\neg Q \rightarrow \neg P)$
T9 $(P \rightarrow \neg Q) \rightarrow (Q \rightarrow \neg P)$
T10 $(\neg P \rightarrow Q) \rightarrow (\neg Q \rightarrow P)$
T11 $\neg \neg P \rightarrow P$
T12 $P \rightarrow \neg \neg P$
T13 $P \rightarrow (\neg P \rightarrow Q)$
T14 $\neg P \rightarrow (P \rightarrow Q)$
T15 $(\neg P \rightarrow P) \rightarrow P$
T16 $(P \rightarrow \neg P) \rightarrow \neg P$
T17 $\neg(P \rightarrow Q) \rightarrow P$
T18 $\neg(P \rightarrow Q) \rightarrow \neg Q$
T19 $((P \rightarrow Q) \rightarrow P) \rightarrow P$
T20 $(\neg P \rightarrow Q) \rightarrow (P \vee Q)$
T21 $\neg(P \rightarrow \neg Q) \rightarrow (P \wedge Q)$
T22 $((P \rightarrow Q) \wedge (Q \rightarrow P)) \rightarrow (P \leftrightarrow Q)$
T23 $P \rightarrow (Q \rightarrow (P \wedge Q))$
T24 $P \rightarrow (Q \rightarrow (Q \wedge P))$
T25 $(P \wedge Q) \rightarrow P$
T26 $(P \wedge Q) \rightarrow Q$
T27 $P \rightarrow (P \vee Q)$
T28 $Q \rightarrow (P \vee Q)$
T29 $(P \leftrightarrow Q) \rightarrow ((P \rightarrow Q) \wedge (Q \rightarrow P))$
T30 $\neg P \vee P$
T31 $P \vee \neg P$
T32 $\neg(P \wedge \neg P)$
T33 $(P \wedge \neg P) \rightarrow Q$
T34 $(P \vee Q) \leftrightarrow (Q \vee P)$ Commutativity of '$\vee$'
T35 $(P \wedge Q) \leftrightarrow (Q \wedge P)$ Commutativity of '$\wedge$'
T36 $P \leftrightarrow \neg \neg P$
T37 $(P \rightarrow Q) \rightarrow ((Q \rightarrow P) \rightarrow (P \leftrightarrow Q))$
T38 $(P \leftrightarrow Q) \rightarrow (P \rightarrow Q)$
T39 $(P \leftrightarrow Q) \rightarrow (Q \rightarrow P)$
T40 $((P \vee \neg Q) \wedge Q) \rightarrow P$
T41 $((P \vee Q) \wedge \neg Q) \rightarrow P$
T42 $((\neg P \vee \neg Q) \wedge Q) \rightarrow \neg P$
T43 $((\neg P \vee Q) \wedge \neg Q) \rightarrow \neg P$
T44 $((P \wedge Q) \rightarrow R) \leftrightarrow (P \rightarrow (Q \rightarrow R))$

T45 $((P \to R) \land (Q \to S)) \to ((P \land Q) \to (R \land S))$
T46 $P \leftrightarrow (P \lor P)$
T47 $(P \to Q) \to ((P \lor Q) \to Q)$
T48 $(P \to Q) \to ((P \lor R) \to (Q \lor R))$
T49 $(P \to Q) \lor (Q \to R)$
T50 $((P \to R) \land (Q \to S)) \to ((P \lor Q) \to (R \lor S))$
T51 $((P \to Q) \land (\neg P \to Q)) \to Q$
T52 $(P \land Q) \to (P \leftrightarrow Q)$
T53 $(\neg P \land \neg Q) \to (P \leftrightarrow Q)$
T54 $((P \land Q) \lor (\neg P \land \neg Q)) \leftrightarrow (P \leftrightarrow Q)$
T55 $(P \lor (Q \lor R)) \leftrightarrow ((P \lor Q) \lor R)$ Associativity of '$\lor$'
T56 $(P \to (Q \lor R)) \leftrightarrow ((P \to Q) \lor (P \to R))$
         The distributive law for '$\to$' over '$\lor$'
T57 $\neg(P \land Q) \leftrightarrow (\neg P \lor \neg Q)$
T58 $\neg(P \lor Q) \leftrightarrow (\neg P \land \neg Q)$
T59 $(P \to Q) \leftrightarrow (\neg P \lor Q)$
T60 $(P \to Q) \leftrightarrow \neg(P \land \neg Q)$
T61 $(P \lor Q) \leftrightarrow (\neg P \to Q)$
T62 $(P \lor Q) \leftrightarrow \neg(\neg P \land \neg Q)$
T63 $(P \land Q) \leftrightarrow \neg(P \to \neg Q)$
T64 $(P \land Q) \leftrightarrow \neg(\neg P \lor \neg Q)$
T65 $\neg(P \to Q) \leftrightarrow (P \land \neg Q)$
T66 $(P \leftrightarrow Q) \leftrightarrow (Q \leftrightarrow P)$ Commutativity of '$\leftrightarrow$'
T67 $\neg(P \leftrightarrow Q) \leftrightarrow (\neg P \leftrightarrow Q)$
T68 $\neg(P \leftrightarrow Q) \leftrightarrow (P \leftrightarrow \neg Q)$
T69 $(\neg P \leftrightarrow Q) \leftrightarrow (P \leftrightarrow \neg Q)$
T70 $((P \to R) \lor (Q \to R)) \leftrightarrow ((P \land Q) \to R)$
T71 $(P \land (Q \lor R)) \leftrightarrow (P \land Q) \lor (P \land R)$
T72 $(P \lor (Q \land R)) \leftrightarrow ((P \lor Q) \land (P \lor R))$
T73 $P \leftrightarrow ((P \land Q) \lor (P \land \neg Q))$
T74 $P \leftrightarrow ((P \lor Q) \land (P \lor \neg Q))$
T75 $(P \to (Q \leftrightarrow R)) \leftrightarrow ((P \to Q) \leftrightarrow (P \to R))$
         distributivity of '$\to$' over '$\leftrightarrow$'
T76 $(P \leftrightarrow Q) \lor (P \leftrightarrow \neg Q)$
T77 $((P \leftrightarrow Q) \to R) \leftrightarrow (((P \land Q) \to R) \land ((\neg P \land \neg Q) \to R))$
T78 $\neg(P \leftrightarrow Q) \leftrightarrow ((P \land \neg Q) \lor (\neg P \land Q))$
T79 $(P \land \neg Q) \to \neg(P \leftrightarrow Q)$
T80 $(\neg P \land Q) \to \neg(P \leftrightarrow Q)$
T81 $(P \leftrightarrow (Q \leftrightarrow R)) \leftrightarrow ((P \leftrightarrow Q) \leftrightarrow R)$
T82 $(P \leftrightarrow Q) \leftrightarrow (\neg P \leftrightarrow \neg Q)$
T83 $(P \leftrightarrow R) \land (Q \leftrightarrow S) \to ((P \leftrightarrow Q) \leftrightarrow (R \leftrightarrow S))$

Some combined theorems from above are:
T84 $(P \to (P \to Q)) \leftrightarrow (P \to Q)$
T85 $(\neg P \to P) \leftrightarrow P$
T86 $(P \to \neg P) \leftrightarrow \neg P$

2.6 DERIVATIONS FROM ASSUMPTIONS

Up to now we have used our rules of inference to derive theorems. We now wish to use them to **derive a sentence from (a set of) assumptions**. A derivation of a theorem, you will recall, is a derivation having that sentence on the last line with <u>no</u> numbers to its right. A **derivation of φ from the set of assumptions** Γ is a derivation with φ on the last line having Γ as its assumptions. Remember: the assumptions of a sentence on a line of a derivation are the sentences corresponding to the numbers to the right of the given sentence. So, for example, here is a derivation of 'Q' from the set of assumptions, $\{P, P \rightarrow Q\}$:

```
1.  P                            1, A
2.  P → Q                        2, A
                                 √SHOW Q
3.  Q                            1, 2, MP
```

In the above derivation line 1 has 'P' derived from $\{P\}$, line 2 has 'P $\rightarrow$ Q' derived from $\{P \rightarrow Q\}$, and line 3 has 'Q' derived from $\{P, P \rightarrow Q\}$. We can also say we have **derived the conclusion 'Q' from the premises, 'P', 'P $\rightarrow$ Q'**. Put another way, we can say that we have **derived the conclusion of this argument from its premises**:

$$P$$
$$P \rightarrow Q$$
$$\overline{\therefore Q}$$

Let us derive the conclusion of this argument from its premises (assumptions):

$$(P \wedge Q) \rightarrow R$$
$$R \rightarrow S$$
$$Q \wedge \neg S$$
$$\overline{\therefore \neg P}$$

```
1.  (P ∧ Q) → R                  1, A
2.  R → S                        2, A
3.  Q ∧ ¬S                       3, A
                                 √SHOW ¬P
4.  P                            4, A
```

5.	Q	3, SIMP
6.	P ∧ Q	3,4, SIMP(4,5)
7.	R	1,3,4, MP(1,6)
8.	S	1,2,3,4, MP(2,7)
9.	¬S	3, SIMP
10.	¬P	1,2,3, CONTRA(8,9)

Normally, in deriving a sentence (conclusion) from (the set of) its assumptions (the premises of an argument), the assumptions (premises) are all written first and the derivation ends with the desired sentence on a line with the numbers 1, 2, . . . , n on the right that correspond to the assumptions.

Here is another example:

$$(P \rightarrow Q) \vee (R \rightarrow S)$$
$$\overline{\therefore (P \rightarrow S) \vee (R \rightarrow Q)}$$

1.	(P → Q) ∨ (R → S)	1, A
	√SHOW (P → S) ∨ (R → Q)	
	√SHOW ¬(P → S) → (R → Q)	
2.	¬(P → S)	2, A
	√SHOW R → Q	
3.	R	3, A
	√SHOW Q	
4.	¬(P → S) ↔ (P ∧ ¬S)	T65 P , Q / P , S
5.	P ∧ ¬S	2, BD(2,4)
6.	¬S	2, SIMP(5)
7.	R ∧ ¬S	2,3, CONJ(3,6)
8.	¬(R → S) ↔ (R ∧ ¬S)	T65 P , Q / R , S
9.	¬(R → S)	2,3, BD(7,8)
10.	P → Q	1,2,3, MTP(1,9)
11.	P	2, SIMP(5)
12.	Q	1,2,3, MP(10,11)
13.	R → Q	1,2, CD(3,12)
14.	¬(P → S) → (R → Q)	1, CD(2,13)
15.	(P → S) ∨ (R → Q)	1, INT(∨)(14)

Exercises:

For each of the arguments below, derive the conclusion from its premises (assumptions):

1. $P \rightarrow Q$
    $(Q \rightarrow P) \rightarrow P$
    ─────────────
    $\therefore Q$

2. $(P \wedge \neg Q) \vee (P \wedge R)$
    $\neg Q \rightarrow \neg P$
    ─────────────
    $\therefore R$

3. $(P \rightarrow Q) \rightarrow Q$
    $(T \rightarrow P) \rightarrow R$
    $(R \rightarrow S) \rightarrow \neg(S \rightarrow Q)$
    ─────────────────────
    $\therefore R$

4. $P \rightarrow \neg S$
    $Q \rightarrow \neg R$
    $T \rightarrow (\neg P \rightarrow Q)$
    $\neg T \rightarrow T$
    ─────────────
    $\therefore \neg(R \wedge S)$

2.7 SUMMARY OF INFERENCE RULES

Basic Rules

A: n. φ n, A

MP: $\varphi \rightarrow \psi$
          φ
          ―――――
          $\therefore \psi$

CD: ψ n. φ n, A
          ――――――― ψ
          $\therefore \varphi \rightarrow \psi$ ―――――――――――
                                                   $\therefore \varphi \rightarrow \psi$ [-n]

CONTRA: ψ n. CONTRA(φ)   n, A
               $\neg \psi$ ψ
          ――――― $\neg \psi$
          $\therefore \varphi$ ――――――――
                                                   $\therefore \varphi$ [-n]

INT: $\varphi \vee \psi$ $\varphi \wedge \psi$ $\varphi \leftrightarrow \psi$
        ═══════════   ═══════════════   ═══════════════════════════════
        $\neg \varphi \rightarrow \psi$ $\neg(\varphi \rightarrow \neg \psi)$ $(\varphi \rightarrow \psi) \wedge (\psi \rightarrow \varphi)$

* The double lines indicate that the inference can go in both directions.
* In each case the sentence below the line (or above and below the double lines) has all the assumption numbers of the sentence(s) above the line.
* '[-n]' means to remove the number n from the assumption numbers.

Derived Rules

T RULE: Any substitution instance of a previous theorem can appear on a line of a derivation with no assumption numbers.

SHORT-CUT RULES

MT: $\varphi \to \psi$
CONTRA(ψ)
∴ CONTRA(φ)

MTP: $\varphi \lor \psi$
CONTRA(ψ)
∴ φ

$\varphi \lor \psi$
CONTRA(φ)
∴ ψ

BC: $\varphi \leftrightarrow \psi$
∴ $\varphi \to \psi$

$\varphi \leftrightarrow \psi$
∴ $\psi \to \varphi$

CTB: $\varphi \to \psi$
$\psi \to \varphi$
∴ $\varphi \leftrightarrow \psi$

DN: φ
$\neg\neg\varphi$

SIMP: $\varphi \land \psi$
∴ φ

$\varphi \land \psi$
∴ ψ

CONJ: φ
ψ
∴ $\varphi \land \psi$

ADJ: φ
∴ $\varphi \lor \psi$

φ
∴ $\psi \lor \varphi$

BD: $\varphi \leftrightarrow \psi$
φ
∴ ψ

$\varphi \leftrightarrow \psi$
ψ
∴ φ

$\varphi \leftrightarrow \psi$
CONTRA(φ)
∴ CONTRA(ψ)

$\varphi \leftrightarrow \psi$
CONTRA(ψ)
∴ CONTRA(φ)

SOC: $\varphi \to \chi$
$\psi \to \theta$
∴ $(\varphi \lor \psi) \to (\chi \lor \theta)$

$\varphi \to \chi$
$\psi \to \theta$
$\varphi \lor \psi$
∴ $\chi \lor \theta$

$\varphi \to \chi$
$\psi \to \chi$
∴ $(\varphi \lor \psi) \to \chi$

$\varphi \to \chi$
$\psi \to \chi$
$\varphi \lor \psi$
∴ χ

2.8 (Optional) AN AXIOMATIC VERSION OF SL

The derivational system presented so far is a kind of natural deduction system. We can also present sentential logic from an axiomatic perspective. To do that, we first specify a set of sentences called "axioms", and then we derive theorems from those axioms using MP as the sole inference rule. Here are the three axioms we are going to use:

A1: $P \to (Q \to P)$
A2: $(P \to (Q \to R)) \to ((P \to Q) \to (P \to R))$
A3: $(\neg Q \to \neg P) \to (P \to Q)$

Only MP can be used in conjunction with A1-A3, and the Theorem Rule permitting (substitution instances of) previously derived theorems can be used as well. We call sentences that can be derived this way 'theorems of Ax-Sent' ('Ax-Sent' is short for "an axiomatic version of sentential logic"). Derive in order the following theorems of Ax-Sent:

TT1: $P \to P$
1. $P \to ((P \to P) \to P)$ A1
2. $[P \to ((P \to P) \to P)] \to [(P \to (P \to P)) \to (P \to P)]$ A2
3. $(P \to (P \to P)) \to (P \to P)$ MP(1,2)
4. $P \to (P \to P)$ A1
5. $P \to P$ MP(3,4)

TT2: $P \to (P \to Q) \to (P \to Q)$ *exercise*

[HINT: Begin with '$(P \to (P \to Q)) \to ((P \to P) \to (P \to Q))$' as an instance of A2. Then use that entire sentence as the antecedent of another instance of A2, whose consequent is '$[(P \to (P \to Q)) \to (P \to P)] \to [(P \to (P \to Q)) \to (P \to Q)]$'. By MP the consequent itself is reached. Its antecedent can be gotten by taking the instance of A1, where 'Q' is replaced by '$P \to (P \to Q)$' and 'P' is replaced by '$P \to P$', and then by using MP together with '$P \to P$' from TT1. Another use of MP then gives the desired conclusion.]

TT3: ¬P → (P → Q)
1. [(¬Q → ¬P) → (P → Q)] → [¬P → ((¬Q → ¬P) →
   (P → Q))]   A1
2. (¬Q → ¬P) → (P → Q)   A3
3. ¬P → ((¬Q → ¬P) → (P → Q))   MP(1,2)
4. [¬P → ((¬Q → ¬P) → (P → Q))] →
   [(¬P → (¬Q → ¬P)) → (¬P → (P → Q))]   A2
5. (¬P → (¬Q → ¬P)) → (¬P → (P → Q))   MP(3,4)
6. ¬P → (¬Q → ¬P)   A1
7. ¬P → (P → Q)   MP(5,6)

TT4: (¬P → P) → (¬P → Q)   *exercise*

TT5: ¬¬P → P
1. ¬¬P → (¬P → ¬¬¬P)   TT3
2. (¬P → ¬¬¬P) → (¬¬P → P)   A3
3. [(¬P → ¬¬¬P) → (¬¬P → P)] →
   [¬¬P → ((¬P → ¬¬¬P) → (¬¬P → P))]   A1
4. ¬¬P → [(¬P → ¬¬¬P) → (¬¬P → P)]   MP(2,3)
5. [¬¬P → ((¬P → ¬¬¬P) → (¬¬P → P))] →
   [(¬¬P → (¬P → ¬¬¬P)) → (¬¬P → (¬¬P → P))]   A2
6. [¬¬P → (¬P → ¬¬¬P)] → [¬¬P → (¬¬P → P)]   MP(4,5)
7. ¬¬P → (¬¬P → P)   MP(1,6)
8. (¬¬P → (¬¬P → P)) → (¬¬P → P)   TT2
9. ¬¬P → P   MP(7,8)

TT6: P → ¬¬P   *exercise*
TT7: [P → (P → Q)] → (P → Q)   *exercise*

Deriving further theorems in Ax-Sent (using only MP and the previous theorems and axioms) can be extremely difficult. In order to make derivations easier, we will first informally demonstrate a procedure called "inserting antecedents". The idea is to demonstrate in a loose (but we hope <u>convincing</u>) sort of way that whenever a sentence appears on a line of a derivation of Ax-Sent, we can get a conditional on a further line with that sentence as consequent and any antecedent we wish. Let's take a specific example. TT1 is 'P → P'. Let's suppose we would like to "insert 'Q'" in front of 'P → P'. Here's what we do:

1. P → P   TT1
2. (P → P) → (Q → (P → P))   A1
3. Q → (P → P)   MP(1,2)

That was a pretty easy example; we just used A1 and MP. Suppose we want to continue with this and place '(R → P)' in front of the sentence on line 3. We continue this way:

    4. $[Q \to (P \to P)] \to [(R \to P) \to (Q \to (P \to P))]$ A1
    5. $(R \to P) \to (Q \to (P \to P))$ MP(3,4)

Before going any further, we will first explain why we're inserting antecedents. We wish to demonstrate that it is okay in Ax-Sent to assume the antecedents of conditionals we want to derive. The reason it is okay is that when we wind up with the consequent we desired, we can insert the antecedents and remove them as assumptions — something like the CD Rule. Let's take an example of a helpful theorem that we would like to be able to derive, but which would be difficult without making any assumptions. On line 1 below we assume 'P' and try to arrive at '¬P → Q':

TT8: $P \to (\neg P \to Q)$
    1. P 1,A
    2. $P \to (\neg P \to P)$ A1
    3. $\neg P \to P$ 1,MP(1,2)
    4. $(\neg P \to P) \to (\neg P \to Q)$ TT3
    5. $\neg P \to Q$ 1,MP(3,4)

Now, we insert 'P' in front of every sentence on every line of the above derivation (except 3 because line 2 is the desired sentence). The derivation below will have no assumptions and be a derivation in Ax-Sent of '$P \to (\neg P \to Q)$':

    1.0. $P \to P$ TT1
    1.1. $[P \to (\neg P \to P)] \to [P \to (P \to (\neg P \to P))]$ A1
    1.2. $P \to (\neg P \to P)$ A1
    2.0. $P \to (P \to (\neg P \to P))$ MP(1.1,1.2)
    3.1. $[(\neg P \to P) \to (\neg P \to Q)] \to [P \to ((\neg P \to P) \to (\neg P \to Q))]$ A1
    3.2. $(\neg P \to P) \to (\neg P \to Q)$ TT3
    4.0. $P \to ((\neg P \to P) \to (\neg P \to Q))$ MP(3.1,3.2)
    4.1. $[P \to ((\neg P \to P) \to (\neg P \to Q))] \to [(P \to (\neg P \to P)) \to (P \to (\neg P \to Q))]$ A2
    4.2. $[P \to (\neg P \to P)] \to [P \to (\neg P \to Q)]$ MP(4.1,4.2)
    4.3. $P \to (\neg P \to P)$ A1
    5.0. $P \to (\neg P \to Q)$ MP(4.2,4.3)

Notice that the assumption 'P' was eliminated in the actual derivation of 'P → (¬P → Q)' when we inserted 'P' as an antecedent. Why? Because inserting 'P' as an antecedent turns the assumption of 'P' into 'P → P', which is a theorem. In this manner we can <u>always</u> assume the antecedents of sentences we wish to derive, and then eliminate those assumptions (once we get the desired consequent), just as we eliminated 'P'.

Here is one version of the theorem we have been describing that authorizes our assuming antecedents of conditionals we wish to derive:

(A) DEDUCTION THEOREM: In Ax-Sent if there is a derivation of ψ when φ is assumed, then $\varphi \to \psi$ is a theorem.

A more symbolic way to say the above is:

(B) DEDUCTION THEOREM:

$$\text{If } \varphi \vdash_{\text{Ax-Sent}} \psi, \text{ then } \vdash_{\text{Ax-Sent}} (\varphi \to \psi)$$

The completely general version of the DEDUCTION THEOREM involves a potentially infinite set of assumptions Γ — though, of course only a finite bunch of them can be used in a given derivation. We take this potentially infinite set of sentences Γ (even though we'll use only finitely many) together with some more assumptions, say $\varphi_1, \varphi_2, \ldots, \varphi_n$ that will become antecedents, and we get this completely general formulation of the:

(C) DEDUCTION THEOREM:[1]

$$\text{If } \Gamma \cup \{\varphi_1, \varphi_2, \ldots, \varphi_n\} \vdash_{\text{Ax-Sent}} \psi, \text{ then}$$
$$\Gamma \vdash_{\text{Ax-Sent}} \varphi_1 \to (\varphi_2 \to \ldots \to (\varphi_n \to \psi) \ldots).$$

(Our claim above that the DEDUCTION THEOREM is correct is based on examples and intuitions. To make the claim

---

[1] The history of the DEDUCTION THEOREM is a bit involved. Normally it is credited to Herbrand and the year most frequently cited is 1930, though Herbrand used it without proof as early as 1928. Tarski indicates, however, that he established it as early as 1921, and used it to produce further results, some of which were done in collaboration with Lindenbaum (in Tarski [1956], p.32, footnote). Herbrand died in 1931 at the age of 23, while climbing in the Alps.

utterly convincing requires a proof [by Complete Induction on the length of the derivation of ψ].)

Returning to our task of arriving at theorems in Ax-Sent, this latest version of the DEDUCTION THEOREM tells us we can assume all the antecedents of a conditional we wish. Once we obtain the desired consequent, the DEDUCTION THEOREM assures us that a derivation without assumptions of the entire conditional can be constructed. If we like, we can actually construct a derivation by using our (informally explained) method of "inserting antecedents".

Let's look at a highly desirable theorem, which expresses the transitivity of the '$\rightarrow$':

TT9: $(P \rightarrow Q) \rightarrow ((Q \rightarrow R) \rightarrow (P \rightarrow R))$

Instead of deriving TT9, we will establish this:

$\{P \rightarrow Q,\ Q \rightarrow R,\ P\} \vdash R$

1.	$P \rightarrow Q$	1, A
2.	$Q \rightarrow R$	2, A
3.	P	3, A
4.	Q	1,3 MP
5.	R	1,2,3 MP(2,4)

The instance of the DEDUCTION THEOREM that assures us that TT9 is a theorem is:

DEDUCTION THEOREM FOR TT9:
$\quad$ If $\emptyset \cup \{P \rightarrow Q,\ Q \rightarrow R,\ P\} \vdash_{\text{Ax-Sent}} R$, then
$\quad \emptyset \vdash_{\text{Ax-Sent}} (P \rightarrow Q) \rightarrow ((Q \rightarrow R) \rightarrow (P \rightarrow R))$

Here, the empty set $\emptyset$ takes the place of Γ in (C), and '$P \rightarrow Q$', '$Q \rightarrow R$', and 'P' are assumptions φ_1, φ_2, and φ_3 respectively.

TT10: $[P \rightarrow (Q \rightarrow R)] \rightarrow [Q \rightarrow (P \rightarrow R)]$
TT11: $(P \rightarrow Q) \rightarrow (\neg Q \rightarrow \neg P)$

1.	$P \rightarrow Q$	1, A
2.	$\neg\neg P \rightarrow P$	TT5
3.	$(\neg\neg P \rightarrow P) \rightarrow [(P \rightarrow Q) \rightarrow (\neg\neg P \rightarrow Q)]$	TT9
4.	$\neg\neg P \rightarrow Q$	1, MP(twice)
5.	$Q \rightarrow \neg\neg Q$	TT6
6.	$(\neg\neg P \rightarrow Q) \rightarrow [(Q \rightarrow \neg\neg Q) \rightarrow (\neg\neg P \rightarrow \neg\neg Q)]$	TT9
7.	$\neg\neg P \rightarrow \neg\neg Q$	1, MP(twice)

8. $(\neg\neg P \to \neg\neg Q) \to (\neg Q \to \neg P)$ A3
9. $\neg Q \to \neg P$ 1, MP(7,8)

TT12: $(P \to \neg Q) \to (Q \to \neg P)$

The next theorem can be derived elegantly without the use of the DEDUCTION THEOREM. See whether you can derive it <u>without</u> looking at the derivation below (if you do look, see whether you can produce a <u>different</u> derivation of TT13, either with or without the use of the DEDUCTION THEOREM):

TT13: $(\neg P \to P) \to P$
1. $(\neg P \to P) \to (\neg P \to \neg(\neg P \to P))$ TT4
2. $[\neg P \to \neg(\neg P \to P)] \to [(\neg P \to P) \to P]$ A3
3. $(\neg P \to P) \to ((\neg P \to P) \to P)$ 1,2 TT9 (MP twice)
4. $(\neg P \to P) \to P$ 4, TT7

TT14: $(P \to \neg P) \to \neg P$
TT15: $(P \to Q) \to ((P \to \neg Q) \to \neg P)$

At this point it should start to become clear that for the language of sentential logic including only the connectives '$\neg$' and '$\to$', Ax-Sent is <u>exactly</u> as strong as our system of rules including A, MP, CD, and CONTRA. If we add the additional connectives to Ax-Sent and we add the INT rules, the two systems contain exactly the same theorems! (The <u>proof</u> that these systems are equivalent is an exercise at the end of Chapter 4.)

You may wish at this point to derive a few more theorems of Ax-Sent, perhaps adding the INT rules, in order to compare it with our earlier system.

CHAPTER 3

THE SEMANTICS OF SENTENTIAL LOGIC

TOPICS: *boolean interpretation, truth for a boolean interpretation, model of a set, satisfiable, tautology, tautological consequence, proving facts about models, similarity between proofs and derivations, strategy for proofs, "informal" vs. "formal" proofs, the "crux" of a proof, the structure of a proof, compactness.*

In Chapter 1 we began describing syntactic features of sentential logic — those features pertaining to the formal arrangement of strings of symbols. In that chapter we first focussed attention on "the sentence". Rules were presented defining which arrangements of symbols were sentences and which were not. For example, the string '(P $\rightarrow$ ¬(Q v R))' *is* an official sentence of L_{SL}, whereas the string of symbols '(P ¬Q ($\rightarrow$ v R))' is *not* a sentence. Later in Chapter 1 a "derivation" was defined to consist of certain sequences of sentences (together with numbers to the right of them). Just as the rules for forming sentences allow us to visually distinguish a sentence from a non-sentence, the rules of inference allow us to distinguish a derivation from a non-derivation, based only on the formal arrangement of sentences and numbers. Chapter 2 extended those rules of inference to make derivations easier and more intuitive.

Those features of sentences and derivations presented in Chapters 1 and 2 are <u>syntactic</u> features — those based only on symbol patterns. Here in this chapter we are interested in the *meanings* of sentences of L_{SL}. So, we investigate the <u>semantic</u> features of sentential logic. In order to give meaning to sentences, we first postulate an entity called "a boolean interpretation" that determines for any sentence of L_{SL} whether it is true or false.

3.1 BOOLEAN INTERPRETATION

A boolean interpretation $\mathcal{B}$ assigns either **T** or **F** (<u>but not both</u>) to each sentence letter of L_{SL}.

For example, take all sentence letters:

$$P, Q, R, S, T, P_1, \ldots, T_1, P_2, \ldots,$$

and define a boolean interpretation $\mathcal{B}_1$ to assign **T** to all letters P, R, T with or without subscripts and to assign **F** to all the other sentence letters with or without subscripts We will make boxes to indicate boolean interpretations, like this:

$\mathcal{B}_1$

P_i, R_i, T_i	: **T**
Q_i, S_i,	: **F**

where i stands for any subscript or no subscript at all.

Thus, for example, $\mathcal{B}_1(P) = \mathbf{T}$, $\mathcal{B}_1(R) = \mathbf{T}$, $\mathcal{B}_1(Q) = \mathbf{F}$, $\mathcal{B}_1(Q_{1000}) = \mathbf{F}$, $\mathcal{B}_1(R_{43}) = \mathbf{T}$, $\mathcal{B}_1(T_9) = \mathbf{T}$.

Exercises:

What values does $\mathcal{B}_1$ assign to:
1. Q_{45} **2.** S_{21} **3.** R_{743} **4.** T_{101} **5.** P_{10}

3.2 TRUTH FOR A BOOLEAN INTERPRETATION

We now define what it means for **any sentence φ to be True (or False) for a boolean interpretation $\mathcal{B}$**. To do this we take the assignment a boolean interpretation $\mathcal{B}$ makes to the sentence letters of L_{SL} and we <u>extend</u> it to cover <u>all the sentences of the language</u>. We know that such an extension can be made so that each sentence will be uniquely **True** or **False**, since each sentence is uniquely readable. If a given sentence letter θ is assigned **T** by a boolean interpretation, say, $\mathcal{B}_{15}$, then we say "θ is **True** for $\mathcal{B}_{15}$" or "$\mathcal{B}_{15}$ of θ is **True**", and we write: $\mathcal{B}_{15}(\theta) = \mathbf{T}$. If for some sentence letter θ', $\mathcal{B}_{15}$ assigns the value **F** to θ', then θ' is **False** for $\mathcal{B}_{15}$.

We now describe the intuitive meaning of the connectives and how the truth of a molecular sentence is related to the truth of its component sentences.

The symbol '¬' is meant to convey "negation". So, since $\mathcal{B}_{15}$ of θ is **True**, $\mathcal{B}_{15}$ of ¬θ is **False**. And, since θ' is **False** for $\mathcal{B}_{15}$, ¬θ' is **True** for $\mathcal{B}_{15}$. The idea is that a negation has the truth value that the <u>un</u>negated sentence does <u>not</u> have, for a given boolean interpretation.

Similarly, since '∧' is intended to convey the idea of "and", we want a conjunction of two sentences to be **True** for a given boolean interpretation just in case both conjuncts are **True** for that boolean interpretation.

'∨' expresses something like "or" in English — at least <u>inclusive</u> "or". We want a disjunction of two sentences to be **True** for a given boolean interpretation just in case <u>either or both</u> of the disjuncts are **True** for that boolean interpretation.

"If and only if" and "just in case", two phrases not often encountered outside of mathematical contexts, are represented by the '↔' symbol. A biconditional is **True** for a given boolean interpretation just in case both sides of the biconditional have the same truth value for that boolean interpretation. That is to say, a biconditional is **True** for a given boolean interpretation if either both sides are **True** for that interpretation or both sides are **False** for it.

We have saved the most problematic case for last, the conditional, reflected by the '→' symbol. The truth conditions for the conditional in L_{SL} correspond somewhat to the truth conditions for "if ..., then _ _ _" statements in ordinary English. But, it is unclear exactly when an "if ..., then _ _ _" statement of ordinary English is true. Two cases seem clear: when the "if" part is true and the "then" part is false, the "if ..., then _ _ _" is <u>false</u> in ordinary English, and when the "if" part is true and the "then" part is also true, then the "if ..., then _ _ _" is <u>true</u>. But, what about the cases in ordinary English when the "if" is false? In L_{SL} we stipulate that the conditional be treated as **True** when the antecedent is **False** (for a given boolean interpretation). (We believe good reasons can be given for accepting the truth of a conditional in ordinary English whose antecedent is false, but we won't give them here.) We define a conditional sentence of L_{SL} to be **True** for a given boolean interpretation just in case either the antecedent and consequent are <u>both</u> **True** for that boolean interpretation, or the antecedent is **False** for that boolean interpretation. Or, an equivalent way to express this is to say that the <u>only</u> case when a conditional is **False** for a given boolean interpretation is when <u>both</u> the antecedent is

True and the consequent is **False** for that boolean interpretation.

In this way, as we said above, by explaining how the truth of compound sentences depends on the truth of its components, we extend the truth assignment (for a given boolean interpretation) from just sentence letters to all the sentences. Now we present again the conditions under which a compound sentence is true, but this time in a more formal way.

3.3 FORMAL DEFINITION OF TRUTH FOR A BOOLEAN INTERPRETATION

The formal definition of what it means for **a given sentence φ of L_{SL} to be True (or False) for a boolean interpretation $\mathcal{B}$** proceeds as follows:

(i) if φ is a sentence letter, θ, then $\mathcal{B}(\theta) = T$ iff $\mathcal{B}$ assigns T to φ.

(ii) if φ is a negation, $\neg\psi$, then $\mathcal{B}(\neg\psi) = T$ iff $\mathcal{B}(\psi) = F$.

(iii) if φ is a conditional, $(\psi \rightarrow \chi)$, then $\mathcal{B}(\psi \rightarrow \chi) = T$ iff either $\mathcal{B}(\psi) = F$ or $\mathcal{B}(\chi) = T$ (or both).

(iv) if φ is a conjunction, $(\psi \wedge \chi)$, then $\mathcal{B}(\psi \wedge \chi) = T$ iff $\mathcal{B}(\psi) = T$ and $\mathcal{B}(\chi) = T$.

(v) if φ is a disjunction, $(\psi \vee \chi)$, then $\mathcal{B}(\psi \vee \chi) = T$ iff either $\mathcal{B}(\psi) = T$ or $\mathcal{B}(\chi) = T$ (or both).

(vi) if φ is a biconditional, $(\psi \leftrightarrow \chi)$, then $\mathcal{B}(\psi \leftrightarrow \chi) = T$ iff $\mathcal{B}(\psi) = \mathcal{B}(\chi)$ (i.e., both are T or both are F).

Examples: Here is $\mathcal{B}_1$ again:

$\mathcal{B}_1$

P_i, R_i, T_i	:	T
Q_i, S_i	:	F

Now we ask whether '$(Q \rightarrow \neg P)$' is **True** for $\mathcal{B}_1$. Since $\mathcal{B}_1(Q) = F$, $\mathcal{B}_1(Q \rightarrow \neg P) = T$ by clause (iii). So, '$(Q \rightarrow \neg P)$' is **True** for $\mathcal{B}_1$.

Is '$(P \rightarrow \neg Q)$' **True** for $\mathcal{B}_1$? Well, $\mathcal{B}_1(Q) = F$. Thus, by (ii), $\mathcal{B}_1(\neg Q) = T$. And, by (iii), $\mathcal{B}_1(P \rightarrow \neg Q) = T$ (since the consequent is T). So, '$(P \rightarrow \neg Q)$' is **True** for $\mathcal{B}_1$.

Is '$(P \rightarrow Q)$' **True** for $\mathcal{B}_1$? Since both $\mathcal{B}_1(P) = T$ and $\mathcal{B}_1(Q) = F$, $\mathcal{B}_1(P \rightarrow Q) = F$, by (iii). So, '$(P \rightarrow Q)$' is **False** for $\mathcal{B}_1$.

Let's consider: $(P \leftrightarrow \neg Q)$. $\mathcal{B}_1(P \leftrightarrow \neg Q) = T$, by (vi), since $\mathcal{B}_1(P) = \mathcal{B}_1(\neg Q)$. So, '$(P \leftrightarrow \neg Q)$' is **True** for $\mathcal{B}_1$.

Exercises:

Evaluate whether the following sentences are **True** or **False** for $\mathcal{B}_1$ above:
1. $(P \land \neg Q)$
2. $(P \lor Q)$
3. $(P \leftrightarrow \neg Q) \land (\neg P \leftrightarrow Q)$
4. $(P \rightarrow \neg Q) \lor (\neg P \leftrightarrow Q)$
5. $(\neg(\neg P \land \neg Q) \land \neg Q)$

3.4 BOOLEAN MODEL OF A SET

If <u>every</u> sentence belonging to a set Γ of sentences of L_{SL} is true for a boolean interpretation $\mathcal{B}$, then $\mathcal{B}$ **is a (boolean) model of** Γ (or $\mathcal{B}$ **models** Γ). Take, for example, the boolean interpretation $\mathcal{B}_1$ defined above. $\mathcal{B}_1(R_3) = T$, $\mathcal{B}_1(P_5) = T$, and $\mathcal{B}_1(T_7) = T$. Thus $\mathcal{B}_1$ models $\{R_3, P_5, T_7\}$. Also, $\mathcal{B}_1(R_3 \land P_5) = T$, by (iv), $\mathcal{B}(R_3 \lor Q) = T$ by (v), and, $\mathcal{B}(Q \rightarrow S) = T$ by (iii). So, $\mathcal{B}_1$ models $\{(R_3 \land P_5), (R_3 \lor Q), (Q \rightarrow S)\}$. But, $\mathcal{B}_1$ does <u>not</u> model $\{R_3, P_5, T_7, (R_3 \land Q)\}$ because $\mathcal{B}_1(R_3 \land Q) = F$.

'$\mathcal{B}(\Gamma) = T$' is short for '$\mathcal{B}$ models Γ'. Also, we will sometimes write '$\mathcal{B}$ **models** φ' to mean $\mathcal{B}$ models $\{\varphi\}$, which means 'φ is true for $\mathcal{B}$' (or '$\mathcal{B}(\varphi) = T$'). Then '$\mathcal{B}$ does not model φ' means 'φ is false for $\mathcal{B}$' ($\mathcal{B}(\varphi) = F$).

If **no** boolean interpretation models a set Γ, then Γ is **not satisfiable**, or **unsatisfiable**. That is, Γ **is satisfiable** if Γ has a model. For example, if $\Gamma = \{R, \neg R\}$, then Γ is <u>unsatisfiable</u>, since <u>no single</u> boolean interpretation can model <u>all</u> the sentences in Γ.

If Γ is a set containing a single sentence, say φ, (i.e., $\Gamma = \{\varphi\}$), we will sometimes say φ **is satisfiable**, meaning that $\{\varphi\}$ (i.e., Γ) is satisfiable, and we will say φ **is not satisfiable (or unsatisfiable)**, meaning that $\{\varphi\}$ is not satisfiable.

Exercises:

For each set of sentences below create a model of that set <u>or</u> say why the set is not satisfiable. [#7 and #10 are solved below.]

1. $\{P, P \to Q, Q\}$
2. $\{P \to Q, \neg Q, \neg P\}$
3. $\{P \to Q, P \to \neg Q\}$
4. $\{(P \to Q) \to P, \neg Q\}$
5. $\{(P \vee Q) \to R, \neg R\}$
6. $\{\neg(P \wedge Q) \wedge \neg(R \wedge T) \wedge (P \wedge R)\}$
7. $\{P \to Q, \neg Q, P\}$
8. $\{P \vee Q, \neg P, \neg Q\}$
9. $\{P \leftrightarrow Q, \neg P, Q\}$
10. $\{P \leftrightarrow Q, \neg P, \neg Q\}$
11. $\{P \wedge Q, \neg(R \wedge S), S \to \neg P, R \to \neg Q\}$
12. $\{\neg(P \leftrightarrow Q), P, \neg Q\}$
13. $\{((P \to Q) \vee (Q \to R)) \to (S \wedge T)\}$

<u>Solutions</u> for **7, 10**:

    7) Any interpretation, $\mathcal{B}$, that models 'P $\to$ Q' must either not model 'P' or model 'Q' by (iii). Since '¬Q' is in the set, $\mathcal{B}$ must model '¬Q'. So, $\mathcal{B}$ does <u>not</u> model 'Q'. Therefore, $\mathcal{B}$ must not model 'P' if $\mathcal{B}$ is to model 'P $\to$ Q'. But, since 'P' is itself in the set, $\mathcal{B}$ must model 'P'. But, $\mathcal{B}$ cannot both model 'P' and not model 'P'. Therefore, since $\mathcal{B}$ was a completely arbitrary boolean interpretation, the set has no model. That is, it is not satisfiable.

    10)

$\mathcal{B}_2$

P, Q :	F
others:	T

Since $\mathcal{B}_2(P) = F$ and $\mathcal{B}_2(Q) = F$, $\mathcal{B}_2(P) = \mathcal{B}_2(Q)$. So, by (vi), $\mathcal{B}_2(P \leftrightarrow Q) = T$.

Therefore, $\mathcal{B}_2$ models $\{P \leftrightarrow Q, \neg P, \neg Q\}$

For any sentence φ of L_{SL}, if φ is true for every boolean interpretation, then φ **is a tautology**, written: $\models \varphi$.

Example 1: 'P $\to$ P' is a tautology. Let us prove this.

PF: Consider an arbitrary boolean interpretation $\mathcal{B}$. Either $\mathcal{B}(P) = \mathbf{T}$ or $\mathcal{B}(P) = \mathbf{F}$, we do not know which. Suppose $\mathcal{B}(P) = \mathbf{T}$. Then, by clause (iii), $\mathcal{B}(P \rightarrow P) = \mathbf{T}$, since the consequent is $\mathbf{T}$. Now, suppose $\mathcal{B}(P) = \mathbf{F}$. By clause (iii) again $\mathcal{B}(P \rightarrow P) = \mathbf{T}$, this time because the antecedent is $\mathbf{F}$. So, whether $\mathcal{B}(P) = \mathbf{T}$ or whether $\mathcal{B}(P) = \mathbf{F}$ (and it must be one or the other), $\mathcal{B}(P \rightarrow P) = \mathbf{T}$. Therefore, 'P $\rightarrow$ P' is true for any boolean interpretation. Hence, 'P $\rightarrow$ P' is a tautology. ∎

For a sentence φ of L_{SL} and a set of sentences Γ, if any boolean interpretation that models Γ also models φ, then φ **is a tautological consequence of** Γ ($\Gamma \models \varphi$). Another way to put this is that φ **is a tautological consequence of** Γ if there is no boolean interpretation that models Γ and does not model φ. These **two definitions of 'tautological consequence'** are equivalent. Sometimes we will use one, sometimes the other, depending on which one seems more appropriate to the context.

Example 2: 'Q' is a tautological consequence of $\{P \rightarrow Q, P\}$ (i.e., $\{P \rightarrow Q, P\} \models Q$). We will prove this by showing that any boolean interpretation that models $\{P \rightarrow Q, P\}$ also models 'Q':

PF: Suppose $\mathcal{B}(P \rightarrow Q) = \mathbf{T}$ and $\mathcal{B}(P) = \mathbf{T}$. By clause (ii), since $\mathcal{B}(P \rightarrow Q) = \mathbf{T}$, either $\mathcal{B}(P) = \mathbf{F}$ or $\mathcal{B}(Q) = \mathbf{T}$ (or both). Since $\mathcal{B}(P) \neq \mathbf{F}$, then it must be the case that $\mathcal{B}(Q) = \mathbf{T}$. This shows $\mathcal{B}$ models 'Q'. ∎

Example 3: Any sentence φ is a tautological consequence of $\{\psi, \neg\psi\}$ for any sentence ψ (i.e., $\{\psi, \neg\psi\} \models \varphi$). We will use our alternative definition of "tautological consequence" to prove this, by showing there is no boolean interpretation that models $\{\psi, \neg\psi\}$ and does not model φ.

PF: Any boolean interpretation that models ψ does not model $\neg\psi$ and vice versa, by clause (ii). Therefore there is no boolean interpretation that models $\{\psi, \neg\psi\}$. Therefore, there is no boolean interpretation that models $\{\psi, \neg\psi\}$ and does not model φ. (This is a "vacuous truth argument". If you do not understand it, return to the section on vacuous truth in the *Mathematical Preliminaries* chapter.) ∎

3.5 PROVING FACTS ABOUT MODELS

The facts about models that we prove below are not very difficult, but they are quite abstract. If you are unaccustomed to abstract mathematical proofs, you should read this section quite carefully and be prepared to do some *exercises* over and over again until you become comfortable with the patterns of reasoning and can apply them in proofs. For every derivational rule, A, MP, CD, CONTRA, and INT, there is a corresponding principle of reasoning used in mathematical proofs. There is also a corresponding principle of mathematical reasoning for every derived rule. Further, the strategies used to obtain derivations correspond to strategies employed in arriving at mathematical proofs. Thus, having produced derivations *in* L_{SL} should help you to understand how to use corresponding principles *outside of* L_{SL} in mathematical proofs.

In the pages that follow we will show how the derivational rules and strategies for deriving sentences of L_{SL} can be helpful in proving mathematical statements. Typically, we exhibit a normal mathematical proof of a given proposition, often followed by an *ANALYSIS* of the proof. Then, we exhibit what we call a **semi-formal proof** of the same proposition that closely resembles a derivation in L_{SL}. A "semi-formal proof" contains much more structure than an "informal proof". A **formal proof**, by contrast, is one that is written in a language whose grammar has been mathematically defined in advance and whose rules of inference have been previously specified for well-defined grammatical units of that language. Our derivations in L_{SL} are, in this sense, formal proofs. But, our semi-formal proofs are <u>not</u> formal, since the language in which they are written is informal (i.e., it is everyday English supplemented by some set-theoretic expressions), and the inference rules have not been fixed in advance. Our semi-formal proofs are merely "dressed up proofs" in the sense of dressing up normal (or informal) proofs in semi-formal attire.

In the theorems that follow, $\mathcal{B}$ is always a boolean interpretation, Γ and Δ are sets of sentences of L_{SL}, φ, ψ, and χ are sentences of L_{SL} and $\emptyset$ is the empty set.

THEOREM 3.0: For any boolean interpretation $\mathcal{B}$,
$\quad\quad\quad\quad\mathcal{B}(\emptyset) = \mathbf{T}$.

PF1: It is true that for any sentence φ belonging to $\emptyset$ that $\mathcal{B}(\varphi) = T$, **BECAUSE THERE ARE NONE** (it's true *vacuously*). ∎

PF2: Suppose the theorem is false. Then it is <u>not</u> true for every boolean interpretation that for any $\varphi \in \emptyset$ that $\mathcal{B}(\varphi) = T$. Therefore, there is some boolean interpretation, $\mathcal{B}_0$, such that $\mathcal{B}_0(\varphi_1) \neq T$, where $\varphi_1 \in \emptyset$. That is, $\mathcal{B}_0(\varphi_1) = F$. But, since there are <u>no</u> sentences in $\emptyset$, there is no sentence φ_1 such that $\mathcal{B}_0(\varphi_1) = F$. Thus, the theorem can't be false. Hence, it is true (by the WCBFMBT principle). ∎

In the statement of many theorems to follow there is an implicit 'For **any** . . .' prefix, where model(s), set(s), and sentences are concerned. For example, THEOREM 3.1 below is written as 'If $\mathcal{B}(\Gamma) = T$ and $\mathcal{B}(\Delta) = T$, then $\mathcal{B}(\Gamma \cup \Delta) = T$', but what it says is: For any boolean interpretation $\mathcal{B}$ and any sets of sentences of L_{SL}, Γ and Δ, if $\mathcal{B}(\Gamma) = T$ and $\mathcal{B}(\Delta) = T$, then $\mathcal{B}(\Gamma \cup \Delta) = T$.

THEOREM 3.1: If $\mathcal{B}(\Gamma) = T$ and $\mathcal{B}(\Delta) = T$, then $\mathcal{B}(\Gamma \cup \Delta) = T$.

PF(INFORMAL): Consider any sentence $\varphi \in (\Gamma \cup \Delta)$. Either $\varphi \in \Gamma$ or $\varphi \in \Delta$. By hypothesis, if $\varphi \in \Gamma$, $\mathcal{B}(\varphi) = T$; and if $\varphi \in \Delta$, then $\mathcal{B}(\varphi) = T$. Thus, in either case $\mathcal{B}(\varphi) = T$. So, $\mathcal{B}(\Gamma \cup \Delta) = T$. ∎

ANALYSIS: The first thing to note is the <u>form</u> of THEOREM 3.1. It is a **hypothetical** sentence. The sentence between the 'if' and the 'then' is the **hypothesis**, and the sentence after the 'then' is the **conclusion**. In deriving conditional sentences of L_{SL}, we normally assumed the antecedent and tried to derive the consequent. In proving hypothetical sentences, *outside of* L_{SL}, we do the same. For example, let's look at a semi-formal proof of THEOREM 3.1. First, we write a TO PROVE line indicating what it is we wish to prove (like our SHOW lines earlier):

TO PROVE: If $\mathcal{B}(\Gamma) = T$ and $\mathcal{B}(\Delta) = T$, then $\mathcal{B}(\Gamma \cup \Delta) = T$.

Then we write down the hypothesis as an assumption:

1. $\mathcal{B}(\Gamma) = T$ and $\mathcal{B}(\Delta) = T$ *hypothesis*

And we write another "TO PROVE" line to indicate that now we wish to prove the conclusion:

   TO PROVE: $\mathcal{B}(\Gamma \cup \Delta) = T$

We next ask what '$\mathcal{B}(\Gamma \cup \Delta) = T$' <u>means</u>. It makes sense to ask this question periodically because understanding the *meaning* of what we're trying to prove will definitely make proving it easier. In this case '$\mathcal{B}(\Gamma \cup \Delta) = T$' <u>means</u> 'for any sentence φ, if $\varphi \in (\Gamma \cup \Delta)$, then $\mathcal{B}(\varphi) = T$.' So, we leave the 'for any sentence φ' implicit, and we write:

   TO PROVE: If $\varphi \in (\Gamma \cup \Delta)$, then $\mathcal{B}(\varphi) = T$.

This leads naturally to our assuming the hypothesis of the TO PROVE sentence (just as we assumed the antecedent of a conditional that we wanted to SHOW in L_{SL}):

   2. $\varphi \in (\Gamma \cup \Delta)$.   *hypothesis*

Then we wish:

   TO PROVE: $\mathcal{B}(\varphi) = T$.

From 1, we can get the following two statements, by an anolog of SIMP:

   3. $\mathcal{B}(\Gamma) = T$ *from 1*
   4. $\mathcal{B}(\Delta) = T$ *from 1*

We then write on separate lines what lines 3 and 4 mean, as we did above:

   5.   If $\varphi \in \Gamma$, then $\mathcal{B}(\varphi) = T$ *def of 3*
   6.   If $\varphi \in \Delta$, then $\mathcal{B}(\varphi) = T$ *def of 4*

Since we know from line 2 that $\varphi \in (\Gamma \cup \Delta)$, we know:

   7. $\varphi \in \Gamma$ or $\varphi \in \Delta$ *2, def of $\cup$*

It now looks like a separation of cases argument will give us what we want:

   8. $\varphi \in \Gamma$ *SOC assumption*
   9. $\mathcal{B}(\varphi) = T$ *5,8*
   10. $\varphi \in \Delta$ *SOC assumption*

11. $\mathcal{B}(\varphi) = T$		6,10
12. If $\varphi \in \Gamma$ or $\varphi \in \Delta$, then $\mathcal{B}(\varphi) = T$		SOC
13. $\mathcal{B}(\varphi) = T$		2,12

Now, we check off the bottom TO PROVE line to show the proof is complete.

Here is what the semi-formal proof of THEOREM 3.1 looks like with all the steps together:

TO PROVE: If $\mathcal{B}(\Gamma) = T$ and $\mathcal{B}(\Delta) = T$, then $\mathcal{B}(\Gamma \cup \Delta) = T$.
1. $\mathcal{B}(\Gamma) = T$ and $\mathcal{B}(\Delta) = T$ hypothesis
        TO PROVE: $\mathcal{B}(\Gamma \cup \Delta) = T$
        TO PROVE: If $\varphi \in (\Gamma \cup \Delta)$, then $\mathcal{B}(\varphi) = T$
2. $\varphi \in (\Gamma \cup \Delta)$ hypothesis
        √TO PROVE: $\mathcal{B}(\varphi) = T$
3. $\mathcal{B}(\Gamma) = T$ from 1
4. $\mathcal{B}(\Delta) = T$ from 1
5. If $\varphi \in \Gamma$, then $\mathcal{B}(\varphi) = T$ def of 3
6. If $\varphi \in \Delta$, then $\mathcal{B}(\varphi) = T$ def of 4
7. $\varphi \in \Gamma$ or $\varphi \in \Delta$ def of 2
8. $\varphi \in \Gamma$ SOC assumption
9. $\mathcal{B}(\varphi) = T$ 5,8
10. $\varphi \in \Delta$ SOC assumption
11. $\mathcal{B}(\varphi) = T$ 6,10
12. If $\varphi \in \Gamma$ or $\varphi \in \Delta$, then $\mathcal{B}(\varphi) = T$ SOC
13. $\mathcal{B}(\varphi) = T$ 2,12

If we look back at the informal proof of THEOREM 3.1, we see that that proof only hints at the longer, semi-formal version we just spelled out. Standard mathematical proofs, like the informal proof of THEOREM 3.1, typically contain few details about their underlying logical organization, because a person reading them is supposed to be able to instinctively grasp their structure even when it has not been made explicit. But sometimes the implicit logical structure of a proof is difficult even for a seasoned mathematician to fathom. So, we will frequently present semi-formal versions of mathematical proofs, in order to clarify the underlying logical principles that are normally left implicit.

In addition to its structure, a proof usually has a "main idea" behind it. When there is just one idea behind the proof, we will speak of "the idea", or "the point", or "the crux" of the proof. For example, the crux of the proof of THEOREM 3.1 is the point that if every element of each of two sets has a certain property, then every element belonging to the union of those sets has the given property. Roughly speaking, there are two aspects to a proof: (1) the main idea behind it, and (2) the separate, individually justifiable statements in its presentation.

Exercises: [#6 is solved below]:

1. If $\mathcal{B}(\Gamma_1) = T$ and $\mathcal{B}(\Gamma_2) = T$ and $\mathcal{B}(\Gamma_3) = T$, then $\mathcal{B}((\Gamma_1 \cup \Gamma_2) \cup \Gamma_3) = T$.
2. If $\varphi \in \Gamma$ and $\mathcal{B}(\Gamma) = T$, then $\mathcal{B}(\varphi) = T$.
3. If $\mathcal{B}(\varphi) = T$ and $\mathcal{B}(\psi) = T$, then $\mathcal{B}(\varphi \wedge \psi) = T$.
4. If $\mathcal{B}(\varphi) = T$, then $\mathcal{B}(\varphi \vee \psi) = T$.
5. If $\mathcal{B}(\psi) = T$, then $\mathcal{B}(\varphi \rightarrow \psi) = T$.
6. If $\mathcal{B}(\varphi) = T$ and $\mathcal{B}(\varphi \rightarrow \psi) = T$, then $\mathcal{B}(\psi) = T$.

Solution to #6:
(I) INFORMALLY:
PF: Since $\mathcal{B}(\varphi \rightarrow \psi) = T$, by hypothesis, *either* $\mathcal{B}(\varphi) = F$ *or* $\mathcal{B}(\psi) = T$ by clause (iii) of our truth definition. But, by hypothesis, $\mathcal{B}(\varphi) \neq F$. Hence $\mathcal{B}(\psi) = T$. ∎

(II) SEMI-FORMALLY:
√TO PROVE: If $\mathcal{B}(\varphi) = T$ and $\mathcal{B}(\varphi \rightarrow \psi) = T$, then $\mathcal{B}(\psi) = T$

1. $\mathcal{B}(\varphi) = T$ and $\mathcal{B}(\varphi \rightarrow \psi) = T$ hypothesis
      √TO PROVE $\mathcal{B}(\psi) = T$
2. $\mathcal{B}(\varphi) = T$ from 1
3. $\mathcal{B}(\varphi \rightarrow \psi) = T$ from 1
4. Either $\mathcal{B}(\varphi) = F$ or $\mathcal{B}(\psi) = T$ from 3 (by truth def.)
5. $\mathcal{B}(\varphi) \neq F$ from 2
6. $\mathcal{B}(\psi) = T$ 4,5
7. If $\mathcal{B}(\varphi) = T$ and $\mathcal{B}(\varphi \rightarrow \psi) = T$, then $\mathcal{B}(\psi) = T$.

We now prove another theorem:

THEOREM 3.2: If $\mathcal{B}(\Delta) = T$ and $\Gamma \subseteq \Delta$, then $\mathcal{B}(\Gamma) = T$

PF: (INFORMAL) We prove for any sentence $\varphi \in \Gamma$ that $\mathcal{B}(\varphi) = T$. Since $\varphi \in \Gamma$ and $\Gamma \subseteq \Delta$, $\varphi \in \Delta$. But $\mathcal{B}(\Delta) = T$ by hypothesis. So, $\mathcal{B}(\varphi) = T$ as well. ∎

PF: (SEMI-FORMAL)
√TO PROVE: If $\mathcal{B}(\Delta) = \mathbf{T}$ and $\Gamma \subseteq \Delta$, then $\mathcal{B}(\Gamma) = \mathbf{T}$
1. $\mathcal{B}(\Delta) = \mathbf{T}$ and $\Gamma \subseteq \Delta$ hypothesis
             √TO PROVE: $\mathcal{B}(\Gamma) = \mathbf{T}$
             √TO PROVE: If $\varphi \in \Gamma$, then $\mathcal{B}(\varphi) = \mathbf{T}$
2. $\varphi \in \Gamma$ Assumption
                     √TO PROVE: $\mathcal{B}(\varphi) = \mathbf{T}$
3. $\mathcal{B}(\Delta) = \mathbf{T}$ from 1
4. $\Gamma \subseteq \Delta$ from 1
5. $\varphi \in \Delta$ 2,4
6. $\mathcal{B}(\varphi) = \mathbf{T}$ 3,5
7. If $\varphi \in \Gamma$, then $\mathcal{B}(\varphi) = \mathbf{T}$ | these three lines
8. $\mathcal{B}(\Gamma) = \mathbf{T}$ | have been added to '√'
9. If $\mathcal{B}(\Delta) = \mathbf{T}$ and $\Gamma \subseteq \Delta$, then $\mathcal{B}(\Gamma) = \mathbf{T}$ | TO PROVE lines above.

ANALYSIS: The crux of the above two proofs is the simple fact that a model of all sentences of a given set is certainly a model of all sentences of a subset of that set. To take a concrete example, suppose $\mathcal{B}(\{P, Q, P \rightarrow R\}) = \mathbf{T}$. Then clearly $\mathcal{B}(\{P, P \rightarrow R\}) = \mathbf{T}$ as well, since $\{P, P \rightarrow R\} \subseteq \{P, Q, P \rightarrow R\}$.

Perhaps the informal proof of THEOREM 3.2 is too short and cryptic, and perhaps the semi-formal one is too long and elaborate. Since proofs that one finds in a math textbook usually resemble informal ones, the student should aim, first, to be able to follow informal proofs, and, later, to be able to construct them.

Long-winded though the above semi-formal proof is, it still contains an area or two of possible mystery. The third TO PROVE line from the top is: TO PROVE: If $\varphi \in \Gamma$, then $\mathcal{B}(\varphi) = \mathbf{T}$. Why does that line follow 'TO PROVE: $\mathcal{B}(\Gamma) = \mathbf{T}$'? The answer is that '$\mathcal{B}(\Gamma) = \mathbf{T}$' is **defined** to mean 'for any sentence φ, if $\varphi \in \Gamma$, then $\mathcal{B}(\varphi) = \mathbf{T}$'. We have stripped the part 'for any sentence φ' in order to focus on the "if ... then _ _ _" nature of that sentence. We hope it is clear that the TO PROVE statement is really about **any** sentence meeting the "if" condition and that line 7 is about **any** sentence belonging to Γ.

Exercises:

Prove the following by exhibiting a semi-formal and an
  informal proof of each:

1. If $(\varphi \rightarrow \psi) \in \Gamma$, $\neg\psi \in \Gamma$ and $\mathcal{B}(\Gamma) = T$, then $\mathcal{B}(\neg\varphi) = T$.
2. If $\mathcal{B}(\Gamma - \{\varphi\}) = T$ and $\mathcal{B}(\varphi) = T$, then $\mathcal{B}(\Gamma) = T$.
3. If $\mathcal{B}(\Gamma \cup \Delta) = T$, then $\mathcal{B}(\Gamma) = T$ and $\mathcal{B}(\Delta) = T$.
4. $\mathcal{B}(\varphi) = F$ iff $\mathcal{B}(\neg\varphi) = T$.
5. Prove one direction of the following and give a counter-example to the other direction:
$$\mathcal{B}(\Gamma \cap \Delta) = T \text{ iff } \mathcal{B}(\Gamma) = T \text{ and } \mathcal{B}(\Delta) = T.$$

THEOREM 3.3: If $\Gamma \models \varphi$ and $\Gamma \subseteq \Delta$, then $\Delta \models \varphi$.

PF: (INFORMAL): Assume that $\Gamma \models \varphi$ and assume that $\Gamma \subseteq \Delta$. We want TO PROVE $\Delta \models \varphi$. That is, we want TO PROVE that for any boolean interpretation $\mathcal{B}$, such that $\mathcal{B}(\Delta) = T$, $\mathcal{B}(\varphi) = T$ as well. Toward that end, we take any $\mathcal{B}$ such that $\mathcal{B}(\Delta) = T$. Since $\Gamma \subseteq \Delta$, THEOREM 3.2 gives us that $\mathcal{B}(\Gamma) = T$. Since $\Gamma \models \varphi$, $\mathcal{B}(\varphi) = T$. So, the proof is finished. ∎

THEOREM 3.2, on which the proof of THEOREM 3.3 is based, says that any boolean model of a set is a boolean model of any of its subsets. THEOREM 3.3 tells us that any consequence of a set is a consequence of any superset of that set. THEOREM 3.3 will come in handy in proving some more complicated-looking theorems:

THEOREM 3.4: If $\Gamma \models \varphi$ and $\Delta \models (\varphi \rightarrow \psi)$, then $(\Gamma \cup \Delta) \models \psi$

First of all, let's analyze the structure of THEOREM 3.4. We know that after assuming the hypothesis, we need TO PROVE: $(\Gamma \cup \Delta) \models \psi$. But, the structure of '$(\Gamma \cup \Delta) \models \psi$' is itself a disguised "if..., then _ _ _", because its definition is: For any $\mathcal{B}$, if $\mathcal{B}(\Gamma \cup \Delta) = T$, then $\mathcal{B}(\psi) = T$. So, to prove that $(\Gamma \cup \Delta) \models \psi$ holds, we take any $\mathcal{B}$ such that $\mathcal{B}(\Gamma \cup \Delta) = T$, and we try to prove that $\mathcal{B}(\psi) = T$. This strategy motivates the following informal proof:

PF (INFORMAL): Take any $\mathcal{B}$ such that $\mathcal{B}(\Gamma \cup \Delta) = T$. By THEOREM 3.2, $\mathcal{B}(\Gamma) = T$ and $\mathcal{B}(\Delta) = T$, since both $\Gamma \subseteq \Gamma \cup \Delta$ and $\Delta \subseteq \Gamma \cup \Delta$. By hypothesis, $\Gamma \models \varphi$ and $\Delta \models \varphi \rightarrow \psi$. So, $\mathcal{B}(\varphi) = T$ and $\mathcal{B}(\varphi \rightarrow \psi) = T$. By solved *exercise* 6 (a few pages back), $\mathcal{B}(\psi) = T$. Hence, we're done. ∎

Here is a semi-formal proof of THEOREM 3.4:

TO PROVE: If $\Gamma \models \varphi$ and $\Delta \models \varphi \to \psi$, then $(\Gamma \cup \Delta) \models \psi$
1. $\Gamma \models \varphi$
2. $\Delta \models \varphi \to \psi$
                      TO PROVE: $(\Gamma \cup \Delta) \models \psi$
                      TO PROVE: If $\mathcal{B}(\Gamma \cup \Delta) = T$,
                                      then $\mathcal{B}(\psi) = T$
3. $\mathcal{B}(\Gamma \cup \Delta) = T$
                √TO PROVE: $\mathcal{B}(\psi) = T$
4. $\mathcal{B}(\Gamma) = T$ 3, THEOREM 3.2
5. $\mathcal{B}(\Delta) = T$ 3, THEOREM 3.2
6. $\mathcal{B}(\varphi) = T$ 1, 4 (+ def. of '$\models$')
7. $\mathcal{B}(\varphi \to \psi) = T$ 2, 5 (+ def. of '$\models$')
8. $\mathcal{B}(\psi) = T$ solved exercise #6

(We stop at line 8 rather than continuing to the bitter end.)

ANALYSIS: The real point to THEOREM 3.4 is contained in solved *exercise* 6 earlier: if $\mathcal{B}(\varphi) = T$ and $\mathcal{B}(\varphi \to \psi) = T$, then $\mathcal{B}(\psi)$. The only additional component to the above proof is provided by THEOREM 3.2, which says that a model of a set models any of its subsets.

In fact, we have many more theorems that seem complicated (like THEOREM 3.4) simply because THEOREM 3.2 or THEOREM 3.3 is built into them. For example, consider the following:

(a) If $\Gamma \models \varphi \to \psi$ and $\Delta \models \neg\psi$, then $(\Gamma \cup \Delta) \models \neg\varphi$
(b) If $\Gamma \models \varphi \to \psi$ and $\Delta \models \varphi \to \neg\psi$, then $(\Gamma \cup \Delta) \models \neg\varphi$

In both cases the set $(\Gamma \cup \Delta)$ before the '$\models$' in the conclusion includes the sets Γ, Δ before the '$\models$' in the hypothesis. Thus, to prove (a) and (b), it is really only necessary to prove:

(a)' If $\mathcal{B}(\varphi \to \psi) = T$ and $\mathcal{B}(\neg\psi) = T$, then $\mathcal{B}(\neg\varphi) = T$
(b)' If $\mathcal{B}(\varphi \to \psi) = T$ and $\mathcal{B}(\varphi \to \neg\psi) = T$, then $\mathcal{B}(\neg\varphi) = T$

since the part of (a) that's not in (a)' is due to THEOREM 3.3, and the same is true of (b)'s relation to (b)'.

Exercises:

Prove the following two theorems either semi-formally or informally (or somewhere in between) by using THEOREM 3.3.

THEOREM 3.5: If $\Gamma \models (\varphi \rightarrow \psi)$ and $\Delta \models \neg\psi$, then $(\Gamma \cup \Delta) \models \neg\varphi$.
THEOREM 3.6: If $\Gamma \models (\varphi \rightarrow \psi)$ and $\Delta \models (\varphi \rightarrow \neg\psi)$, then $(\Gamma \cup \Delta) \models \neg\varphi$.

THEOREM 3.7: $\Gamma \models \varphi$ iff $\Gamma \models \neg\neg\varphi$

ANALYSIS: In order to prove "left side iff right side" we first prove "if left side, then right side", and then we prove "if right side, then left side". This was our basic strategy in deriving biconditionals, and "iff" is very much like '$\leftrightarrow$'.

So, we first reduce THEOREM 3.7 to:

(a) If $\Gamma \models \varphi$, then $\Gamma \models \neg\neg\varphi$, and
(b) If $\Gamma \models \neg\neg\varphi$, then $\Gamma \models \varphi$.

Our next step is to notice that since $\Gamma \subseteq \Gamma$, THEOREM 3.3 applies. That is, we can further reduce THEOREM 3.7 to:

(a)' If $\mathcal{B}(\varphi) = \mathbf{T}$, then $\mathcal{B}(\neg\neg\varphi) = \mathbf{T}$
(b)' If $\mathcal{B}(\neg\neg\varphi) = \mathbf{T}$, then $\mathcal{B}(\varphi) = \mathbf{T}$

Exercises:

Prove (a)' and (b)', and explain how THEOREM 3.3 is used to give us (a) and (b).

Here are some more theorems that should be proved semi-formally as *exercises*. (There is a brief *ANALYSIS* of THEOREM 3.8, and much discussion and proofs of THEOREM 3.9 below.)

THEOREM 3.8: $\Gamma \models (\neg\varphi \rightarrow \psi)$ iff $\Gamma \models (\varphi \vee \psi)$
THEOREM 3.9: $(\Gamma \cup \{\varphi\}) \models \psi$ iff $\Gamma \models (\varphi \rightarrow \psi)$
THEOREM 3.10: If $\Gamma \models (\varphi \wedge \neg\varphi)$ then $(\Gamma - \{\neg\varphi\}) \models \varphi$

ANALYSIS of THEOREM 3.8: Break it into two conditionals, use THEOREM 3.3 to further reduce it, and the rest is easy, using the appropriate clauses from the truth definition.

Here is a semi-formal proof of THEOREM 3.9:

TO PROVE (THEOREM 3.9): $\Gamma \cup \{\varphi\} \models \psi$ iff $\Gamma \models (\varphi \rightarrow \psi)$.
                TO PROVE: If $\Gamma \cup \{\varphi\} \models \psi$,
                          then $\Gamma \models (\varphi \rightarrow \psi)$

1. $(\Gamma \cup \{\varphi\}) \models \psi$
                TO PROVE: $\Gamma \models (\varphi \rightarrow \psi)$
                TO PROVE: If $\mathcal{B}(\Gamma) = \mathbf{T}$,
                          then $\mathcal{B}(\varphi \rightarrow \psi) = \mathbf{T}$
2. $\mathcal{B}(\Gamma) = \mathbf{T}$
                TO PROVE: $\mathcal{B}(\varphi \rightarrow \psi) = \mathbf{T}$
                TO PROVE: either $\mathcal{B}(\varphi) = \mathbf{F}$ or
                                    $\mathcal{B}(\psi) = \mathbf{T}$
3. $\mathcal{B}(\varphi) \neq \mathbf{F}$
                √TO PROVE: $\mathcal{B}(\psi) = \mathbf{T}$
4. $\mathcal{B}(\varphi) = \mathbf{T}$ 3
5. $\mathcal{B}(\Gamma) = \mathbf{T}$ and $\mathcal{B}(\varphi) = \mathbf{T}$ 2,4
6. $\mathcal{B}(\psi) = \mathbf{T}$ 1,5

(We skip the steps leading to the completion of this direction.)
∴ If $\Gamma \cup \{\varphi\} \models \varphi$, then $\Gamma \models \varphi \rightarrow \psi$.

Now, we prove the other direction:

                      TO PROVE: If $\Gamma \models (\varphi \rightarrow \psi)$,
                               then $\Gamma \cup \{\varphi\} \models \psi$
7. $\Gamma \models (\varphi \rightarrow \psi)$
                TO PROVE: $\Gamma \cup \{\varphi\} \models \psi$
                TO PROVE: If $\mathcal{B}(\Gamma \cup \{\varphi\}) = \mathbf{T}$,
                            then $\mathcal{B}(\psi) = \mathbf{T}$
8. $\mathcal{B}(\Gamma \cup \{\varphi\}) = \mathbf{T}$
                √TO PROVE: $\mathcal{B}(\psi) = \mathbf{T}$
9. $\mathcal{B}(\Gamma) = \mathbf{T}$ and $\mathcal{B}(\varphi) = \mathbf{T}$ 8
10. $\mathcal{B}(\Gamma) = \mathbf{T}$ 9
11. $\mathcal{B}(\varphi) = \mathbf{T}$ 9
12. $\mathcal{B}(\varphi \rightarrow \psi) = \mathbf{T}$ 7,10
13. $\mathcal{B}(\psi) = \mathbf{T}$ 11,12

Here is an informal proof of the left-to-right direction of:

THEOREM 3.9: $(\Gamma \cup \{\varphi\}) \models \psi$ iff $\Gamma \models (\varphi \rightarrow \psi)$.

PF (=>): We assume the hypothesis and suppose that the conclusion is <u>not</u> true. So, we assume $(\Gamma \cup \{\varphi\}) \models \psi$, and suppose $\mathcal{B}(\Gamma) = \mathbf{T}$ but $\mathcal{B}(\varphi \rightarrow \psi) \neq \mathbf{T}$. Since $\mathcal{B}(\varphi \rightarrow \psi) \neq \mathbf{T}$, then $\mathcal{B}(\varphi) = \mathbf{T}$ and $\mathcal{B}(\psi) = \mathbf{F}$. Since $\mathcal{B}(\Gamma) = \mathbf{T}$ and $\mathcal{B}(\varphi) = \mathbf{T}$, $\mathcal{B}(\Gamma \cup \{\varphi\}) = \mathbf{T}$. Thus, by hypothesis, $\mathcal{B}(\psi) = \mathbf{T}$, which contradicts $\mathcal{B}(\psi) = \mathbf{F}$, and we're done. ∎

Closer inspection of this informal proof reveals that it is not "essentially" a proof by contradiction, not a WCBFMBT argument. When deriving conditional sentences of L_{SL}, we learned that sometimes it was easier to derive the contrapositive first. That is, the derivation would go something like this:

√SHOW $\varphi \rightarrow \psi$

                             √SHOW $\neg\psi \rightarrow \neg\varphi$

1. $\neg\psi$

                             √SHOW $\neg\varphi$

    ⋮

n. $\neg\varphi$
n+1. $\neg\psi \rightarrow \neg\varphi$
n+2. $(\neg\psi \rightarrow \neg\varphi) \leftrightarrow (\varphi \rightarrow \psi)$
n+3. $\varphi \rightarrow \psi$

Similarly, when proving 'if left side, then right side' outside of L_{SL}, we may sometimes prove 'if not right side, then not left side', the contrapositive of what we wish to prove. The only difference is that in a proof, we do not include the counterparts to lines n+1 through n+3. The reason these extra lines are not included is that the equivalence of the contrapositives is assumed to have been already accepted.

We now prove the left-to-right direction of THEOREM 3.9 by proving its contrapositive. First we prove it semi-formally and then informally. Notice that we get more elegant proofs in both cases:

TO PROVE: If $\Gamma \not\models \varphi \rightarrow \psi$, then $\Gamma \cup \{\varphi\} \not\models \psi$
1. $\Gamma \not\models \varphi \rightarrow \psi$

                  TO PROVE: $\Gamma \cup \{\varphi\} \not\models \psi$
                 √TO PROVE: $\mathcal{B}(\Gamma \cup \{\varphi\}) = \mathbf{T}$ and
                                      $\mathcal{B}(\psi) = \mathbf{F}$ (for some $\mathcal{B}$)
2. $\mathcal{B}(\Gamma) = \mathbf{T}$ and $\mathcal{B}(\varphi \rightarrow \psi) = \mathbf{F}$ (for some $\mathcal{B}$)   1
3. $\mathcal{B}(\Gamma) = \mathbf{T}$ 2
4. $\mathcal{B}(\varphi \rightarrow \psi) = \mathbf{F}$ 2
5. $\mathcal{B}(\varphi) = \mathbf{T}$ and $\mathcal{B}(\psi) = \mathbf{F}$ 4

6.	$\mathcal{B}(\varphi) = \mathbf{T}$	5
7.	$\mathcal{B}(\Gamma \cup \{\varphi\}) = \mathbf{T}$	3,6
8.	$\mathcal{B}(\psi) = \mathbf{F}$	5
9.	$\mathcal{B}(\Gamma \cup \{\varphi\})$ and $\mathcal{B}(\psi) = \mathbf{F}$	7,8

Notice that the above semi-formal proof of the left-to-right direction of THEOREM 3.9 — the proof by contraposition — is much more straightforward than the earlier semi-formal proof.

Now, let's look at an informal proof of the contrapositive of the left-to-right direction of THEOREM 3.9.

TO PROVE: If $\Gamma \not\models \varphi \rightarrow \psi$, then $\Gamma \cup \{\varphi\} \not\models \psi$

PF: Suppose $\Gamma \not\models \varphi \rightarrow \psi$. Then there is a boolean interpretation $\mathcal{B}$ such that $\mathcal{B}(\Gamma) = \mathbf{T}$ and $\mathcal{B}(\varphi \rightarrow \psi) = \mathbf{F}$. Thus, $\mathcal{B}(\varphi) = \mathbf{T}$ and $\mathcal{B}(\psi) = \mathbf{F}$. Hence, $\mathcal{B}(\Gamma \cup \{\varphi\}) = \mathbf{T}$ and $\mathcal{B}(\psi) = \mathbf{F}$, completing the proof. ∎

The above informal proof of the contrapositive of the left-to- right direction of THEOREM 3.9 shows that once the "idea of the proof" is clear, there is not really much to do.

We still have the right-to-left direction of THEOREM 3.9 to prove, namely:

TO PROVE: If $\Gamma \models (\varphi \rightarrow \psi)$, then $\Gamma \cup \{\varphi\} \models \psi$.

To prove the contrapositive of the above statement, one needs

TO PROVE: If $\Gamma \cup \{\varphi\} \not\models \psi$, then $\Gamma \not\models \varphi \rightarrow \psi$

Exercise:

Give a semi-formal and an informal proof of the right-to-left direction of THEOREM 3.9 by proving its contrapositive.

Now, this next theorem is much trickier:

THEOREM 3.10: If $\Gamma \models (\psi \wedge \neg\psi)$, then $\Gamma - \{\neg\varphi\} \models \varphi$

Exercises: Prove THEOREM 3.10 two ways:

1. by contradiction, and
2. by contraposition (see below for a "third" way). In both cases, either give a semi-formal or an informal proof

[Hints: For **1.** by contradiction, if Γ has no model and removing $\neg\varphi$ yields at least one model (of $\Gamma - \{\neg\varphi\}$) in which $\neg\varphi$ is true, then Γ (with $\neg\varphi$ possibly in Γ) must have had a model to begin with. For **2.** by contraposition, if a model of $\Gamma - \{\neg\varphi\}$ does not model φ, then Γ itself has a model.] A "third" proof of 3.10 begins as a direct one, but at a certain stage in the reasoning becomes an indirect proof:

Here is the third way to prove THEOREM 3.10, mentioned in *exercise* **2.** above:

TO PROVE: If $\Gamma \models \psi \wedge \neg\psi$, then $\Gamma - \{\neg\varphi\} \models \varphi$
1. $\Gamma \models \psi \wedge \neg\psi$
2. There is no $\mathcal{B}$ such that $\mathcal{B}(\Gamma) = T$
                                           TO PROVE: $\Gamma - \{\neg\varphi\} \models \varphi$
                                           TO PROVE: If $\mathcal{B}(\Gamma - \{\neg\varphi\}) = T$, then $\mathcal{B}(\varphi) = T$
3. $\mathcal{B}(\Gamma - \{\neg\varphi\}) = T$
                                           TO PROVE: $\mathcal{B}(\varphi) = T$
4. $\mathcal{B}(\varphi) \neq T$
                                          √TO PROVE: A contradiction
5. $\mathcal{B}(\neg\varphi) = T$
6. $\mathcal{B}((\Gamma - \{\neg\varphi\}) \cup \{\neg\varphi\}) = T$ 3,5
7. $\Gamma \subseteq (\Gamma - \{\neg\varphi\}) \cup \{\neg\varphi\}$ by set theory
8. $\mathcal{B}(\Gamma) = T$ T2
(*contradicting line 2*)

ANALYSIS: The above proof starts out being a direct proof. That is, there is no immediate assumption of the hypothesis and the *negation* of the conclusion. The proof proceeds directly until: TO PROVE: $\mathcal{B}(\varphi) = T$. At that point, it is clear that the negation of '$\mathcal{B}(\varphi) = T$' needs to be assumed. The rest of the proof is straightforward, except for the set-theoretical fact on line 7. For a proof of this, consult Chapter 0, *Mathematical Preliminaries*.

Exercises:
Prove the following, either semi-formally or informally:
1. If $\Gamma \models (\psi \wedge \neg\psi)$ then $(\Gamma - \{\varphi\}) \models \neg\varphi$
2. $\{\varphi, \neg\varphi\} \models \psi$
3. $\{\varphi, (\varphi \rightarrow \psi)\} \models \psi$
4. $\Gamma \models (\varphi \rightarrow \varphi)$

5. $(\Gamma \cup \{\varphi\}) \models (\varphi \vee \psi)$

THEOREM 3.11: If $\Gamma \models \psi$, then $\Gamma - \{\varphi\} \models (\varphi \rightarrow \psi)$

PF: Assume $\Gamma \models \psi$. Suppose additionally that $\varphi \notin \Gamma$. Then $\Gamma \cup \{\varphi\} \models \psi$ by THEOREM 3.3. Hence, by THEOREM 3.9, $\Gamma \models (\varphi \rightarrow \psi)$. In this case, $\Gamma - \{\varphi\} = \Gamma$. So: $\Gamma - \{\varphi\} \models (\varphi \rightarrow \psi)$.

Suppose instead that $\varphi \in \Gamma$. Let $\Delta = \Gamma - \{\varphi\}$. By hypothesis, $\Delta \cup \{\varphi\} \models \psi$ (since $\Delta \cup \{\varphi\} = \Gamma$). Hence, by THEOREM 3.9 again, $\Delta \models \varphi \rightarrow \psi$. Since $\Gamma - \{\varphi\} = \Delta$, $\Gamma - \{\varphi\} \models \varphi \rightarrow \psi$.

Thus, whether $\varphi \notin \Gamma$ or $\varphi \in \Gamma$, $\Gamma - \{\varphi\} \models \varphi \rightarrow \psi$. ∎

Exercise:

1. Give a counter-example to the right-to-left direction of 3.11, showing why that direction is <u>not</u> a theorem.
2. Prove 3.11 *without* using THEOREMS 3.9 and 3.3, and note the similarity between that proof and your proof of 3.9.

Here are two more theorems closely related to THEOREM 3.9. The first can be proved using THEOREM 3.9, and the second by almost duplicating the proof of THEOREM 3.9.

THEOREM 3.12: If $\Gamma \models \varphi$ and $\Delta \models \neg\varphi$, then $(\Gamma \cup \Delta) - \{\neg\varphi\} \models \varphi$
THEOREM 3.13: If $\Gamma \models \varphi$ and $\Delta \models \neg\varphi$, then $(\Gamma \cup \Delta) - \{\varphi\} \models \neg\varphi$

THEOREMS 3.14 and 3.15 tell us that conditional sentences of two specified forms are tautologies.

THEOREM 3.14: $\models (\varphi \rightarrow (\psi \rightarrow \varphi))$
THEOREM 3.15: $\models (\neg\psi \rightarrow (\psi \rightarrow \varphi))$

Writing 'φ is a tautology' is the same as writing '$\models \varphi$', and we should expect that φ is a tautology iff φ is a tautological consequence of the empty set of sentences. In symbols,

THEOREM 3.16: $\models \varphi$ iff $\varnothing \models \varphi$

Similarly, φ is a tautological consequence of every set of sentences Γ iff φ is a tautology. In symbols again,

THEOREM 3.17: For every Γ, $\Gamma \models \varphi$ iff $\models \varphi$
THEOREM 3.18: $\models \varphi$ iff $\{\neg\varphi\}$ is not satisfiable
THEOREM 3.19: For every sentence φ, $\Gamma \models \varphi$ iff Γ is not satisfiable.

3.6 COMPACTNESS

Compactness is an important model-theoretic principle that we should discuss before concluding this chapter. Let's start out with an example. Suppose you have an infinite set of sentences, Γ such that $\Gamma = \{\varphi_1, \varphi_2, \varphi_3, \ldots\}$. Now, let's suppose that *every finite subset* of Γ has a model. Not necessarily the same model; each may have a different model. Now we raise the following question: does Γ itself have a model?

Before you answer the above question, let's look at a related, but different question. Suppose, as before, that $\Gamma = \{\varphi_1, \varphi_2, \varphi_3, \ldots\}$. This time suppose that each unit set has a model. Let's say $\mathcal{B}_1$ models $\{\varphi_1\}$, $\mathcal{B}_2$ models $\{\varphi_2\}$, $\mathcal{B}_3$ models $\{\varphi_3\}$, ... Now, does it follow that Γ itself must have a model?

No, it does not. Let φ_1 be 'P' and φ_2 be '¬P', for example. Then, even though $\mathcal{B}_1$ models {P}, and $\mathcal{B}_2$ models {¬P}, there is no model for {P, ¬P}.

Well, since that didn't work, let's try something else. Suppose that every set of *pairs* of elements from Γ has a model. Does that imply that Γ itself has a model?

The answer is no, again.

Exercises:

1. Give an example of a three-element set of sentences that is *un*satisfiable though each two-element subset of that set *is* satisfiable.
2. Give an example of an unsatisfiable four-element set of sentences, where each three-element subset is satisfiable.
3. Show how to get an unsatisfiable k-element set of sentences, where each (k-1)-element subset is satisfiable (for k ≥ 3).
4. Prove that if an infinite set Γ has a model, then every finite subset of Γ has a model.

By *exercise* **3** above, we know it is possible to construct an unsatisfiable set of any finite number of sentences such that the removal of any one of the sentences makes the resultant set satisfiable. And, by **4** we know that <u>if</u> an infinite set Γ has a model, <u>then</u> any finite subset of Γ also has a model. But, our original question remains: if every finite subset of an infinite set of sentences has a model, does the infinite set <u>itself</u> have to have a model? The answer is yes, and that is Compactness:

Compactness: An infinite set of sentences Γ has a model if every finite subset of Γ has a model.

Compactness is a purely model-theoretic concept containing no implicit references to any proof-theoretic notions (i.e., notions pertaining to derivatons). Thus, it should be provable by purely model-theoretic means alone, as were the other theorems of this chapter. Compactness *is* provable model-theoretically, but it is much easier to prove it from a result that mixes proof-theoretic and model-theoretic notions. In the next chapter we give that proof. Later on, in the chapter *Models for First-Order Theories*, we present an outline of a beautiful model-theoretic proof of Compactness for a much richer language.

LIST OF THEOREMS

THEOREM 3.0: For any boolean interpretation $\mathcal{B}$,
  $\mathcal{B}(\emptyset) = \mathbf{T}$.
THEOREM 3.1: If $\mathcal{B}(\Gamma) = \mathbf{T}$ and $\mathcal{B}(\Delta) = \mathbf{T}$, then $\mathcal{B}(\Gamma \cup \Delta) = \mathbf{T}$.
THEOREM 3.2: If $\mathcal{B}(\Delta) = \mathbf{T}$ and $\Gamma \subseteq \Delta$, then $\mathcal{B}(\Gamma) = \mathbf{T}$.
THEOREM 3.3: If $\Gamma \models \varphi$ and $\Gamma \subseteq \Delta$, then $\Delta \models \varphi$.
THEOREM 3.4: If $\Gamma \models \varphi$ and $\Delta \models (\varphi \rightarrow \psi)$, then $(\Gamma \cup \Delta) \models \psi$
THEOREM 3.5: If $\Gamma \models (\varphi \rightarrow \psi)$ and $\Delta \models \neg\psi$,
  then $(\Gamma \cup \Delta) \models \neg\varphi$.
THEOREM 3.6: If $\Gamma \models (\varphi \rightarrow \psi)$ and $\Delta \models (\varphi \rightarrow \neg\psi)$,
  then $(\Gamma \cup \Delta) \models \neg\varphi$.
THEOREM 3.7: $\Gamma \models \varphi$ iff $\Gamma \models \neg\neg\varphi$
THEOREM 3.8: $\Gamma \models (\neg\varphi \rightarrow \psi)$ iff $\Gamma \models (\varphi \vee \psi)$
THEOREM 3.9: $(\Gamma \cup \{\varphi\}) \models \psi$ iff $\Gamma \models (\varphi \rightarrow \psi)$
THEOREM 3.10: If $\Gamma \models (\varphi \wedge \neg\varphi)$ then $(\Gamma - \{\neg\varphi\}) \models \varphi$
THEOREM 3.11: If $\Gamma \models \psi$, then $\Gamma - \{\varphi\} \models (\varphi \rightarrow \psi)$
THEOREM 3.12: If $\Gamma \models \varphi$ and $\Delta \models \neg\varphi$,
  then $((\Gamma \cup \Delta) - \{\neg\varphi\}) \models \varphi$

THEOREM 3.13: If $\Gamma \models \varphi$ and $\Delta \models \neg\varphi$,
then $((\Gamma \cup \Delta) - \{\varphi\}) \models \neg\varphi$
THEOREM 3.14: $\models (\varphi \rightarrow (\psi \rightarrow \varphi))$
THEOREM 3.15: $\models (\neg\psi \rightarrow (\psi \rightarrow \varphi))$
THEOREM 3.16: $\models \varphi$ iff $\varnothing \models \varphi$
THEOREM 3.17: For every Γ, $\Gamma \models \varphi$ iff $\models \varphi$
THEOREM 3.18: $\models \varphi$ iff $\{\neg\varphi\}$ is not satisfiable
THEOREM 3.19: For every sentence φ, $\Gamma \models \varphi$ iff Γ is not satisfiable.

Chapter Exercises:

1. Let φ be a sentence of L_{SL} containing only the binary connective '$\leftrightarrow$' and possibly the unary connective '$\neg$'. Then:

    $\models \varphi$ iff the number of occurrences of both '$\leftrightarrow$' and '$\neg$' is even.

2. Let φ be a sentence of L_{SL} not containing '$\neg$'. Then if 'P $\rightarrow$ P' is substituted for each occurrence of each sentence letter in φ, the resultant sentence is a tautology.

3. If $\neg\varphi$ is a tautology, then φ contains at least one occurrence of '$\neg$'.

4. If φ is a sentence of L_{SL} containing only the connectives '$\wedge$', '$\vee$', and '$\neg$' and φ^* results from interchanging '$\vee$' and '$\wedge$' and replacing each sentence letter in φ by its negation, then

    $\models \neg\varphi \leftrightarrow \varphi^*$.

CHAPTER 4

CONNECTING THE SYNTAX AND SEMANTICS OF SENTENTIAL LOGIC

(SOUNDNESS, CONSISTENCY & COMPLETENESS)

4.1 SYNTAX vs. SEMANTICS

In this chapter we combine the syntax with the semantics of L_{SL} and prove several metatheorems that connect the two independent concepts.

First, let us recall some of the differences between syntactic and semantic concepts. Syntactic concepts concern identifiable patterns of strings of symbols. For example, the question whether a given sentence has an occurrence of 'P' in it is syntactic. The inference rules are all syntactic. Take Modus Ponens, for example. According to MP, if a sentence with the shape $(\varphi \rightarrow \psi)$ appears on one line of a derivation and a sentence with the shape φ also appears on another line of that derivation, then ψ may be placed on a line below the other two. Notice that there is no mention of what sentences "mean" — there is reference only to the way sentences look. So, for example, we could have a rule, say MIDDLE, that allows the "middle" sentence in a sentence of the form $\varphi \rightarrow (\psi \rightarrow \chi)$ to be placed on a later line (than the line on which $\varphi \rightarrow (\psi \rightarrow \chi)$ occurred). Again, to apply MIDDLE correctly would involve only the ability to visually inspect the sentence's form. Thus, the rule MIDDLE would be purely syntactic.

The concept of "truth", on the other hand, pertains to the meaning of sentences, and hence is semantic. Similarly, a boolean interpretation, since it makes truth assignments, is a semantic entity. In short, Chapters 1 and 2, both of which cover derivations, are syntactic. Chapter 3 covers boolean interpretations, and is thus semantic.

At this point, however, we wish to show that semantic and syntactic concepts — though completely different in kind — are connected to each other in certain interesting and important ways. First, though, we explain derivations from a more technical point of view.

4.2 TECHNICAL DERIVATIONS

In an ordinary derivation the assumption numbers appearing to the right of a sentence on a line stand for sentences that have been assumed on those lines (i.e., sentences introduced by the Assumption Rule).

For example, the first line of every derivation is always an assumption. The derivation of T1 began:

     1. P   1,A

The 'A' is superfluous, but the 1 (more strictly: '1') stands for the sentence that is assumed on that line, which in this case is 'P' itself. Since 'P' is the only assumption of line 1, {P} is the set of all assumptions of that line. We can thus rewrite the line as an ordered pair consisting of the set of assumptions and the sentence on that line, respectively:

     ⟨{P}, P⟩

To take another example, look at line 4 of the derivation of T7. That line is:

     4. ¬Q   1,3,MP

(where, again, the 'MP' is superfluous). In this derivation the assumption number 1 stands for the sentence assumed on line 1, which is '¬P → ¬Q'. Assumption number 3 stands for '¬P', since '¬P' was assumed on line 3. So, the set of assumptions for the given line is {¬P → ¬Q, ¬P}. Rewriting that line in ordered pair notation, we get:

     ⟨{¬P → ¬Q, ¬P}, ¬Q⟩

In this way, we can rewrite any derivation so that each line is in ordered pair notation. Consider the following derivation of '¬P' from {P → Q, Q → R, ¬R}:

  1. P → Q   1,A
  2. Q → R   2,A
  3. ¬R   3,A
  4. ¬Q   2,3,MT
  5. ¬P   1,2,3,MT(1,4)

We can rewrite it in ordered pair notation as follows:

$$\langle \{P \to Q\}, P \to Q \rangle$$
$$\langle \{Q \to R\}, Q \to R \rangle$$
$$\langle \{\neg R\}, \neg R \rangle$$
$$\langle \{Q \to R, \neg R\}, \neg Q \rangle$$
$$\langle \{Q \to R, \neg R, P \to Q\}, \neg P \rangle$$

We call a derivation written in ordered pair notation a **technical derivation**. More formally, **a technical derivation of a sentence of L_{SL}** is a finite (non-empty) sequence of ordered pairs $\langle \Gamma', \varphi \rangle$, where φ is a sentence of L_{SL} and Γ' is a finite set of assumptions for φ in accordance with rules A, CD, MP, CONTRA, and INT (and any short-cut rules, since they can all be eliminated in favor of our five basic rules):

Exercises:

Write technical derivations of 'P $\to$ ¬¬P', 'P $\to$ (Q $\to$ P)', and 'P $\to$ (P v Q)'.

A sentence φ of L_{SL} is derivable from a set of sentences Γ if and only if $\langle \Gamma', \varphi \rangle$ is an element of a technical derivation, where Γ' is finite and $\Gamma' \subseteq \Gamma$. We write '$\Gamma \vdash \varphi$' to mean 'φ is derivable from Γ using the derivational rules for L_{SL}'. '$\vdash \varphi$' is short for '$\emptyset \vdash \varphi$', which means that φ is derivable from the empty set of assumptions, which also means that φ is a theorem of sentential logic.

4.3 SYNTACTIC (META)THEOREMS

The following two proof-theoretic facts are immediate consequences of the definition of derivability:

THEOREM 4.1: If $\Gamma \vdash \varphi$ and $\Gamma \subseteq \Delta$, then $\Delta \vdash \varphi$.

PF:
1. $\Gamma \vdash \varphi$ hypothesis
2. $\langle \Gamma', \varphi \rangle$ is an element of a technical derivation, where Γ' is finite and $\Gamma' \subseteq \Gamma$.   1, by def.
3. $\Gamma \subseteq \Delta$ hypothesis
4. $\langle \Gamma', \varphi \rangle$ is an element of a technical derivation, where Γ' is finite and $\Gamma' \subseteq \Delta$.   2,3
5. $\Delta \vdash \varphi$ 4, by def.

THEOREM 4.2: If $\Gamma \vdash \varphi$, then there is some finite $\Gamma' \subseteq \Gamma$ such that $\Gamma' \vdash \varphi$.

PF:
1. $\Gamma \vdash \varphi$ hypothesis
2. $\langle \Gamma', \varphi \rangle$ is an element of a technical derivation, where Γ' is finite and $\Gamma' \subseteq \Gamma$.   1, def.
3. $\langle \Gamma', \varphi \rangle$ is an element of a technical derivation, where Γ' is finite and $\Gamma' \subseteq \Gamma'$.   2
4. $\Gamma' \vdash \varphi$ 3, def.

Corresponding to each rule of derivation there is a straightforward principle that is easily proved. For example, THEOREM 4.3 "says" that if a sentence is a member of a set of sentences Γ, then that sentence is derivable from Γ. Another way to look at the theorem is that it "says" that any assumption is derivable. Hence, THEOREM 4.3 states a principle based on the Assumption Rule.

THEOREM 4.3: If $\varphi \in \Gamma$, then $\Gamma \vdash \varphi$

PF:
1. $\varphi \in \Gamma$ hypothesis
2. $\{\varphi\} \subseteq \Gamma$, and $\{\varphi\}$ is finite   facts about sets
3. $\langle \{\varphi\}, \varphi \rangle$ is an element of a technical derivation
                                                       by the Assumption Rule.
4. $\Gamma \vdash \varphi$ by 1,2, def.

Similarly, the following principle is based on the Conditional Derivation Rule:

THEOREM 4.4: If $\Gamma \vdash \psi$, then $(\Gamma - \{\varphi\}) \vdash (\varphi \rightarrow \psi)$

PF:
1. $\Gamma \vdash \psi$ hypothesis
2. $\langle \Gamma', \psi \rangle$ is an element of a technical derivation, where Γ' is finite and $\Gamma' \subseteq \Gamma$.   1, def.
3. $\langle \{\varphi\}, \varphi \rangle$ is an element of a technical derivation
                                                       by Assumption Rule
4. $\langle \Gamma' - \{\varphi\}, \varphi \rightarrow \psi \rangle$ is an element of a technical derivation.   from 2,3,CD
5. $(\Gamma' - \{\varphi\}) \subseteq (\Gamma - \{\varphi\})$ from 2, (since $\Gamma' \subseteq \Gamma$)
6. $(\Gamma - \{\varphi\}) \vdash (\varphi \rightarrow \psi)$ 5, def.

The next theorem is based on Modus Ponens, and the two after that are based on CONTRA:

THEOREM 4.5: If $\Gamma \vdash (\varphi \rightarrow \psi)$, then $\Gamma \cup \{\varphi\} \vdash \psi$.
THEOREM 4.6: If $\Gamma \vdash \psi$ and $\Gamma \vdash \neg\psi$, then $\Gamma - \{\neg\varphi\} \vdash \varphi$.
THEOREM 4.7: If $\Gamma \vdash \psi$ and $\Gamma \vdash \neg\psi$, then $\Gamma - \{\varphi\} \vdash \neg\varphi$.

Exercises:

Prove THEOREM 4.5 and one of THEOREM 4.6 and THEOREM 4.7.

Once you prove a few of the above theorems, you will see two repetitive steps. Each proof begins with a sentence φ being deriv<u>able</u> from Γ, where Γ is possibly infinite. And, then the definition of derivability is used to give us the finite set Γ' from which φ is actually deriv<u>ed</u>, where $\Gamma' \subseteq \Gamma$. Toward the end of each proof, the definition of derivability is used again to return to a possibly infinite set again. These two steps, shifting from a possibly infinite set to a finite subset and back again to a possibly infinite set are repeated in almost an identical fashion in virtually every proof. Thus, we will stop calling attention to them and focus instead on the unique justification of each theorem. For example, THEOREM 4.3 is justified by the Assumption Rule, 4.4 by CD, 4.5 by MP, and both 4.6 and 4.7 by CONTRA.

THEOREM 4.8: If $\Gamma \vdash \varphi$ and $\Delta \vdash \varphi \rightarrow \psi$, then $\Gamma \cup \Delta \vdash \psi$.

Dispensing with what we call "The Repetitive Argument" (using the definition of derivability to go from a possibly infinite set to a finite subset and then using derivability to return to a possibly infinite set again), we prove THEOREM 4.8 like this:

PF:
1. $\Gamma \vdash \varphi$ hypothesis
2. $\Delta \vdash \varphi \rightarrow \psi$ hypothesis
3. $\Gamma \cup \Delta \vdash \varphi$ THEOREM 4.1, 1
4. $\Gamma \cup \Delta \vdash \psi \rightarrow \psi$ THEOREM 4.1, 2
3. $\Gamma \cup \Delta \vdash \psi$ MP (3,4)

Exercise:

Prove that THEOREM 4.5 is a corollary of THEOREM 4.8.

The rest of these theorems are now easily seen to be straightforward applications of INT:

THEOREM 4.9:
a) $\Gamma \vdash (\varphi \vee \psi)$ iff $\Gamma \vdash \neg\varphi \rightarrow \psi$
b) $\Gamma \vdash (\varphi \wedge \psi)$ iff $\Gamma \vdash \neg(\varphi \rightarrow \neg\psi)$
c) $\Gamma \vdash (\varphi \leftrightarrow \psi)$ iff $\Gamma \vdash (\varphi \rightarrow \psi)$ and $\Gamma \vdash (\psi \rightarrow \varphi)$.

And then, of course, all of the derived rules have corresponding theorems. For example,

$$\Gamma \vdash (\varphi \wedge \psi) \text{ iff } \Gamma \vdash \varphi \text{ and } \Gamma \vdash \psi,$$

where one direction is by SIMP and the other by CONJ.

Since an ordinary derivation and a technical derivation can easily and mechanically be transformed into the other, there is no essential reason to distinguish them, except for example, to refer to a "line" of an ordinary derivation, or to refer to the ordered pair belonging to a given technical derivation. Hence, unless some specific feature is called for, we will often use the word 'derivation' to cover both.

4.4 (STRONG) SOUNDNESS

Soundness connects the syntax of sentential logic and its semantics in an important way. Recall that our derivational rules are purely syntactic. The rules make reference to only the shapes of sentences, not to their truth or falsity. Our intention in framing those rules, though, was for them to preserve truth; we don't want our rules to lead us from a true sentence to a false one. Preserving truth, though, is a semantic concern making no reference to syntactic matters pertaining to the shapes of sentences. What we need to do is bridge the gap between our semantic desire that the rules be truth-preserving and the fact that our system of derivations is solely syntactic. This gap is bridged by our proving a metatheorem that shows our rules to be (Strongly) Sound.

A system of rules of derivation is **(Strongly) Sound** iff:

$$\text{if } \Gamma \vdash \varphi \text{ then } \Gamma \models \varphi.$$

In words, **a system of derivational rules of sentential logic is (strongly) sound** if and only if whenever φ is derivable from Γ, φ is a tautological consequence of Γ.

Why Bother With Soundness?

It may seem as though soundness for our rules of derivation is a kind of "given". That is, some may think, "Of course our rules are good ones. Why wouldn't they be sound?" To stimulate our thinking about soundness, let us modify one of our rules slightly.

We will define a new rule, CD', that is only "slightly" different from CD, except that we make CD' somewhat more permissive. Its exact statement is:

> CD': If a sentence ψ is on a line, then $\varphi \rightarrow \psi$ may be placed on a later line. As assumption numbers of $\varphi \rightarrow \psi$ take all the assumption numbers of ψ, but remove the assumption number of a line on which φ appears if φ has only one assumption number.

CD, you'll recall, permits a single assumption number to be removed <u>only when the antecedent, φ, is an assumption</u>. CD' now permits the removal of a single assumption number whenever the antecedent φ has only a single assumption number, <u>even when φ is not itself an assumption</u>. Here is an example of the use of CD':

1.	P		1, A
2.	¬¬P		1, DN
3.	¬¬P ∨ Q		1, ADJ
4.	¬¬P → (¬¬P ∨ Q)		CD'

On line 4, CD' was used to eliminate assumption number 1 even though '¬¬P' was <u>not</u> introduced as an assumption — only 'P' was assumed. CD' <u>also</u> permits eliminating the assumption number of an assumption, as CD does. But, only CD' permits deleting the sole assumption number of an antecedent when the antecedent has <u>not</u> been assumed.

Incidentally, many students often want to liberalize CD in just this way, <u>and</u> such liberalizations have sometimes appeared in logic books as well. Unfortunately, substituting CD' for CD in our derivational system renders it **UNSOUND!**

Here is a proof that in the resulting system that includes CD' **EVERY SENTENCE IS A THEOREM**. Let 'Q' stand for any sentence at all and let's derive 'Q' as a theorem:

```
              1.   P ∧ ¬P              1,A
              2.   Q ∨ ¬Q              1,SL(from 1)
              3.   Q                   1,SL(also from 1)
       X      4.   (Q ∨ ¬Q) → Q    X   CD'           X
              5.   Q ∨ ¬Q             SL
              6.   Q                  MP(4,5)
```

Exercise:

Suppose we permit "Backwards Modus Ponens," (Back-MP) as a derivational rule, which is:

Back-MP: If ψ and $\varphi \to \psi$ are on two lines, then φ may be placed on a later line with the assumption numbers of ψ and $\varphi \to \psi$ taken together.

1. Derive any sentence 'Q' as a theorem using Back-MP, together with the other derivational rules (<u>not</u> using CD').

In light of the previous examples of unsound rules, it should be clear why we wish to <u>prove</u> our system of rules is strongly sound. In the absence of a proof, there can always be a lingering suspicion that our rules are not faithful to the truth. We now nip this suspicion in the bud.

STRONG SOUNDNESS METATHEOREM: The system of rules consisting of A, CD, MP, CONTRA, and INT is (strongly) sound. In symbols (for any set of sentences Γ of L_{SL} and any sentence φ of L_{SL}):

$$\text{If } \Gamma \vdash \varphi \text{ then } \Gamma \vDash \varphi.$$

PF: The proof will be carried out using Complete Induction on the length (number of ordered pairs) of a technical derivation. A technical derivation can have 1 or more elements. Thus, the induction is performed on the set of positive integers. (Turn to the *Mathematical Preliminaries* chapter, if needed, for more detail about proofs using Complete Induction.)

Case 1: The derivation consists of a single line (more technically: a single ordered pair). This is our "base case" for the induction, when the length of the derivation is 1. The only way a sentence φ can appear on line 1 of a derivation is by using rule A. Thus, the ordered pair for that line is $\langle\{\varphi\}, \varphi\rangle$. Clearly, every boolean interpretation $\mathcal{B}$ that models $\{\varphi\}$ models φ.

Case k + 1: Our Inductive Hypothesis is that the Metatheorem holds for all derivations of lengths 1, 2,..., k. More explicitly, IH (the Inductive Hypothesis) is (for any set of sentences Γ of L_{SL} and any sentence ψ of L_{SL} whose derivation is of length k or less):

IH: If $\Gamma \vdash \psi$ then $\Gamma \models \psi$.

We will prove that it must also hold for derivations of length k + 1:

Subcase A: If rule A is used for line k + 1, the same argument given in Case 1 works.

Subcase MP: If rule MP is used, then we need to prove that if $\Gamma \models \varphi$ and $\Delta \models (\varphi \rightarrow \psi)$, then $(\Gamma \cup \Delta) \models \psi$. This has been proved already as THEOREM 3 in the previous chapter.

Subcase CONTRA: We need to prove two things: (1) if $\Gamma \models \psi$, then $((\Gamma \cup \Delta) - \{\neg\varphi\}) \models \varphi$, and (2) if $\Gamma \models \psi$ and $\Delta \models \neg\psi$, then $((\Gamma \cup \Delta) - \{\varphi\}) \models \neg\varphi$. These are justified, respectively, by THEOREM 4.11 and THEOREM 4.12.

Subcase CD: Show that if $\Gamma \models \psi$, then $\Gamma - \{\varphi\} \models (\varphi \rightarrow \psi)$. By left-to-right direction of THEOREM 4.10.

Subcase INT: Three THEOREMS need to be proved:

(1) $\Gamma \models (\varphi \lor \psi)$ iff $\Gamma \models (\neg\varphi \rightarrow \psi)$,
(2) $\Gamma \models (\varphi \land \psi)$ iff $\Gamma \models \neg(\varphi \rightarrow \neg\psi)$,
(3) $\Gamma \models (\varphi \leftrightarrow \psi)$ iff $\Gamma \models ((\varphi \rightarrow \psi) \land (\psi \rightarrow \varphi))$.

It suffices (thanks to THEOREM 3.3 of the previous chapter) to prove three easier-looking THEOREMS. (For any boolean interpretation $\mathcal{B}$):

(1)' $\mathcal{B}(\varphi \vee \psi) = T$ iff $\mathcal{B}(\neg \varphi \rightarrow \psi) = T$,
(2)' $\mathcal{B}(\varphi \wedge \psi) = T$ iff $\mathcal{B}(\neg(\varphi \rightarrow \neg\psi)) = T$,
(3)' $\mathcal{B}(\varphi \leftrightarrow \psi) = T$ iff $\mathcal{B}((\varphi \rightarrow \psi) \wedge (\psi \rightarrow \varphi)) = T$.

Some of these theorems have been proved in the previous chapter. The rest are left as *exercises*.

This proves that whenever the Metatheorem holds for derivations of up to k lines, it holds for line k + 1, where the sentence φ on line k + 1 was entered by A, CD, MP, CONTRA, or INT. Since we have also proven that it holds for all derivations of only 1 line, we have proven by Complete Induction that our system of rules is strongly sound. ■

Now that we *know* our rules of inference for sentential logic are (strongly) sound, we may begin to wonder about their consistency.

While soundness is a mixed concept, that is, it involves both syntax (as reflected by derivations) and semantics (referring to truth), consistency (as we define it) is purely syntactic. We want to know whether for some sentence φ, both φ and $\neg\varphi$ are theorems. That is, a system of logic is inconsistent if contradictory theorems can be derived (i.e., if $\vdash \varphi$ and $\vdash \neg\varphi$). Obviously we want our system of sentential logic to be consistent.

4.5 CONSISTENCY (Proof-Theoretic)

A system of rules is consistent iff for no sentence φ, is it possible to derive φ and $\neg\varphi$ from the empty set of assumptions. In symbols, a system is consistent iff it is <u>not</u> the case that there is a sentence φ such that <u>both</u> $\vdash \varphi$ and $\vdash \neg\varphi$. That is, not both $\varnothing \vdash \varphi$ <u>and</u> $\varnothing \vdash \neg\varphi$, if we wish to explicitly mention the empty set.

Corollary (of Strong Soundness): The system of rules consisting of A, CD, MP, CONTRA, and INT is consistent.

PF: Suppose the system were <u>in</u>consistent (i.e., suppose $\vdash \varphi$ and $\vdash \neg\varphi$ for some sentence φ). By Soundness, $\varnothing \models \varphi$ and $\varnothing \models \neg\varphi$. Thus, every boolean interpretation $\mathcal{B}$ is such that $\mathcal{B}(\varphi) = T$ and $\mathcal{B}(\neg\varphi) = T$. But, no $\mathcal{B}$ models φ and $\neg\varphi$. ■

We say **a set of sentences, Γ, is consistent** iff for no sentence φ is it the case that $\Gamma \vdash \varphi$ and $\Gamma \vdash \neg\varphi$. Thus, **a set of sentences, Γ, is inconsistent** iff, for some sentence φ, $\Gamma \vdash \varphi$ and $\Gamma \vdash \neg\varphi$. (So, saying that our rules are consistent is really a special case of saying that a set Γ is consistent — when Γ is the null set.)

THEOREM 4.10: Γ is consistent iff for no sentence φ is it the case that $\Gamma \vdash (\varphi \wedge \neg\varphi)$.

THEOREM 4.11: Γ is consistent iff there is at least one sentence φ such that $\Gamma \nvdash \varphi$.

PF(=>): Assume Γ is consistent and suppose (by way of contradiction) that every sentence φ is derivable from Γ. Then, in particular $\Gamma \vdash \varphi$ and $\Gamma \vdash \neg\varphi$ (for every φ), contradicting the consistency of Γ.

(<=) Assume there is some sentence φ such that $\Gamma \nvdash \varphi$. Suppose (by way of contradiction) that Γ is not consistent. Thus, $\Gamma \vdash (\psi \wedge \neg\psi)$ for some sentence ψ (by THEOREM 4.10)). By T33, $\Gamma \vdash ((\psi \wedge \neg\psi) \to \varphi)$. Thus, by MP, $\Gamma \vdash \varphi$, contradicting our assumption. ∎

THEOREM 4.12: $\Gamma \vdash \varphi$ iff $\Gamma \cup \{\neg\varphi\}$ is inconsistent

PF(=>): Assume $\Gamma \vdash \varphi$. Then, by THEOREM 4.1, $\Gamma \cup \{\neg\varphi\} \vdash \varphi$. But, $\Gamma \cup \{\neg\varphi\} \vdash \neg\varphi$, since there is a derivation of $\neg\varphi$ from $\{\neg\varphi\}$ (by Rule A), and $\{\neg\varphi\} \subseteq \Gamma \cup \{\neg\varphi\}$. Since $\Gamma \cup \{\neg\varphi\} \vdash \varphi$ and $\Gamma \cup \{\neg\varphi\} \vdash \neg\varphi$, $\Gamma \cup \{\neg\varphi\}$ is inconsistent.

(<=) Assume $\Gamma \cup \{\neg\varphi\}$ is inconsistent. Then every sentence is derivable from $\Gamma \cup \{\neg\varphi\}$ by THEOREM 4.11. So, in particular:

$\Gamma \cup \{\neg\varphi\} \vdash \varphi$
$\Gamma \vdash \neg\varphi \to \varphi$ by THEOREM 4.4 (or CD)
$\varnothing \vdash (\neg\varphi \to \varphi) \to \varphi$ T15
$\Gamma \vdash \varphi$ ∎

THEOREM 4.13: $\Gamma \vdash \neg\varphi$ iff $\Gamma \cup \{\varphi\}$ is inconsistent

Exercise: Prove THEOREM 4.13.

A system of rules of derivation is **weakly sound** iff: if $\vdash \varphi$, then $\vDash \varphi$. It is easy to prove strong soundness implies weak soundness. (Just take Γ to be the empty set, $\varnothing$.)

But, it is not so easy to prove the converse direction at this time.

In proving strong soundness, we have proven that for our system of rules every line of every derivation is a consequence of the assumptions of that line. This implies that every initial segment of a derivation is also a derivation. That is, for a given derivation of n lines, lines 1 through k (for k < n) is also a derivation. There are many systems of derivation that do <u>not</u> have this property (e.g., that of Kalish & Montague). That is, in such systems some derivational lines do <u>not</u> follow from the assumptions of those lines.

Corresponding to strong and weak soundness, there is strong and weak completeness. Intuitively, soundness tells us our rules are "good"; we cannot derive sentences that we should not be able to derive. Another (very loose) way to put this is to say that our rules will never lead us from truth to falsity.

But, we want something further of our rules. Not only do we want them to allow us to "reach" <u>only</u> truths, but also to allow us to reach <u>all</u> truths. That is, we want all truths to be derivable. More technically, if a sentence φ is a tautological consequence of a set of sentences Γ, we want to be able to derive φ from Γ by using our rules. This is strong completeness.

"Strong Completeness" does <u>not</u> imply that if φ is a tautological consequence of Γ that someone will be able to construct a derivation. Constructing a derivation is somewhat of an art. There may be a way to derive φ from Γ, but that does <u>not</u> mean you or I or anyone else will discover it. Strong Completeness means only that there is a way.

A system of rules of derivation for L_{SL} is **strongly complete iff:**

$$\text{if } \Gamma \models \varphi, \text{ then } \Gamma \vdash \varphi.$$

In words, the rules of derivation for L_{SL} are **strongly complete** if and only if whenever φ is a tautological consequence of Γ, φ is derivable from Γ.

4.6 STRONG COMPLETENESS[1]

The system of rules consisting of A, CD, MP, CONTRA, and INT is Strongly Complete. That is, for any sentence φ and set of sentences Γ of L_{SL}, if $\Gamma \models \varphi$, then $\Gamma \vdash \varphi$.

Before proving Strong Completeness for sentential logic, let us observe that there is a sense in which sentential logic is not complete. Take the sentence 'P' and ask yourself whether 'P' is a theorem (i.e., $\vdash$ P) or whether '¬P' is a theorem (i.e., $\vdash$ ¬P). The answer to both questions is 'no', neither 'P' nor '¬P' is a theorem. This holds for non-atomic sentences as well. '(P → Q)' cannot be derived, and neither can its negation, '¬(P → Q)'. If, for each sentence φ of L_{SL}, either $\vdash \varphi$ or $\vdash \neg\varphi$, then sentential logic would be a **Complete Theory** and be called **Mathematically Complete**. Sentential logic is therefore not a Complete Theory; sentential logic is not Mathematically Complete.

Even though sentential logic is not Mathematically Complete, we can create a set of sentences that is Mathematically Complete by selecting exactly one of each sentence letter or its negation, and putting that sentence in the set. Let us use the term **unit** to describe a sentence letter or a negation of a sentence letter. Then, we are claiming that if the set Δ contains one unit for each sentence letter (either it or its negation, but not both), then Δ is Mathematically Complete in the sense that for any sentence φ (non-units as well as units), either $\Delta \vdash \varphi$ or $\Delta \vdash \neg\varphi$.

In order to gain a sense of the above claim, let us get units for just the letters 'P' and 'Q', and let's make the set Δ be: {¬P, Q}. The more restricted claim in this case is that for any sentence φ having no occurrences of sentence letters other than 'P' and 'Q', either $\Delta \vdash \varphi$ or $\Delta \vdash \neg\varphi$. Here are a few *exercises* to make this claim seem plausible.

---

[1] For those already familiar with the Henkin construction, the basic modification is that we add only sentence letters, not sentences in general, to an initial set of sentences. The resultant set, then, is not "maximally consistent" in the sense of containing all the sentences it can while retaining its consistency. It is, nonetheless, a "Complete Theory", which enables us to define a sentence as true iff it is derivable from the resultant set (rather than iff it belongs to it).

Exercises:

For each sentence below, either derive it or its negation from Δ, where Δ = {¬P, Q}:

1. P → Q
2. Q → P
3. ¬(¬Q → P)
4. ((Q ∧ P) ∨ P)
5. (Q → P) ↔ P)
6. ¬(Q → (P ∧ ¬P))

After doing the above *exercises*, it should seem plausible that if Δ is a set containing one unit for every sentence letter, then Δ is a Complete Theory (i.e., is Mathematically Complete). There is another fact that should begin to seem plausible from the above *exercises*. Consider *exercise* 2. You should have found that {¬P, Q} ⊬ (Q → P), but that {¬P, Q} ⊢ ¬(Q → P). Hence, {¬P, Q} ∪ {(Q → P)} is inconsistent. Looking at it from another direction, if we already have the set {(Q → P), ¬P}, then adding the sentence letter 'Q' to that set yields an inconsistent set. We know that set would be inconsistent because {¬P, Q} ⊢ ¬(Q → P) by *exercise* 2. Since adding 'Q' to {(Q → P), ¬P} yields an inconsistent set, it must be that {(Q → P), ¬P} ⊢ ¬Q, which is the case, by MT. Thus, when we begin with a consistent set of sentences (such as {(Q → P)}, ¬P) and we try and <u>fail</u> to add a sentence letter θ to that set, the negation of that sentence letter ¬θ is derivable from the set.

The above reasoning leads to the following method for proving Strong Completeness. We actually prove the contrapositive of Strong Completeness which is:

STRONG COMPLETENESS (contrapositive): If Γ ⊬ φ, then Γ ⊭ φ.

We take a set of sentences Γ and any sentence φ, such that Γ ⊬ φ, and we show that Γ ⊭ φ. The way we show Γ ⊭ φ is to exhibit a boolean interpretation of Γ ∪ {¬φ}. That interpretation makes all of Γ true and φ false (since it makes ¬φ true). Now, here's where the reasoning in the above paragraph comes into play: Γ ∪ {¬φ} is consistent (or else, Γ ⊢ φ). So, we add sentence letters to Γ ∪ {¬φ}, one letter at a time, when adding them does not destroy consistency. Adding all these letters to Γ ∪ {¬φ} yields a set, Δ, that includes Γ ∪ {¬φ} and is still consistent. We then interpret all the letters in Δ as **True**, and all the letters <u>not</u> in Δ as **False**. This interpretation makes all the rest of the sentences in Δ true as well. Since Γ ∪ {¬φ} is included in

Δ, all sentences in Γ are **True** and φ is **False** for this boolean interpretation. Thus, we will have shown that $\Gamma \not\models \varphi$, which is what we wished to show. Hence: sentential logic is Strongly Complete.

Lindenbaum's Lemma

We now furnish the formal details for the argument outlined in the paragraph above. First, we establish the following version of Lindenbaum's Lemma.

LINDENBAUM'S LEMMA (LIN): For any consistent set Δ_0 of sentences of L_{SL}, there is a set $\Delta = \Delta_0 \cup \{\chi_1, \chi_2, \ldots\}$, where each χ_i is a sentence letter, such that:

1) Δ is consistent, and
2) Δ is a Complete Theory (Mathematically Complete).

We begin by taking Δ_0 to be any consistent set of sentences, as in the hypothesis of the theorem. We now show by recursion how to get the <u>next</u> consistent set, Δ_{n+1}, from the <u>previous</u> set, Δ_n, where $n \geq 0$. We take the nth sentence letter, θ_n, of our alphabetical listing of sentence letters: $\theta_0, \theta_1, \theta_2, \ldots, \theta_n, \ldots$, where $n \geq 0$, and we attempt to add θ_n to the set Δ_n in the following way:

$$\Delta_{n+1} = \begin{cases} \Delta_n \cup \{\theta_n\} & \text{if this set is consistent} \\ \Delta_n & \text{otherwise} \end{cases}$$

That is, if the union of $\{\theta_n\}$ and Δ_n is consistent, Δ_{n+1} is equal to the union of the two sets. If the union of $\{\theta_n\}$ and Δ_n is <u>not</u> consistent, then Δ_{n+1} is Δ_n itself. So, for each n, $\Delta_n \subseteq \Delta_{n+1}$, and each Δ_n is consistent. For example, if $\Delta_0 \cup \{P\}$ is consistent, then $\Delta_1 = \Delta_0 \cup \{P\}$. If $\Delta_0 \cup \{P\}$ is <u>not</u> consistent, then $\Delta_1 = \Delta_0$. (Remember, 'P' is the first sentence letter in the alphabetical list of sentence letters. So 'P' = θ_0.)

We then let Δ be the set-theoretic union of all the Δ_n's. Formally, the set-theoretic union of all the Δ_n's is written this way: $\Delta = \bigcup_{n \in \mathbb{N}} \Delta_n$, where $\bigcup_{n \in \mathbb{N}} \Delta_n = \{\psi | \exists n (n \in \mathbb{N} \ \& \ \psi \in \Delta_n)\}$. That is, Δ is the set of all sentences ψ such that ψ is an element of <u>some</u> Δ_n. Clearly, every Δ_n is consistent because the construction maintains their consistency.

In order to prove both 1) and 2) of the theorem, we prove this statement:

$\qquad\qquad \Delta \vdash \neg\psi \quad$ iff $\quad \Delta \nvDash \psi$.

can be broken down into its two conditional statements as follows:

$\qquad$ #1: If $\Delta \vdash \neg\psi$, then $\Delta \nvdash \psi$
$\qquad$ #2: If $\Delta \nvdash \psi$, then $\Delta \vdash \neg\psi$

$\qquad$ According to **##**, exactly one of $\Delta \vdash \neg\psi$, and $\Delta \vdash \psi$ holds. By #1, at most one holds. For, if $\Delta \vdash \neg\psi$, then ψ is not derivable from Δ (i.e., $\Delta \nvdash \psi$). And, if it is not the case that $\Delta \nvdash \psi$ (in other words, $\Delta \vdash \psi$), then it is not the case that $\Delta \vdash \neg\psi$ (i.e., $\Delta \nvdash \neg\psi$).

$\qquad$ By #2, at least one of $\Delta \vdash \neg\psi$ and $\Delta \vdash \psi$ holds.

Exercise:

Argue, using #2, that at least one of $\Delta \vdash \neg\psi$ and $\Delta \vdash \psi$ holds. The reasoning will be very similar to the argument that at most one of $\Delta \vdash \neg\psi$ and $\Delta \vdash \psi$ holds, which used #1.

$\qquad$ Therefore, by #1 and #2, exactly one of $\Delta \vdash \neg\psi$ and $\Delta \vdash \psi$ holds. Hence, proving **##** suffices to prove LINDENBAUM'S LEMMA (LIN).

Exercise:

Explain in detail why **##** establishes LIN (LINDENBAUM'S LEMMA).

$\qquad$ We now prove **##**, which establishes LIN.

PF of #1: Suppose $\Delta \vdash \neg\psi$ and $\Delta \vdash \psi$. We will show that a contradiction follows from this supposition. By THEOREM 4.2, there are finite sets, Δ' and Δ'' such that $\Delta' \vdash \neg\psi$ and $\Delta'' \vdash \neg\psi$. By THEOREM 4.1, $(\Delta' \cup \Delta'') \vdash \neg\psi$ and $(\Delta' \cup \Delta'') \vdash \psi$. So, $(\Delta' \cup \Delta'')$ is inconsistent. Since each of Δ' and Δ'' is finite, $\Delta' \cup \Delta''$ is finite. Also, $(\Delta' \cup \Delta'') \subseteq \Delta$. Suppose the sentence letter in $(\Delta' \cup \Delta'')$ that is furthest from the beginning of the alphabetical list of sentence letters is θ_k. Then, we make the following **CLAIM**:

$$(\Delta' \cup \Delta'') \subseteq \Delta_{k+1}$$

One way to verify this claim is to consider any other sentence letter in $(\Delta' \cup \Delta'')$, say θ_j, where θ_j precedes θ_k in the alphabetical list of sentence letters. Then, supposing $\theta_j \notin \Delta_0$, the only way for θ_j to be in Δ (which must be the case for θ_j to be in $(\Delta' \cup \Delta'')$) is for θ_j to be added to Δ_j in the formation of Δ_{j+1}. Since Δ_{k+1} is formed later than Δ_{j+1} in the above construction, $\Delta_{j+1} \subseteq \Delta_{k+1}$. So, $\theta_j \in \Delta_{k+1}$. And, for any sentence letter in $(\Delta' \cup \Delta'')$ that <u>does</u> belong to Δ_0, that sentence letter must also belong to Δ_{k+1}. Since our construction does not add any sentences that are not sentence letters to sets, the only way a non-atomic sentence can belong to Δ is for it to belong to Δ_0. Clearly, $\Delta_0 \subseteq \Delta_{k+1}$, by construction. So, for any sentence in $(\Delta' \cup \Delta'')$, whether it is atomic or non-atomic, must belong to Δ_{k+1}. Since $\Delta' \cup \Delta''$ is inconsistent and $(\Delta' \cup \Delta'') \subseteq \Delta_{k+1}$, then Δ_{k+1} must be inconsistent as well (by THEOREM 4.1). This contradicts the fact that each Δ_n is consistent. Therefore, #1 holds: If $\Delta \vdash \neg\psi$, then $\Delta \nvdash \psi$. So, <u>at most one</u> of $\Delta \vdash \psi$ and $\Delta \vdash \neg\psi$ holds. ∎

To establish #2, that <u>at least one</u> of $\Delta \vdash \psi$ and $\Delta \vdash \neg\psi$ holds, is more difficult, requiring a proof by Complete Induction. (If you are not already familiar with this technique, you may wish to return to the *Mathematical Preliminaries* chapter.) The induction is performed on the total number of connectives ('¬', '→', '∨', '∧', '↔') occurring in ψ. This was called the **height** of ψ in *Mathematical Preliminaries* (though sentences in language L of that chapter had fewer connectives). Since the height of an atomic sentence (a sentence letter) is 0, the induction is performed on the set of natural numbers, $\mathbb{N}$ ($\mathbb{N} = \{0, 1, 2, \ldots\}$).

In some of the cases below it is easier to prove the contrapositive of #2. Here is #2 followed by its contrapositive:

#2: If $\Delta \nvdash \psi$, then $\Delta \vdash \neg\psi$

CONTRAPOS. of #2: If $\Delta \nvdash \neg\psi$, then $\Delta \vdash \psi$

Both **#2** and its contrapositive express the same idea, namely: if one of ψ, $\neg\psi$, can not be derived from Δ, then the other one can be derived from Δ. So, in each case below, we assume one of the two cannot be derived and we **prove** that the other can be derived from Δ.

PF **#2**:
Case 1. Suppose ψ is a sentence letter, θ_k, having no connectives whatsoever. This is our "base case" for the induction argument, where the height of ψ is 0. Assume $\Delta \not\vdash \neg\theta_k$. By THEOREM 4.13, $\Delta \cup \{\theta_k\}$ is consistent. Thus, $\Delta_k \cup \{\theta_k\}$ is consistent. So, by our construction, $\Delta_{k+1} = \Delta_k \cup \{\theta_k\}$. Thus, $\Delta_{k+1} \vdash \theta_k$, and $\Delta \vdash \theta_k$. This concludes the case where ψ is a sentence letter. ∎

Exercises:

These all pertain to Case 1, above:
1. Explain how THEOREM 4.13 establishes that $\Delta \cup \{\theta_k\}$ is consistent.
2. Why does the consistency of $\Delta \cup \{\theta_k\}$ imply the consistency of $\Delta_k \cup \{\theta_k\}$?
3. Why is it that $\Delta_{k+1} \vdash \theta_k$?
4. Explain why $\Delta_{k+1} \vdash \theta_k$ implies $\Delta \vdash \theta_k$.

Case 2. ψ is not a sentence letter. This case splits into several subcases. But, before examining these subcases one by one, let's look at our Induction Hypothesis. We assume that **#2** holds for all sentences whose height is less than the height of sentence ψ. So, if the height of ψ is k, then we are assuming that for any sentence φ whose height is less than k:

IH: If $\Delta \not\vdash \varphi$, then $\Delta \vdash \neg\varphi$

Subcase 2a. ψ is of the form $\neg\chi$. Assume $\Delta \not\vdash \neg\psi$. That is, we are assuming $\Delta \not\vdash \neg\neg\chi$. By DN, $\Delta \not\vdash \chi$. Since χ has one fewer connective than ψ, our Inductive Hypothesis for χ is: if $\Delta \not\vdash \chi$, then $\Delta \vdash \neg\chi$. By IH and the fact that $\Delta \not\vdash \chi$, we get $\Delta \vdash \neg\chi$. But $\neg\chi$ is ψ. Therefore, $\Delta \vdash \psi$. ∎

Subcase 2b. ψ is of the form $(\chi \to \theta)$. Assume $\Delta \not\vdash \psi$ (i.e., we are assuming $\Delta \not\vdash (\chi \to \theta)$). Then $\Delta \not\vdash \theta$ and $\Delta \not\vdash \neg\chi$ (because if $\Delta \vdash \theta$, then $\Delta \vdash \chi \to \theta$ by CD, and if $\Delta \vdash \neg\chi$, then $\Delta \vdash \chi \to \theta$ by T14). Both θ and χ contain fewer connectives than ψ. So, by IH, $\Delta \vdash \neg\theta$ and $\Delta \vdash \chi$. Thus, $\Delta \vdash (\chi \land \neg\theta)$ by CONJ and, by T65 $\Delta \vdash \neg(\chi \to \theta)$. ∎

Subcase 2c. ψ is of the form $(\chi \land \theta)$. Assume $\Delta \not\vdash \psi$ (i.e., assume $\Delta \not\vdash (\chi \land \theta)$). Then, not <u>both</u> $\Delta \vdash \chi$ and $\Delta \vdash \theta$. Suppose $\Delta \not\vdash \chi$. Since χ has fewer connectives than ψ, IH holds. So, $\Delta \vdash \neg\chi$. Any instance of $(\chi \land \theta) \to \chi$ is a theorem by T25. Then, by MT we get $\Delta \vdash \neg(\chi \land \theta)$. If, instead of supposing $\Delta \not\vdash \chi$, we suppose that $\Delta \not\vdash \theta$, then by IH again we have that $\Delta \vdash \neg\theta$. And again, we get $\Delta \vdash \neg(\chi \land \theta)$, by the same reasoning and T26. Since ψ is $(\chi \land \theta)$, we have proven $\Delta \vdash \neg\psi$. ∎

Exercises:

Prove #2 for the remaining subcases:
1. ψ is of the form $(\chi \lor \theta)$.
2. ψ is of the form $(\chi \leftrightarrow \theta)$.

This concludes the proof of LIN (LINDENBAUM'S LEMMA). ∎

Overall Reasoning for Strong Completeness

Now that we have established (a modified form of) LINDENBAUM'S LEMMA, Strong Completeness is not very difficult. The overall reasoning in the proof that follows is this: we prove the contrapositive of Strong Completeness, which is: if $\Gamma \not\vdash \varphi$, then $\Gamma \not\models \varphi$. To do this, we first assume $\Gamma \not\vdash \varphi$. Then, by THEOREM 4.12, $\Gamma \cup \{\neg\varphi\}$ is consistent. Then, $\Gamma \cup \{\neg\varphi\}$ is taken as set Δ_0 in LIN. Then, as in that lemma, Δ is a set that includes $(\Gamma \cup \{\neg\varphi\})$, and Δ has property ##. The only thing left to do is to define a boolean interpretation $\mathcal{B}$ for which the sentences of $\Gamma \cup \{\neg\varphi\}$ are all **True**. Hence, for interpretation $\mathcal{B}$, $\mathcal{B}(\Gamma) = \mathbf{T}$ and $\mathcal{B}(\varphi) = \mathbf{F}$. Thus, $\Gamma \not\models \varphi$, concluding the proof.
    Where do we find the interpretation $\mathcal{B}$ that enables us to conclude the proof? It is obtained from the set Δ constructed in the proof of LINDENBAUM'S LEMMA (where $\Gamma \cup \{\neg\varphi\} = \Delta_0$). In fact, the whole reason for constructing Δ is that Δ "imposes" an interpretation via the notion of

derivability. That is, there is a boolean-valued interpretation making any sentence derivable from Δ **True**, and making any sentence <u>not</u> derivable from Δ **False**. We now prove that derivability from Δ gives us such an interpretation, and then the proof of Strong Completeness will be finished.

The Interpretation Lemma

Define a function $f:\{\text{sentences of } L_{SL}\} \rightarrow \{T, F\}$ such that $f(\psi) = T$ iff (for Δ in LIN) $\Delta \vdash \psi$. In words, f is a function that assigns to any sentence ψ of our language the truth value **T**, provided that $\Delta \vdash \psi$. If $\Delta \nvdash \psi$, then f assigns **F** to sentence ψ.

INTERP LEMMA: f is a boolean interpretation.

f is by definition a function from sentences to truth values. But, we need to prove that f assigns the <u>correct</u> value to each sentence as spelled out by our definition of the truth of a sentence for a given boolean interpretation. Let us ask how f could <u>fail</u> to be a boolean interpretation? Well, if f assigned **T** to a given sentence ψ, and also assigned **F** to that sentence, then f would fail to be a boolean interpretation, since an interpretation assigns <u>exactly</u> <u>one</u> truth value to a given sentence. What else could go wrong? Here's another example of the way f could <u>fail</u> to be a boolean interpretation. Suppose f assigns **T** to the sentence $(\varphi \rightarrow \chi)$, but does <u>not</u> assign **F** to φ, nor assigns **T** to χ. In that case, f would <u>fail</u> to obey the truth condition for conditional sentences. So, to prove that f <u>is</u> a boolean interpretation, we need to show that f <u>does</u> obey <u>all</u> the truth conditions for a boolean interpretation.

We will now prove that f is a boolean interpretation by showing that f assigns the correct truth value to each sentence ψ, as spelled out by our truth definition. The proof is by Complete Induction on the height of ψ (i.e., the number of connectives in ψ):

> PF: Base Case: ψ is a sentence letter θ_i. By LIN, <u>exactly</u> <u>one</u> of $\Delta \vdash \theta_i$ or $\Delta \nvdash \theta_i$ holds. Thus, each sentence letter is assigned either **T** or **F**, but not both, by f. (It happens because of our manner of constructing Δ that for a sentence letter θ_i, $\Delta \vdash \theta_i$ iff $\theta_i \in \Delta$, but this fact is not essential to the proof.)

Case 1: Suppose ψ is of the form $\neg\chi$. Then, we need to prove for clause (ii) of our truth definition that $f(\neg\chi) = T$ iff $f(\chi) = F$. This follows by LIN and the definition of f.

Case 2: Suppose ψ is a conditional, $\chi \to \theta$. We need to prove $f(\chi \to \theta) = T$ iff either $f(\chi) = F$ or $f(\theta) = T$.

($\Rightarrow$) Suppose $f(\chi \to \theta) = T$ and assume $f(\chi) = T$ and $f(\theta) = F$. Then, $\Delta \vdash \chi$ and since $\Delta \nvdash \theta$, $\Delta \vdash \neg\theta$ by LIN. Hence, $\Delta \vdash (\chi \land \neg\theta)$. Therefore, $\Delta \vdash \neg(\chi \to \theta)$, by T65. But, since $f(\chi \to \theta) = T$, $\Delta \vdash \chi \to \theta$. So, our assumption that $f(\chi) = T$ and $f(\theta) = F$ is wrong. Hence either $f(\chi) = F$ or $f(\theta) = T$.

($\Leftarrow$) Suppose $f(\chi) = F$ or $f(\theta) = T$. Then $\Delta \nvdash \chi$ or $\Delta \vdash \theta$. If $\Delta \nvdash \chi$, then, by Lemma 2, $\Delta \vdash \neg\chi$. By T14, $\Delta \vdash \chi \to \theta$. And, if $\Delta \vdash \theta$, then $\Delta \vdash \chi \to \theta$ by T6.

Exercises:

1. Prove the claim made in the Base Case that for any sentence letter θ_i, $\Delta \vdash \theta_i$ iff $\theta_i \in \Delta$.
2. Finish the proof for the rest of the cases, when:
   a) ψ is $(\chi \lor \theta)$,
   b) ψ is $(\chi \land \theta)$,
   c) ψ is $(\chi \leftrightarrow \theta)$.

This completes the proof of the INTERP LEMMA. ∎

Our boolean interpretation f assigns the value T to all sentences in Γ and assigns F to φ. Thus, $\Gamma \nvDash \varphi$.

Hence, **we have established Strong Completeness!** ∎

Exercises:

1. Explain why f establishes $\Gamma \nvDash \varphi$.
2. Justify each step of the following summary of the Strong Completeness proof by citing theorems proved above:

   √TO PROVE: If $\Gamma \vDash \varphi$, then $\Gamma \vdash \varphi$
                 √TO PROVE: If $\Gamma \nvdash \varphi$, then $\Gamma \nvDash \varphi$
   1. $\Gamma \nvdash \varphi$ _____
   2. $\Gamma \cup \{\neg\varphi\}$ is consistent. _____
   3. There is a set Δ such that $\Gamma \cup \{\neg\varphi\} \subseteq \Delta$, and Δ is both consistent and mathematically complete. _____

4. If $f(\psi) = T$ iff $\Delta \vdash \psi$, then
   f is a boolean interpretation. _____
5. $f(\Gamma \cup \{\neg\varphi\}) = T$ _____
6. $\Gamma \not\models \varphi$ _____

4.7 COMPACTNESS (Revisited)

Before defining "compactness", let us ask a few questions. First, let's suppose every <u>finite</u> subset of a given <u>infinite</u> set of sentences, Γ, has a model. And then let's ask whether that implies that Γ itself has a model. Should it be true if there are <u>infinitely many</u> models, one for <u>each</u> of the finite subsets of Γ, that a model exists for all of Γ itself?

Here is a question that turns out to be equivalent to the one we just asked: if a sentence is a tautological consequence of an <u>infinite</u> set of sentences Γ, is it a tautological consequence of some <u>finite</u> subset of Γ? Think of Γ as information. If some <u>infinite</u> amount of information establishes something to be true beyond a shadow of a doubt, does that mean that some <u>finite</u> amount of information is sufficient to establish the same thing?

Corresponding to the two questions posed are two forms of compactness, Compact$_1$ and Compact$_2$:

Compact$_1$: Γ has a model iff every finite subset of Γ has a model.

Compact$_2$: $\Gamma \models \varphi$ iff there is <u>some</u> finite subset $\Gamma' \subseteq \Gamma$, such that $\Gamma' \models \varphi$.

Each version of compactness has an "easy direction" and a "hard direction".

Exercises:

Prove the following "easy directions" of compactness:
1. If Γ has a model, then every finite subset of Γ has a model.
2. If there is some finite subset $\Gamma' \subseteq \Gamma$ such that $\Gamma' \models \varphi$, then $\Gamma \models \varphi$.

These are, then, the "hard directions" of compactness:

Comp$_1$: If every finite subset of Γ has a model, then Γ itself has a model.

Comp$_2$: If $\Gamma \models \varphi$, then there is <u>some</u> finite subset $\Gamma' \subseteq \Gamma$, such that $\Gamma' \models \varphi$.

As mentioned above, the hard directions are equivalent.

THEOREM: Comp$_2 \Longrightarrow$ Comp$_1$.
PF:
1. Assume: if $\Gamma \models \varphi$, then there is a finite subset $\Gamma' \subseteq \Gamma$, such that $\Gamma' \models \varphi$.
   TO PROVE: Comp$_1$
2. Assume: every finite subset $\Gamma' \subseteq \Gamma$, is such that Γ' has a model.
   TO PROVE: Γ itself has a model.
3. Suppose Γ has <u>no</u> model (to give us a contradiction).
4. Then $\Gamma \models \varphi$, for any φ.
5. Take '$P \land \neg P$' to be φ.
6. So, $\Gamma \models P \land \neg P$.
7. By line 1, there is a finite subset $\Gamma' \subseteq \Gamma$, such that $\Gamma' \models P \land \neg P$.
8. Since <u>every</u> finite subset of Γ has a model (by assumption), Γ' has a model, say $\mathcal{B}$.
9. Since $\Gamma' \models P \land \neg P$, $\mathcal{B}(P \land \neg P) = \mathbf{T}$, which is impossible. ■

Exercises:

1. Explain why line 4 in the above proof is true.
2. Why is line 9 true, and how does that give us the desired result?

THEOREM: Comp$_1 \Longrightarrow$ Comp$_2$.

Exercise:

Prove the above metatheorem. [Hint: after assuming Comp$_1$ and the antecedent of Comp$_2$, suppose that what you wish to prove is false. That is, suppose that for <u>every</u> finite subset $\Gamma' \subseteq \Gamma$, $\Gamma' \not\models \varphi$.]

Though Compactness is a purely semantic concept, it is most easily proved as a consequence of Strong Completeness and (Strong) Soundness.

Exercises:
1. Prove $Comp_1$.
   Weak Completeness is the following: for any sentence ψ, if $\vdash \psi$, then $\models \psi$.
2. Prove: Weak Completeness + Compactness $\implies$ Strong Completeness.
3. Prove that the following statement is equivalent to Strong Completeness:

   Every consistent set of sentences has a model.

It is possible to give a purely semantic proof of Compactness, rather than using syntactic properties of derivations, as we have. In *Model Theory for First Order Theories*, we outline a beautiful semantic proof of Compactness for a much richer language.

Chapter Exercises:
Prove the following statements.
1. If Γ is satisfiable, Γ is consistent.
2. If Γ is consistent, then Γ is satisfiable
3. Consider any sentence φ of L_{SL} with sentence letters $\theta_1, \theta_2, \ldots, \theta_k$, and any boolean interpretation $\mathcal{B}$.
   If $\mathcal{B}(\varphi) = T$, let $\varphi' = \varphi$. If $\mathcal{B}(\varphi) = F$, let $\varphi' = \neg\varphi$.
   Similarly, for each θ_i such that $1 \leq i \leq k$, let $\theta'_i = \theta_i$ if $\mathcal{B}(\theta_i) = T$, and let $\theta'_i = \neg\theta_i$ if $\mathcal{B}(\theta_i) = F$.
   Then:
   $$\{\theta'_1, \theta'_2, \ldots, \theta'_k\} \vdash \varphi'.$$

4. Suppose φ is a sentence of L_{SL} that is not a tautology. Let $\Gamma = \{\psi : \psi \text{ is a substitution instance of } \varphi\}$. Then:

   Γ is inconsistent.

5. Form the inference system SL' by replacing the rule CONTRA by this single version of Modus Tollens (MT'):

$$\frac{\neg\varphi \rightarrow \neg\psi}{\therefore \varphi} \quad \text{MT'}$$

SL' now consists of A, CD, MP, MT', and INT, whereas SL consists of A, CD, MP, CONTRA, and INT. Then, for any sentence φ and set of sentences Γ of L_{SL},

$$\Gamma \vdash_{SL} \varphi \text{ iff } \Gamma \vdash_{SL'} \varphi.$$

6. (a) Prove the deduction theorem (version c) for Ax-Sent, the system of sentential logic presented at the end of Chapter 2.

   (b) Prove what we'll call 'the CONTRA theorem' for Ax-Sent (the language includes only the connectives '$\neg$' and '$\rightarrow$'):

   CONTRA THEOREM: If $\Gamma \vdash_{Ax-Sent} \psi$ and $\Gamma \vdash_{Ax-Sent} \neg\psi$, then $\Gamma - \{\neg\varphi\} \vdash_{Ax-Sent} \varphi$

   (c) Prove that every theorem of sentential logic that has only the connectives '$\neg$' and '$\rightarrow$' is a theorem of Ax-Sent. In symbols, prove (for this restricted language):

   If $\vdash_{SL} \varphi$, then $\vdash_{Ax-Sent} \varphi$

   [Hint: use (a) and (b)]

   (d) Prove the converse of (c), which, together with (c), shows:

   $\vdash_{SL} \varphi$ iff $\vdash_{Ax-Sent} \varphi$

   (That is, the two systems have exactly the same theorems for the language of sentential logic with the connectives '$\neg$' and '$\rightarrow$'.)

CHAPTER 5

THE SYNTAX OF FIRST-ORDER PREDICATE LOGIC

TOPICS: *logical symbol, non-logical symbol, individual variable, logical name, quantifier, relation letter, individual symbol, formula, scope of a quantifier, bound vs. free variable, sentence, universal, existential, molecular formula, name that is new to a formula or to a derivation, rules of derivation, informal "meanings" of sentences and intuitions behind rules, instance, UG strategy, Existential Switch rule and strategy, derivations from assumptions.*

The preceding three chapters have all been about sentential logic, which is the logic of "whole sentences". According to sentential logic, if 'P' is true (for a given boolean interpretation), and if 'P $\rightarrow$ Q' is also true (for that interpretation), then 'Q' must be true as well. Since the meaning of '$\rightarrow$' is modeled on 'if... then_ _ _' of ordinary English, we can paraphrase 'P $\rightarrow$ Q' as saying 'if P then Q'. Then, the inference becomes: (1) 'P' is true; and (2) 'if P then Q' is true; (3) therefore, 'Q' is true.

We now wish to create a more expressive language than L_{SL}, capable of modeling more extensive inferences. The new language will enable us to model inferences like the following: (1) any even number less than 10^{23} is equal to the sum of two primes; (2) $10^{23} - 2$ is an even number less than 10^{23}; (3) therefore, $10^{23} - 2$ is equal to the sum of two primes.

The key to being able to model arguments like the above is having the capability to express concepts like "for all" and "for some". To do this, we introduce two **quantifiers**, '$\forall$' and '$\exists$' in a language that we call "the language of first-order predicate logic", abbreviated 'L_{PL}'.

We now present the syntax of first-order predicate logic, which consists of the grammar for the language L_{PL} and its rules of inference. First, the syntax is presented very formally, in terms of strings of symbols that are arranged in certain precisely-specified patterns. Afterwards, we explain the informal "meanings" for such a language and develop the intuitions behind the rules of inference in an informal way.

5.1 THE FIRST-ORDER LANGUAGE L_{PL}

The language of first-order predicate logic, L_{PL}, is much richer than L_{SL}. The symbols of L_{PL} are split into two categories: (1) **logical symbols**, those pertaining purely to logic; and (2) **non-logical symbols**, those reflecting content that is outside the realm of logic.

The **logical symbols** of L_{PL} are **individual variables**:

$$x, y, z, w, x_1, y_1, z_1, w_1, x_2, \ldots,$$

logical names:

$$a, b, c, d, a_1, b_1, \ldots \quad (\text{in alphabetic order}),$$

and two **logical constants (quantifiers):**

$$\forall, \text{ and } \exists.$$

The **non-logical symbols** are **relation letters** (where k and j are ≥ 0):

$$F_j^k, G_j^k, H_j^k, I_j^k, J_j^k, K_j^k, L_j^k.$$

An **individual symbol** is an individual variable or a logical name.

An **atomic formula** of L_{PL} is a k-ary relation letter R^k followed by a string of length k of individual symbols: $\delta_1 \delta_2 \ldots \delta_k$, where k can be 0. The result is the atomic formula: $R^k \delta_1 \delta_2 \ldots \delta_k$. There may or may not be a subscript j attached to a relation letter. Subscripts are used simply to distinguish otherwise identical relation letters.

Examples: 'F^0', 'F_1^0', '$F^4 xya_3 b_4$', '$K^3 a_{72} w_2 c_{100}$', are atomic formulas. '$L^5 xyzw$', '$P^2 xy$', '$F^0 a$' are not (why?).

Exercise:

Which strings below are atomic formulas of L_{PL}, and which ones are not:

$$P_1^0, \; F_3^2 xy, \; H^2 PQ, \; H_0^4 axax, \; (P \wedge Q).$$

If a relation letter R^k has 0 for superscript k, then R^k is a **sentence letter** of L_{PL}.

A **formula of L_{PL}** is given by the following inductive definition:

(i)   An atomic formula is a formula;
(ii) If ψ is a formula, then $\neg\psi$ is a formula;
(iii) If ψ and χ are both formulas, then the following are all formulas:

   (a) $(\psi \rightarrow \chi)$,
   (b) $(\psi \wedge \chi)$,
   (c) $(\psi \vee \chi)$,
   (d) $(\psi \leftrightarrow \chi)$;

(iv) If α is a variable and ψ is a formula, then both of the following are formulas:

   (a) $\forall\alpha\psi$,
   (b) $\exists\alpha\psi$.

Nothing is a formula unless it satisfies one of the four clauses (i)-(iv).

Here are some examples of formulas that are not atomic:

$\forall x F^1 x$, $\exists y \forall x F^1_1 a$, $\exists x \exists y F^2 xy$, $\exists x \forall y (F^1 x \vee \neg G^3 ayb)$.

In sentences of the form $\forall\alpha\varphi$ and $\exists\alpha\varphi$, φ is the **scope** of the quantifiers $\forall$ and $\exists$. An occurrence of a variable α is **bound** in a formula iff it immediately follows a quantifier ($\forall$ or $\exists$) or it occurs within the scope of a quantifier that has α as its variable. Otherwise, the occurrence is **free**.

Example: In the formula '$(\exists x \forall y F^3 xay \vee \forall z G^2 yy)$' the first and second occurrences of 'x' are bound, the first and second occurrences of 'y' are bound, the third and fourth occurrences of 'y' are free, and the one occurrence of 'z' is bound.

Exercise:

State which occurrences of the variables 'x', 'y', and 'z' are bound and which are free in the following formula:

$\exists z(\forall y \exists x F^3 yxa \rightarrow \neg \exists y G^2 xz)$.

A **sentence of** L_{PL} is a formula in which no variable is free.

A formula that begins with a '$\forall$' is a **universal**. A formula that begins with '$\exists$' is an **existential**. A formula that is universal or existential is a **quantified formula**. Thus, '$\exists x(F^1 x \rightarrow G^1 x)$' is a quantified formula (since it <u>begins</u> <u>with</u> a quantifier), whereas '$(\exists x F^1 x \rightarrow \exists x G^1 x)$' is <u>not</u> a quantified formula, it is a conditional (it <u>begins</u> with a left parenthesis).

A formula whose left-most symbol is either a '$\neg$' or a left parenthesis is a **molecular formula**.

Thus, every formula of L_{PL} is either atomic or, if not atomic, molecular or quantified. If it is quantified, it is either universal or existential.

Exercises:

Classify each formula below as atomic, molecular or quantified. If it is quantified, state whether it is universal or existential:
F^0, $G_5^3 abz$, $(\exists x F^1 x \rightarrow \exists y G^1 y)$, $\forall x(\exists y G^1 y \rightarrow F^1 z)$, $\neg \exists x F^1 x$, $\exists x \neg F^2 xa$, $\exists z(\forall x F^0 \rightarrow F_2^0)$.

A formula all of whose relation letters have 0 or 1 as a superscript is called a **monadic formula**.

The formulas of L_{PL} have the **unique readability** property that held of the sentences of L_{SL}, namely that each sentence of L_{PL} can be read in one and only one way (see Chapter 1). (This result can be proved rigorously by Complete Induction on the number of connectives and/or quantifiers — the "height" — of a given formula.)

For any formula φ, an individual variable α, and a logical name β, $\varphi \alpha/\beta$ is identical with φ except for having occurrences of β wherever φ has free occurrences of α.

Examples:

φ	α	β	$\varphi\, \alpha/\beta$
$F_1^3 xax$	x	b	$F_1^3 bab$

φ	α	β	$\varphi\ \alpha/\beta$
$\exists x F^2_0 xy$	y	a	$\exists x F^2_0 xa$
$\forall x F^1_1 x$	x	c	$\forall x F^1_1 x$
$(\exists x F^1_1 x \lor F^1_1 x)$	x	a	$(\exists x F^1_1 x \lor F^1_1 a)$

Exercises:

Fill in the blanks:

φ	α	β	$\varphi\ \alpha/\beta$
1. $F^2_3 xa$	x	a	_____
2. _____	x	b	F^0_5
3. $\forall z(F^1_1 y \lor \neg \exists y G^1_1 y)$	y	b	_____
4. $(\exists y F^2 cc \lor \neg \forall x(F^1 x \to G^2 yy))$	y	c	_____
5. $(G^4 zazx \to \neg \forall z F^2 zb)$	z	__	$(G^4 aaax \to \neg \forall z F^2 zb)$

Informal Conventions

Again, we will usually *not* write external parentheses. And, we usually drop superscripts and subscripts of relation letters when no ambiguity results. Remember that the "official" formulas have external parentheses where required. So, the formula '$\forall x F^1 x \to \forall x G^1 x$' does not begin with a quantifier, since the official formula is '$(\forall x F^1 x \to \forall x G^1 x)$'. (It begins with a left parenthesis.) We also write 'P', 'Q', etc., as shorthand for 0-ary relation letters (though they are not official relation letters at all).

Unless otherwise indicated, the letters 'Γ' and 'Δ' (with perhaps primes or subscripts) always stand for sets of sentences; 'φ', 'ψ', 'χ', and 'θ' stand for sentences (sometimes 'θ' stands for an atomic sentence); 'α' always stands for a variable; and 'β', 'δ', 'λ', and 'γ' usually stand for logical names. In each context where these symbols are used, clarifying remarks will be made to eliminate possible ambiguities.

CHAPTER 5

In the following exercises $\forall\alpha\varphi$ is a universal sentence of L_{PL}, α is a variable, φ is a formula, β is a name, and $\varphi\alpha/\beta$ is the result of substituting an occurrence of β for every free occurrence of α in φ.

Exercises:

Fill in the blanks:

$\forall\alpha\varphi$	α	φ	β	$\varphi\alpha/\beta$
1. $\forall xFx$	x	Fx	a	___
2. $\forall x(Fx \rightarrow Gx)$	___	___	d	___
3. $\forall x(Fx \rightarrow \forall xGxb)$	___	___	b	___
4. $\forall yGya$	___	___	a	___
5. $\forall xGabc$	___	___	d	___
6. $\forall z\exists yGazb$	___	___	b	___
7. $\forall y\forall xFxx$	___	___	c	___
8. $\forall z(\exists xFxaz \lor \exists zFzax)$	___	___	b	___
9. ___	x	Fax	a	___
10. ___	z	$\forall x\exists yFxya$	a	$\forall x\exists yFxya$
11. ___	z	$\forall x\exists yFxyz$	a	$\forall x\exists yFxya$

We will often say that a logical name β **is new to a formula** if β does not occur in that formula. β **is new to a derivation** at a certain line in the derivation if β does not occur anywhere in that derivation above that line, <u>including SHOW lines</u>.

Exercises:

In which of the above cases is β new to φ?

5.2 RULES OF DERIVATION FOR L_{PL}

We will now formally define inference rules for first-order predicate logic. In the next section, we discuss these rules more informally and try to provide useful intuitions for understanding and applying them. The rules are as follows:

(1) A, MP, CD, CONTRA, and DEF, as well as all derived rules, including the Theorem Rule (putting any substitution instance of a previous theorem on a line of a later derivation). The derived rules are: DN, MT, MTP, SOC, T, CONJ, SIMP, ADJ, and CTB. Informally, we now permit **SL** to be

written to the right of a line as an abbreviation of Sentential Logic, provided that there is an <u>obvious</u> derivation of the sentence on that line using rules of sentential logic alone and that the correct assumption numbers are written to the right of the line.

(2) Universal Instantiation (UI). If $\forall \alpha \varphi$ appears on an earlier line and β is a name, $\varphi \alpha / \beta$ may be entered on a later line with the same assumption numbers as $\forall \alpha \varphi$.

(3) Universal Generalization (UG). If $\varphi \alpha / \beta$ appears on a line, where β is a name that is new to φ and to all the assumptions of $\varphi \alpha / \beta$, then $\forall \alpha \varphi$ may be entered on a later line with the same assumption numbers as $\varphi \alpha / \beta$.

(4) Quantifier Exchange (QE). Either of $\neg \forall \alpha \varphi$ and $\exists \alpha \neg \varphi$ may be placed on a line if the other one appears on an earlier line. Similarly, either of $\neg \exists \alpha \varphi$ and $\forall \alpha \neg \varphi$ may be exchanged for the other. The assumption numbers for the later line are the same as those for the earlier one.

$\neg \forall \alpha \varphi$ for $\exists \alpha \neg \varphi$
$\neg \exists \alpha \varphi$ for $\forall \alpha \neg \varphi$

Again, when a sentence "appears", "occurs", "is placed on", or is simply "on" a line, that means that <u>only</u> that sentence is on the line, <u>not</u> as a proper part of another sentence. The definitions of a derivation, of derivable, and a theorem are all as before, except that derivations are now in L_{PL}.

We need to note that the notion of a substitution instance of a sentence in sentential logic is now widened to permit sentences of first-order logic to be substituted for sentence letters of sentential logic. Thus, 'Fa $\rightarrow$ Fa', and '$\forall$xGxb $\rightarrow$ $\forall$xGxb' are both substitution instances of T5, 'P $\rightarrow$ P'. Hence, we can write on any line of a derivation:

k.   Fa $\rightarrow$ Fa   T5 $\dfrac{P}{Fa \rightarrow Fa}$

and:

m. $\forall$xGxb $\rightarrow$ $\forall$xGxb   T5 $\dfrac{P}{\forall xGxb}$

We call UI, UG, QE, and all the sentential rules **basic rules** of derivation in L_{PL} to distinguish them from **derived rules**, which we will obtain from the basic ones.

5.3 INFORMAL "MEANINGS" & THE INFERENCE RULES

A sentence of the form $\forall \alpha \varphi$ "says" (we will be more precise about this later) "for all α, φ". So, the sentence '$\forall x Fx$' says "for all x, x has property F". Or, more simply: everything is F. Sometimes, to make a sentence like '$\forall x Fx$' more intelligible, it is helpful to have a "domain of discourse" in mind that the quantifier '$\forall$' ranges over. Suppose we pick the positive integers as a domain of discourse. Then, '$\forall x Fx$' "says" (in terms of our domain of discourse) that all positive integers have property F. To be more specific, we can invent a property of the positive integers for which the given sentence is true. Suppose we let property F be the property of being greater than or equal to 1. Now (given this interpretation of the symbols), '$\forall x Fx$' "says" that all positive integers are greater than or equal to 1.

Now, let's test the Universal Instantiation (UI) Rule for this interpretation of the symbols. But, before doing that, let's let the logical name 'a' name some object of our domain of discourse. Let's let 'a' name the number 2. Now, to test our UI Rule, suppose '$\forall x Fx$' appears on a line of a derivation. According to UI, 'Fa' can be placed on a later line. In this special case, we can see that the inference is justified, since if <u>all</u> positive integers are greater than or equal to 1, 2 is greater than or equal to 1 also. We can see more generally that the inference is correct, regardless of the domain of discourse and how we interpret 'F' and 'a'. If a universal sentence is true, then any **instance** of it must be true as well. So, 'Fb' must be true and so must 'Fc', 'Fd', 'Fa$_1$', etc. In short, the UI Rule expresses the fact that if *every object* has a given property, then *any object in particular* has that property. So, if the universal generalization $\forall \alpha \varphi$ appears on a line, UI permits us to place any instance $\varphi \alpha / \beta$ on a later line.

The Universal Generalization Rule (UG) proceeds in the opposite direction of the UI Rule. According to the UG Rule, if the instance $\varphi \alpha / \beta$ is on a line of a derivation, the generalization, $\forall \alpha \varphi$, of that instance may be placed on a later line, provided certain conditions are met. The idea behind one of the conditions is to ensure that the instance is "arbitrary".

For example, suppose for some domain of objects 'Ga' is true. To be concrete, assume, as before, that the domain is the positive integers, and suppose 'G' stands for the property of being an even number. Let us further suppose that 'a' names the number 2. So, given these informal meanings, 'Ga' is true, since it "says" that 2 is an even number. Does this mean that '∀xGx' is true? No. It doesn't follow that because 2 is even, <u>every</u> number is even. But, there <u>are</u> special cases when an instance of a generalization is "arbitrary", and then it <u>does</u> lead to the truth of the generalization itself.

For example, recall that in high school geometry class you proved facts about <u>all triangles</u> by looking at the properties of a <u>single triangle</u>. You may have started a proof about all triangles by saying something like, "Let triangle ABC have the following properties...". Then, you continued with the proof until you established the desired fact about triangle ABC. Once you did this, the proof was over. Even though you showed that only triangle ABC had the desired property, the proof was perfectly general and <u>pertained to all triangles</u>. The reason for this — probably not mentioned by your geometry teacher — was that triangle ABC was "completely arbitrary". Though ABC seemed to be only one <u>instance</u> of a triangle, it served as a representative for <u>all triangles</u>. In mathematics, when you prove some fact about a "completely arbitrary instance" of a universal generalization, you show that the universal generalization itself holds. In first-order logic, one condition on the UG Rule ensures that an instance is "arbitrary" before permitting the inference from $\varphi\alpha/\beta$ to $\forall\alpha\varphi$. We will say more about this later.

The Quantifier Exchange Rule (QE) can also be explained by our informal meanings of the quantifiers. Consider $\neg\forall\alpha\varphi$, which "says" that not all α's have property φ. If not all α's have φ, then there must be some α's that <u>don't</u> have φ. That is, $\exists\alpha\neg\varphi$. And, vice versa, if some α's don't have φ, then not all α's <u>do</u> have φ. So, informally, that's why we can switch $\neg\forall\alpha\varphi$ and $\exists\alpha\neg\varphi$.

Exercise:

Explain, using our informal meanings for the quantifiers, why we can switch $\neg\exists\alpha\varphi$ and $\forall\alpha\neg\varphi$.

The fact that we are working hard to cultivate correct intuitions about the "meanings" of the rules of derivation and the sentences of L_{PL} raises an important question about the relationship between these "meanings" and the formal system presented. Formally speaking, what constitutes a derivation in L_{PL} is completely determined by the syntactic rules above, describing conditions for manipulating the specific strings of symbols that are sentences of L_{PL}. On the other hand, the "meanings" of the sentences of L_{PL} are what motivated the construction of the rules in the first place. So, having a feel for the underlying meanings will make it easier to understand and to use the derivational rules, even though the rules themselves are completely syntactic.

To decide whether a string of symbols of L_{PL} constitutes a sentence or whether a series of sentences constitutes a derivation is a purely mechanical task that can be carried out by a machine following the rules for identifying a sentence or a derivation. Further, machines can be created to list sentences in some order, and they can also be created to provide a list of derivations. In both cases the machines simply follow rules. A sentence-listing machine follows the rules for forming sentences and a derivation-listing machine follows the rules for forming derivations.

The "intended meanings", lurking behind the derivational rules are of no use to a machine, but they can help us considerably. The meaning of a sentence of L_{PL} is somewhat similar to the meaning of a related sentence in Mathematical English — the English used to make mathematical statements. And a derivation of a sentence within L_{PL} is very much like a proof of the Mathematical English version of that sentence outside of L_{PL}. Thus, associating informal meanings to sentences and derivations can help to deepen our understanding of proofs in mathematics.

More immediately, associating informal meanings to the formal symbols and rules of first-order logic will help us to better understand what each sentence of L_{PL} "says", and if what it says is always true, then its informal meaning will help us to construct a derivation of it.

In short, the formal system of L_{PL} consisting of its language and rules of derivation is purely syntactic, composed simply of string patterns. Derivations can be cranked out by a machine following mechanical rules describing only the permissible symbol manipulations. Attaching informal meanings to these symbol patterns, nonetheless, has two important benefits: (1) it enriches our understanding of the relationships between the derivations in first-order logic and mathematical proofs, and (2) it improves our proficiency in discovering and constructing derivations and mathematical proofs.

5.4 DERIVATIONS OF THEOREMS IN L_{PL}

We now use the rules of inference to construct derivations. Check each line of the following derivation to assure yourself that it has been entered correctly in accordance with the appropriate rule:

T5.1 $\forall x(Fx \rightarrow Gx) \rightarrow (\forall xFx \rightarrow \forall xGx)$

    SHOW $\forall x(Fx \rightarrow Gx) \rightarrow (\forall xFx \rightarrow \forall xGx)$
1. $\forall x(Fx \rightarrow Gx)$ 1,A
                                   SHOW $\forall xFx \rightarrow \forall xGx$
2. $\forall xFx$ 2,A
                                   SHOW $\forall xGx$
                                   SHOW Ga
3.   Fa $\rightarrow$ Ga   1,UI
4.   Fa   2,UI
5.   Ga   1,2,MP(3,4)

Note that at this point in the derivation we can check off the line 'SHOW Ga' and add the remaining SHOW lines in reverse order, justifying each one by the appropriate rule, and we are done:

6. $\forall xGx$ 1,2,UG(5)
7. $\forall xFx \rightarrow \forall xGx$ 1,CD
8. $\forall x(Fx \rightarrow Gx) \rightarrow (\forall xFx \rightarrow \forall xGx)$ CD

Please double-check line 6 to assure yourself that the restrictions on Rule UG have been adhered to. According to the UG restrictions, the name β in φα/β must not occur in assumptions 1 or 2, or in φ. β is 'a' in this case, and φα/β is 'Ga'. So, 'a' does <u>not</u> occur in lines 1 or 2, or in φ, which is 'Gx'. Thus, the conditions on UG have been met, and line 6 is correct. Notice that immediately after the SHOW line 'SHOW $\forall xGx$', we introduced another SHOW line: SHOW Ga.

See if you can understand how 'SHOW Ga' functions in the above derivation <u>before</u> it is explained.

(As before, you are to derive all theorems as *exercises* when their derivations have not been provided.)

T5.2 $(\forall x(Fx \rightarrow Gx) \land \forall x(Gx \rightarrow Hx)) \rightarrow \forall x(Fx \rightarrow Hx)$
T5.3 $(\forall x(Fx \rightarrow Gx) \land \forall x(Fx \rightarrow Hx)) \rightarrow \forall x(Fx \rightarrow (Gx \land Hx))$

5.5   INFORMAL JUSTIFICATION OF THE UG RULE

You may have started wondering more about the correctness of the UG Rule. We said earlier that the purpose of one restriction on UG was to ensure that the instance generalized on was "completely arbitrary". Let's think of sentences containing names as carrying "specific information" about those named instances. For example, the sentence 'Fa' gives us specific information about the object named by 'a'. It tells us that that object has property F. Suppose our domain of discourse is, as before, the set of positive integers, and suppose 'F' stands for the property of being a prime number. Now, suppose in a derivation we assume 'Fa'. Then the assumption 'Fa' gives us the specific piece of information that the number named by 'a' is prime. To infer '∀xFx', that <u>all</u> numbers are prime in this case, is to jump (incorrectly) from the <u>assumption</u> that one specific number has property F to the general conclusion that all numbers have F. So, if we <u>assume</u> something specific about an instance, we are <u>not</u> permitted to generalize on it.

Suppose we begin a new chain of reasoning by assuming '∀x(Fx ∧ Gx)' is true. Suppose the domain of discourse consists of all multiples of 5 greater than 0. In other words, the domain of discourse is {5, 10, 15, ...}. Now, let F be the property of being evenly divisible by 5, and let G be the property of being greater than or equal to 5. For this case, '∀x(Fx ∧ Gx)' is true, since it "says" that every element of the domain of discourse is evenly divisible by 5 and is greater than or equal to 5. Notice that '∀x(Fx ∧ Gx)' contains <u>no</u> "specific information" about any named instance, since there are no names occurring in '∀x(Fx ∧ Gx)'. Suppose then, we infer 'Fa ∧ Ga' from '∀x(Fx ∧ Gx)'. Formally, we infer 'Fa ∧ Ga' by UI, given the assumption '∀x(Fx ∧ Gx)'. Since the assumption carries no specific information about the object that 'a' names, we know that the inference 'Fa ∧ Ga' is not based on knowing anything specific about object-a. Thus, if we infer 'Fa' alone (formally by SIMP), we know that we have still not <u>assumed</u> anything specific about a. Since no specific information has been provided about object-a, we know that the sentence 'Fa' is really not only <u>about</u> object-a, it's about <u>any object</u> at all. Thus, the inference to '∀xFx' from 'Fa' is correct in this case.

So, one UG restriction is meant to distinguish those cases when specific information is assumed (about a named object) from those cases when no specific information has been assumed. Let us now look at the other important restriction on UG.

We do not want the generalization itself to contain any specific information about the instance. That is, we do not want ∀αφ to contain any specific information about object-β. The reason for requiring that ∀αφ itself carry no specific information about object-β is that we also wish to block this particular brand of faulty reasoning:

1. For any natural number x, x ≤ x
2. So, for the number a, a ≤ a
3. Hence, for number a, ∀x x ≤ a   X FAULTY REASONING X

Line 3 says <u>every number</u> is less than the number a, which is <u>false</u>. We can reproduce this faulty reasoning in L_{PL} as follows:

1. ∀xFxx   1,A
2. Faa   2,UI
3. ∀xFxa   X FAULTY use of UG rule X

When 'Faa' occurs on a line with no occurrence of 'a' in any assumption of that line (i.e., the truth of 'Faa' is not based on any specific information about object-a), then it is <u>correct</u> to infer '∀xFxx' by UG since '∀xFxx' contains no occurrence of 'a' (i.e., '∀xFxx' retains no specific information about object-a). On the other hand, it is a violation of the UG Rule to infer '∀xFxa' because '∀xFxa' contains an occurrence of 'a' (i.e., '∀xFxa' retains some specific information about object-a).

5.6 UG STRATEGY

After a SHOW line of the form 'SHOW ∀αφ', write the SHOW line 'SHOW φα/β' where β is a name that is new to the derivation, including preceding SHOW lines. Then, if you can show φα/β <u>without</u> using any new assumptions containing β, you may use UG to get ∀αφ. To see an example of this, look back at the derivation of T5.1, where 'SHOW Ga' was used.

We show the right-to-left direction of T5.4 and leave the left-to-right direction for you:

T5.4   ∀x(Fx ∧ Gx) ↔ (∀xFx ∧ ∀xGx)

    √SHOW (∀xFx ∧ ∀xGx) → ∀x(Fx ∧ Gx)
    1. ∀xFx ∧ ∀xGx   1,A
                                      √SHOW ∀x(Fx ∧ Gx)
                                      √SHOW Fa ∧ Ga
    2. ∀xFx   1,SIMP
    3. ∀xGx   1,SIMP
    4. Fa   1,UI(2)
    5. Ga   1,UI(3)
    6. Fa ∧ Ga   1,CONJ(4,5)
    7. ∀x(Fx ∧ Gx)   1,UG
    8. (∀xFx ∧ ∀xGx) → ∀x(Fx ∧ Gx)   CD

Exercise:

Derive the left-to-right direction of T5.4.

T5.5 ∀x(P → Fx) ↔ (P → ∀xFx)
T5.6 (∀xFx ∨ ∀xGx) → ∀x(Fx ∨ Gx)   [Hint: Use SOC]

The theorems up to this point illustrate the use of UI and UG for universal sentences and their instances. Since we do not have corresponding rules for existential sentences at this point, we have to convert existentials to universals, using the Quantifier Exchange Rule (QE) when trying to derive an existential sentence:

T5.7 ∀xFx → ∃xFx

    SHOW ∀xFx → ∃xFx
    1. ∀xFx   1,A
                                    SHOW ∃xFx
    2. ¬∃xFx   2,A
                                    √SHOW A CONTRADICTION
    3. ∀x¬Fx   2,QE
    4. ¬Fa   2,UI(3)
    5. Fa   1,UI

T5.8 $\forall x(Fx \rightarrow Gx) \rightarrow (\exists xFx \rightarrow \exists xGx)$
T5.9 $(\forall xFx \wedge \exists xGx) \rightarrow \exists x(Fx \wedge Gx)$
T5.10 $(\forall x(Fx \rightarrow Gx) \wedge \exists x(Fx \wedge Hx)) \rightarrow \exists x(Gx \wedge Hx)$

It soon becomes unwieldy to always use a CONTRA strategy together with QE in deriving an existential. Besides, we would like our rules to mirror our intuitions that tell us why these sentences must be true. When our rules mirror our intuitions, our intuitions will lead us to derivations. And the derivations, in turn, will help us to refine our intuitions. For example, take T5.7. It "says" that if everything is F, then something is F. Here is the informal argument for T5.7:

INFORMAL ARGUMENT:
    Since everything is F, then object-a is F.
    Since a is F, <u>something</u> is F (namely: a).   Q.E.D.

Here is another derivation of T5.7 based on the <u>INFORMAL ARGUMENT</u>, using a new rule called **Existential Generalization (EG)**:

		SHOW $\forall xFx \rightarrow \exists xFx$
1.	$\forall xFx$	1, A
		√SHOW $\exists xFx$
2.	Fa	1, UI
3.	$\exists xFx$	1, EG

Existential Generalization (EG): When a sentence of the form $\varphi\alpha/\beta$ is on a line of a derivation, you may place $\exists\alpha\varphi$ on a later line with the same assumptions as $\varphi\alpha/\beta$.

There is not much to say about EG, since its use is natural and obvious. Also, it is virtually impossible to use EG incorrectly.

5.7 THE EXISTENTIAL SWITCH ARGUMENT

Another very helpful rule is **Existential Switch (ES)**, which is based on a pattern of reasoning frequently found in mathematical proofs. In the following example of a mathematical proof, 'a|b' is defined as meaning "there is an x such that a multiplied by x equals b", or, more informally, 'a|b' means "a divides b evenly" (with no remainder). '·' means multiplication.

TO PROVE: If $a|b$, then $a|b\cdot c$.
1. Assume $a|b$.
                          TO PROVE: $a|(b\cdot c)$
                        √TO PROVE: $\exists x$ such that $a\cdot x = b\cdot c$
2. $\exists x$ such that $a\cdot x = b$ 1, def. of $|$
3. Call one such x: d.   2, naming an x
4. $a\cdot d = b$ 2,3
5. $(a\cdot d)\cdot c = b\cdot c$ 4, multiplying by c
6. $a\cdot(d\cdot c) = b\cdot c$ 5, associativity of $\cdot$
7. Let $e = d\cdot c$.   6, naming $d\cdot c$
8. $a\cdot e = b\cdot c$ 6,7 substitution
9. $\exists x$ such that $a\cdot x = b\cdot c$ 8

    In the argument above, the name 'd' was chosen on line 3 to stand for a number that yields b when multiplied by a. We knew from line 2 that there was <u>some</u> <u>number</u> for which this was true, and 'd' was an arbitrary name to use to stand for that number. How do we know that 'd' was an "arbitrary" choice of names? One way to tell is to pick a letter that <u>never</u> <u>occurs</u> in the above proof, say 'k', and to replace 'd' by 'k' on line 3 where 'd' was introduced. Then, if the proof works for 'k' as well as it did for 'd', without requiring a new assumption with 'k' in it, then the original choice of 'd' was completely arbitrary. It is arbitrary because we have just shown that any <u>other</u> letter would work as well.
    In short, an ES argument is correct when the statement that <u>some</u> <u>element</u> has a given property is replaced by the statement that β has that property, where β is an arbitrary name. To ensure that β is arbitrary in a proof requiring an ES argument, mathematicians often select the first unused β from some alphabetical list of letters implicitly reserved for that purpose. For example, sometimes 'x_0', 'x_1', 'x_2',... are used as arbitrary names in this way.
    Though mathematicians use ES fluidly and effortlessly in their proofs, we will be slow and deliberate in our use of it in derivations, especially at first. While EG is <u>difficult</u> to use incorrectly, ES is <u>easy</u> to use incorrectly. Thus, we will be very careful in applying ES.
    Before formalizing the mathematician's use of ES, let us look at an informal argument for T5.8:

T5.8 $\forall x(Fx \rightarrow Gx) \rightarrow (\exists xFx \rightarrow \exists xGx)$.

INFORMAL ARGUMENT:
  Assume: Every F is G.
  Assume: Something is F.
  Call a thing that is F by the name 'a'.
  Then, since <u>every</u> F is G, a is G as well.
  Thus, <u>something</u> is G (namely a).   Q.E.D.

Here is the new derivation:

SHOW ∀x(Fx → Gx) → (∃xFx → ∃xGx)
1.   ∀x(Fx → Gx)   1,A
                SHOW ∃xFx → ∃xGx
2.   ∃xFx   2,A
                √SHOW ∃xGx
3.   Fa   3,EA ("Existential Assumption")
4.   Fa → Ga   1,UI
5.   Ga   1,3 MP(3,4)
6.   ∃xGx   1,3 EG(5)
7.   ∃xGx   1,2 ES(2,3,6)

The "switch" occurs on line 7 when we switch from the **"Existential Assumption"** on line 3 to the assumptions of the sentence '∃xFx' on line 2.

5.8 EXISTENTIAL SWITCH STRATEGY

When a sentence of the form ∃α𝜑 is on a line of a derivation and you wish to show ψ, assume 𝜑α/β where β is a name that is new to the derivation and new to ψ. We call such an assumption an **Existential Assumption**. If the desired sentence ψ is later obtained on a line and 𝜑α/β is the <u>only</u> assumption of that line containing β, then you may perform an **Existential Switch (ES)**, which means you may write ψ again on the next line with the assumptions of ∃α𝜑 replacing the Existential Assumption 𝜑α/β.

Let us look at another example of an Existential Switch. T5.9, '(∀xFx ∧ ∃xGx) → ∃x(Fx ∧ Gx)', "says" if everything is F and something is G, then something is both F and G. Here is the informal argument: Assume everything is F and something is G. Call a thing that is G: a. Since everything is F, a is F. So a is <u>both</u> F and G. Therefore, <u>something</u> is both F and G.

Here is the formal counterpart to that argument using our new ES Rule:

```
SHOW (∀xFx ∧ ∃xGx) → ∃x(Fx ∧ Gx)
1.   ∀xFx                    1,A
2.   ∃xGx                    2,A
            √SHOW ∃x(Fx ∧ Gx)
3.   Ga                      3,EA("Existential Assumption")
4.   Fa                      1,UI
5.   Fa ∧ Ga                 1,3 CONJ
6.   ∃x(Fx ∧ Gx)             1,3 EG
7.   ∃x(Fx ∧ Gx)             1,2 ES
```

Since no other assumption of line 6 contains 'a' except the Existential Assumption of line 3, we re-write '∃x(Fx ∧ Gx)' on line 7 and Existentially Switch assumption 3 (the only assumption of '∃xGx' on line 2) for Existential Assumption 2.

Our examples so far illustrate switching the single assumption of the existential sentence ∃αφ for the Existential Assumption φα/β. In more complicated cases the existential sentence may have several assumptions. When this happens, <u>all</u> of the assumptions of ∃αφ must replace the <u>single</u> Existential Assumption.

Existential Switch and Existential Generalization make derivations very natural and intuitive. But, it is easy to make a mistake with Existential Switch. You have to be careful about two things: (1) the name β in the Existential Assumption <u>must</u> be new and also *must not* occur in ψ, the sentence you are trying to show; and (2) when ψ has been shown, the <u>only</u> assumption of ψ that may contain β is the existential one.

Exercise:

Modeled on the two derivations above, rederive T5.10:
   (∀x(Fx → Gx) ∧ ∃x(Fx ∧ Hx)) → ∃x(Gx ∧ Hx).

T5.11: ∃x(Fx ∧ Gx) → (∃xFx ∧ ∃xGx)
T5.12: (∀x(Fx → Gx) ∧ ∃xFx) → ∃x(Fx ∧ Gx)
T5.13: ∃x(Fx ∨ Gx) ↔ (∃xFx ∨ ∃xGx)

You may wonder why EG and ES were introduced <u>after</u> the other quantifier rules rather than at the same time, especially since EG and ES are so useful. The reason is that they can be <u>derived</u> from the basic rules. Thus, they are theoretically dispensable, though practically quite useful (we will derive them later).

T5.14: $\neg\forall x(Fx \to Gx) \leftrightarrow \exists x(Fx \land \neg Gx)$
T5.15: $\neg\exists xFx \leftrightarrow (\forall x(Fx \to Gx) \land \forall x(Fx \to \neg Gx))$

T5.15 is our formal, first-order counterpart to the vacuous truth arguments mentioned in the chapter *Mathematical Preliminaries*. T5.15 "says" that it is exactly when nothing is F that everything that is F is G and everything that is F is not-G also.

Here are two "Laws of Confinement":

T5.16 $(\exists xFx \to P) \leftrightarrow \forall x(Fx \to P)$
T5.17 $\exists x(Fx \to P) \leftrightarrow (\forall xFx \to P)$

We show the right-to-left direction of T5.16

```
SHOW ∀x(Fx → P) → (∃xFx → P)
  1.  ∀x(Fx → P)                         1,A
      SHOW ∃xFx → P
  2.  ∃xFx                               2,A
      √SHOW P
  3.  Fa                                 3,EA('a' is new)
  4.  Fa → P                             1,UI
  5.  P
```

Deriving the left-to-right direction of T5.16 is fairly straightforward, as is the derivation of the left-to-right direction of T5.17. Deriving the right-to-left direction of T5.17 is more challenging, however.

T5.18 $(\forall xFx \to \forall xGx) \to \exists x(Fx \to Gx)$
T5.19 $(\forall xFx \to \forall xGx) \leftrightarrow \exists x\forall y(Fx \to Gy)$

```
SHOW (∀xFx → ∀xGx) → ∃x∀y(Fx → Gy)
  1.  ∀xFx → ∀xGx                        1,A
      SHOW ∃x∀y(Fx → Gy)
  2.  ¬∃x∀y(Fx → Gy)                     2,A
      √SHOW A CONTRADICTION
  3.  ∀x¬∀y(Fx → Gy)                     2,QE
  4.  ¬∀y(Fa → Gy)                       2,UI
  5.  ∃y¬(Fa → Gy)                       2,QE
  6.  ¬(Fa → Gb)                         6,EA('b' is new)
  7.  Fa                                 6,SL
```

8.	Fa	2,ES(7)
9.	∀xFx	2,UG(8)
10.	∀xGx	1,2,MP(1,9)
11.	Gb	1,2,UI(10)
12.	¬Gb	6,SL

The crux of this last derivation is to notice that line 2, '¬∃x∀y(Fx → Gy)', yields '∀xFx'. Since '¬(Fa → Gb)' yields 'Fa', which does not contain 'b', we can use ES to switch back to assumption 2, which then permits us to Universally Generalize. If we could not separate 'Fa' from the name 'b' introduced in the Existential Assumption on line 6, this derivation would have been trickier. Consider the following:

T5.20   ∀x∃y(Fx ∧ Gy) → ∃y∀x(Fx ∧ Gy)

	SHOW ∀x∃y(Fx ∧ Gy) → ∃y∀x(Fx ∧ Gy)	
1.	∀x∃y(Fx ∧ Gy)	1,A
2.	∃y(Fa ∧ Gy)	1,UI
3.	Fa ∧ Gb	3 EA('b' is new)

At this point in the derivation, we would like to be able to Universally Generalize to get:

∀x(Fx ∧ Gb)

But, we are *not permitted* to perform UG on line 3, because 'Fa ∧ Gb' is an assumption. Now, there is a sense in which we <u>should</u> be able to Universally Generalize on the 'a' on line 3 because we know that the part of that line that is the assumption concerns only the 'b'. The 'a' came from UI so we <u>should</u> be able to Universally Generalize on <u>it</u> and "not worry about the 'b'." Since the 'a' is in one subsentence on line 3, 'Fa', and the 'b' is in another, 'Gb', a way to continue the derivation would be to *separate* these two sentences.

A tricky way to separate the universal part and the existential part of line 3 is to make *another* Existential Assumption after Universally Instantiating '∀x∃y(Fx ∧ Gy)' to a <u>new</u> <u>name</u> and then mixing the conjuncts:

4.	∃y(Fc ∧ Gy)	1,UI
5.	Fc ∧ Gd	5,EA('d' is new)
6.	Fc ∧ Gb	3,5,SL("mixing")
7.	Fc ∧ Gb	1,3,ES

Now we <u>can</u> Universally Generalize on 'c' since it does not occur in line 1 or 3, and the rest of the derivation proceeds smoothly.

A second, more straightforward way to separate the universal content from the existential part is to begin a new derivation and to derive both '∀xFx' and '∃xGx' first. Then it is easy to get '∃y∀x(Fx ∧ Gy)'.

A <u>third</u> way to reach '∃y∀x(Fx ∧ Gy)' from '∀x∃y(Fx ∧ Gy)' is to not worry about separating universal from existential content and to begin a CONTRA strategy immediately — though this way can still be somewhat involved.

Exercises:

Derive T5.20 all three of the suggested ways and say which derivation is <u>least</u> satisfying and why. Which is <u>most</u> satisfying?

T5.21   ∀x∃y(Fx → Gy) → ∃y∀x(Fx → Gy)
T5.22   ∀x∃y(Fx ∨ Gy) → ∃y∀x(Fx ∨ Gy)

The reason that T5.20, T5.21, and T5.22 are only one-directional is that the other directions of them are all special cases of this theorem:

T5.23   ∃y∀xFxy → ∀x∃yFxy

```
       SHOW  ∃y∀xFxy → ∀x∃yFxy
       1.    ∃y∀xFxy                          1,A
                                       SHOW  ∀x∃yFxy
                                       √SHOW ∃yFay
       2.    ∀xFxb                            2,EA('b' is new)
       3.    Fab                              2,UI
       4.    ∃yFay                            2,EG
```

T5.24   ∃x∃yFxy ↔ ∃y∃xFxy
T5.25   ∀x∀yFxy ↔ ∀y∀xFxy
T5.26   ∀y∃x(Fx ↔ ¬Fy) ↔ (∃xFx ∧ ∃x¬Fx)

In order to extend the Theorem Rule from L_{SL} to L_{PL}, we first define which formulas may be substituted for each other. A sentence φ of L_{PL} is a **substitution instance** of ψ iff φ is obtained by uniformly replacing atomic formulas in ψ by formulas in φ with exactly the same free variables as the formulas they replace.[1] We give some examples of substitution instances:

(1) $\forall x Fx \dfrac{Fx}{Fx \wedge Ga} = \forall x(Fx \wedge Ga)$

(2) $\exists x Fx \dfrac{Fx}{Fx \rightarrow Gx} = \exists x(Fx \rightarrow Gx)$

(3) $P \rightarrow Q \dfrac{P, Q}{\exists x Fx, \forall x Fx} = \exists x Fx \rightarrow \forall x Fx$

These are <u>not</u> correct examples of substitution instances:

✗ (1) $\forall x Fx \rightarrow \exists x Fx \dfrac{\forall x Fx \rightarrow \exists x Fx}{P \rightarrow P} = P \rightarrow P$ ✗

✗ (2) $\forall x \forall y Fxy \dfrac{Fxy}{Gxyz} = \forall x \forall y \forall z Gxyz$ ✗

✗ (3) $\exists x Fx \dfrac{Fx}{P} = \exists x P$ ✗

Now we can update our Theorem Rule: any substitution instance of a previous theorem of L_{PL} can be placed on a line (without an assumption number).

T5.27 $\forall y \exists x(Fx \rightarrow Fy)$
T5.28 $\forall x \exists y(Fx \rightarrow Fy)$
[HINT: Begin both derivations with 'Fa $\rightarrow$ Fa', which is a substitution instance of the sentential theorem 'P $\rightarrow$ P'.]

Given our updated Theorem Rule, T5.27 and T5.28 can also be derived by taking the appropriate instance of T5.23 together with T5.29 and T5.30 respectively:

T5.29 $\exists x \forall y(Fx \rightarrow Fy)$
T5.30 $\exists y \forall x(Fx \rightarrow Fy)$
   [HINT for T5.29: Begin a CONTRA strategy, arrive at
   'Fa $\wedge$ ¬Fb', then get 'Fb $\wedge$ ¬Fc' by universally
   instantiating on 'b'. From 'Fb $\wedge$ ¬Fb' get 'P $\wedge$ ¬P' by SL
   and then use ES twice. T5.30 is similar.]

---

[1] A more extensive account of a substitution instance can be given, but only at the cost of much additional complexity.

SYNTAX OF FIRST-ORDER   155

T5.31 ∃x(Fx → ∀yFy)
T5.32 ∃y(∃xFx → Fy)
T5.33 ∀x∀y(Fx → Fy) ↔ ¬∃xFx ∨ ∀xFx

```
       SHOW  ∀x∀y(Fx → Fy) ↔ (¬∃xFx ∨ ∀xFx)
                         SHOW ∀x∀y(Fx → Fy) → (¬∃xFx ∨ ∀xFx)
   1.  ∀x∀y(Fx → Fy)         1,A
                         SHOW ¬∃xFx ∨ ∀xFx
                         SHOW ∃xFx → ∀xFx
   2.  ∃xFx                  2,A
                         SHOW ∀xFx
                        √SHOW Fa
   3.  Fb                    3,EA('b' is new to SHOW lines too)
   4.  ∀y(Fb → Fy)           1,UI
   5.  Fb → Fa               1,UI(4)
   6.  Fa                    1,3,MP(3,5)
   7.  Fa                    1,2,ES
                         SHOW (¬∃xFx ∨ ∀xFx) → ∀x∀y(Fx → Fy)
  13.  ¬∃xFx ∨ ∀xFx          1,A
                         SHOW ∀x∀y(Fx → Fy)
                         SHOW ∀y(Fa → Fy)
                         SHOW Fa → Fb
  14.  Fa                                       14,A
                        √SHOW Fb
  15.  ¬∃xFx                 15,A
  16.  ∀x¬Fx                 15,QE
  17.  ¬Fa                   15,UI(16)
  18.  Fb                    14,15,CONTRA
  19.  ¬∃xFx → Fb            14,CD
  20.  ∀xFx                  20,A
  21.  Fb                    20,UI
  22.  ∀xFx → Fb             CD(21)
  23.  Fb                    13,14,SOC(13,19,20)
```

T5.34 (∀xFx ↔ ∃xGx) ↔ ∃x∃y∀z∀w((Fx → Gy) ∧ (Gz → Fw))
T5.35 (∃xFx → (∃xGx → ∀xHx)) ↔ ∀x∀y∀z(Fx ∧ Gy → Hz)
T5.36 ∀x∀y∃z(Fx ∧ Gy → Hz) ↔ ∀y∃z∀x(Fx ∧ Gy → Hz)
T5.37 ¬∃y∀x(Fxy ↔ ¬Fxx)
T5.38 ∀z∃y∀x(Fxy ↔ (Fxz ∧ ¬Fxx)) → ¬∃y∀xFxy

T5.37 and T5.38 are both examples of "paradoxes". T5.37 is a first-order rendition of "Russell's Paradox". If you suppose 'F' stands for the '∈' of set theory, then T5.37 "says" that there is no set of all the sets not belonging to themselves. T5.38 says that there is no universal set, given the antecedent of the sentence. Suppose (by way of contradiction) that there *is* a universal set, *U*. If we take *U* to be the 'z' in the antecedent, then it says there's a set Y whose members are those elements of the universal set *U* that do not belong to themselves. But, what about Y itself? Y belongs to the universal set *U* because everything does (that's what makes it "universal"). So, Y belongs to itself if and only if it does not. Q.E.D.

Exercise:

Provide a derivation of T5.38 in accord with the intuitive proof of the previous paragraph.

THEOREMS OF FIRST-ORDER LOGIC (WITHOUT IDENTITY)

T5.1 $\forall x(Fx \to Gx) \to (\forall xFx \to \forall xGx)$
T5.2 $(\forall x(Fx \to Gx) \land \forall x(Gx \to Hx)) \to \forall x(Fx \to Hx)$
T5.3 $(\forall x(Fx \to Gx) \land \forall x(Fx \to Hx)) \to \forall x(Fx \to (Gx \land Hx))$
T5.4 $\forall x(Fx \land Gx) \leftrightarrow (\forall xFx \land \forall xGx)$
T5.5 $\forall x(P \to Fx) \leftrightarrow (P \to \forall xFx)$
T5.6 $(\forall xFx \lor \forall xGx) \to \forall x(Fx \lor Gx)$
T5.7 $\forall xFx \to \exists xFx$
T5.8 $\forall x(Fx \to Gx) \to (\exists xFx \to \exists xGx)$
T5.9 $(\forall xFx \land \exists xGx) \to \exists x(Fx \land Gx)$
T5.10 $(\forall x(Fx \to Gx) \land \exists x(Fx \land Hx)) \to \exists x(Gx \land Hx)$
T5.11 $\exists x(Fx \land Gx) \to (\exists xFx \land \exists xGx)$
T5.12 $(\forall x(Fx \to Gx) \land \exists xFx) \to \exists x(Fx \land Gx)$
T5.13 $\exists x(Fx \lor Gx) \leftrightarrow (\exists xFx \lor \exists xGx)$
T5.14 $\neg \forall x(Fx \to Gx) \leftrightarrow \exists x(Fx \land \neg Gx)$
T5.15 $\neg \exists xFx \leftrightarrow (\forall x(Fx \to Gx) \land \forall x(Fx \to \neg Gx))$
T5.16 $(\exists xFx \to P) \leftrightarrow \forall x(Fx \to P)$
T5.17 $\exists x(Fx \to P) \leftrightarrow (\forall xFx \to P)$
T5.18 $(\forall xFx \to \forall xGx) \to \exists x(Fx \to Gx)$
T5.19 $(\forall xFx \to \forall xGx) \leftrightarrow \exists x \forall y(Fx \to Gy)$
T5.20 $\forall x \exists y(Fx \land Gy) \to \exists y \forall x(Fx \land Gy)$
T5.21 $\forall x \exists y(Fx \to Gy) \to \exists y \forall x(Fx \to Gy)$
T5.22 $\forall x \exists y(Fx \lor Gy) \to \exists y \forall x(Fx \lor Gy)$
T5.23 $\exists y \forall xFxy \to \forall x \exists yFxy$
T5.24 $\exists x \exists yFxy \leftrightarrow \exists y \exists xFxy$
T5.25 $\forall x \forall yFxy \leftrightarrow \forall y \forall xFxy$
T5.26 $\forall y \exists x(Fx \leftrightarrow \neg Fy) \leftrightarrow (\exists xFx \land \exists x \neg Fx)$
T5.27 $\forall y \exists x(Fx \to Fy)$
T5.28 $\forall x \exists y(Fx \to Fy)$
T5.29 $\exists x \forall y(Fx \to Fy)$
T5.30 $\exists y \forall x(Fx \to Fy)$
T5.31 $\exists x(Fx \to \forall yFy)$
T5.32 $\exists y(\exists xFx \to Fy)$
T5.33 $\forall x \forall y(Fx \to Fy) \leftrightarrow \neg \exists xFx \lor \forall xFx$
T5.34 $(\forall xFx \leftrightarrow \exists xGx) \leftrightarrow \exists x \exists y \forall z \forall w((Fx \to Gy) \land (Gz \to Fw))$
T5.35 $(\exists xFx \to (\exists xGx \to \forall xHx)) \leftrightarrow \forall x \forall y \forall z(Fx \land Gy \to Hz)$
T5.36 $\forall x \forall y \exists z(Fx \land Gy \to Hz) \leftrightarrow \forall y \exists z \forall x(Fx \land Gy \to Hz)$
T5.37 $\neg \exists y \forall x(Fxy \leftrightarrow \neg Fxx)$
T5.38 $\forall z \exists y \forall x(Fxy \leftrightarrow (Fxz \land \neg Fxx)) \to \neg \exists y \forall xFxy$

5.9 DERIVATIONS FROM ASSUMPTIONS

As in the end of Chapter 2, *Short Cut Rules for Sentential Logic*, we now start to **derive a sentence of L_{PL} from (a set of) assumptions**. The notions here are exactly the same as those in Chapter 2. Only the language is different. When assumptions are tentative and not intended to be relied upon in any significant way or for long duration, they are usually referred to as *premises*. However, when assumed sentences are intended as providing a basis for concepts in a theory, they are usually called *axioms* (of that theory). The theory itself can then be identified with the set of theorems derivable from those axioms (i.e., those assumed sentences). (A more thorough and precise description of first-order theories appears in Chapters 8 & 10.)

Here we offer some sentences that show how some of the concepts pertaining to "less-than or equal" (L), "equal" (E), and "strictly less-than" (S) relate to each other. The assumed sentences that follow can be thought of as axioms for the beginning of a "theory of less-than":

L1 $\forall x \forall y \forall z (Lxy \land Lyz \rightarrow Lxz)$
L2 $\forall x \forall y (Lxy \lor Lyx)$
L3 $\forall x \forall y (Exy \leftrightarrow Lxy \land Lyx)$
L4 $\forall x \forall y (Sxy \leftrightarrow \neg Lyx)$
L5 $\exists x \forall y (Lxy)$
L6 $\forall x \exists y (Sxy)$

Exercises: Use L1-L6 as assumptions in deriving the following sentences (which can be seen as theorems of the theory of less-than):

(1) $\forall x (Exx)$
(2) $\forall x \forall y (Exy \rightarrow Eyx)$
(3) $\forall x \forall y \forall z (Exy \land Eyz \rightarrow Exz)$
(4) $\forall x \forall y (Sxy \rightarrow \neg Syx)$
(5) $\forall x \forall y \forall z (Sxy \land Syz \rightarrow Sxz)$
(6) $\forall x \forall y (Exy \rightarrow \neg (Sxy \lor Syx))$
(7) $\forall x \forall y (Sxy \rightarrow \neg (Exy \lor Lyx))$
(8) $\forall x (\neg Sxx)$
(9) $\neg \exists x \forall y Sxy$
(10) $\exists x \forall y (\neg Exy \rightarrow Sxy)$
(11) $\neg \exists y \forall x Lxy$
(12) $\forall x \forall y \forall z (Exy \land Syz \rightarrow Sxz)$
(13) $\forall x \forall y \forall z (Exy \land Szx \rightarrow Szy)$

[HINTS: In deriving (1) - (13), consider what each sentence "says," given the intended meaning of 'L', 'E', and 'S'. The derivations become much easier and more informative once the content of each sentence has been clearly understood. For example:(1) "says" everything equals itself; (6) says that if x equals y, then neither is x strictly less than y nor is y strictly less than x; but L3 says that x equals y just in case each is less than or equal to the other. Notice that L5 says there is an element that is less than or equal to any element, but (11) says there's not an element such that any element is less than or equal to it. Why is that true? And, why is it true that there's no element strictly less than any element (9)? Good luck.]

CHAPTER 6

SEMANTICS OF FIRST-ORDER PREDICATE LOGIC

TOPICS: *naming functions, domain of discourse, assignment tree, β-change, β-value, thorough (naming function), interpretation, true, false, new, naming interpretation, bracket notation, partial interpretation, model, satisfiability, validity, invalidity, particularization, logical consequence.*

In this chapter we begin to investigate the "meanings" of expressions of L_{PL}. In the previous chapter we often described informally what a given sentence "says". But, in the present chapter we show formally how expressions of L_{PL} acquire "meanings".

6.1 NAMING FUNCTIONS

A **naming function** n assigns a value from an underlying **domain of discourse**, $\mathcal{D}$, to each logical name of L_{PL}. That is, n is a function from the set of logical names in L_{PL} to $\mathcal{D}$. When n assigns values from a given domain $\mathcal{D}$ we say that n is a naming function **for** $\mathcal{D}$. A domain of discourse gives us objects to "talk about", so we require each domain of discourse to be nonempty. For example, let us specify a domain $\mathcal{D}$ to be the set $\{0,1\}$, and let us suppose that the naming function n assigns 1 to each logical name. Then, each logical name **stands for** or **denotes** or **names** 1, for naming function n. So, in this case, 'a', 'b', 'c', 'd', 'a_1'... all stand for, denote, or name 1, for n.

We can also say, for this particular example, that 1 is the **denotation** of 'c' for n, or that 1 is the **'c'-value** of n. More generally: 1 is the β-**value** of n, where β is 'c'. If n is implicit in the context, we will say that 'c' names 1.

A naming function n is **thorough** for a given domain of discourse if every element from the domain has a logical name. That is, a naming function n is **thorough** for $\mathcal{D}$ if n is onto. So, for our example of n, where $\mathcal{D} = \{0,1\}$ and for any name β, $n(β) = 1$, n is not thorough because 0 has no name.

Suppose we are given a naming function n for a certain domain of discourse and we want another naming function n' that differs from n on at most the denotation of a single name. For example, let $\mathcal{D} = \{0,1\}$ and define n such that for logical name 'a', $n(a) = 0$, but for any other name β, $n(\beta) = 1$. Now, let's define n' such that n' is exactly like n except that n' assigns 1 to 'a'. Hence, for any name β other than 'a', $n'(\beta) = n(\beta)$. But, $n(a) = 0$ and $n'(a) = 1$. We call n' an **'a'-change** of n since they differ on 'a' but are otherwise identical. In this case the 'a'-value of n is 0, whereas the 'a'-value of n' is 1. For any other name β, the β-value of n and n' are identical. We will write '$n' = n(a \to 1)$' to indicate that n' is obtained from n by changing the 'a'-value of n to 1. Just as $n' = n(a \to 1)$, $n = n'(a \to 0)$, since n can be obtained from n' by changing the 'a'-value of n' to 0.

We also write '$n(a) = 0$' and '$n'(a) = 1$', employing standard functional notation. Thus, $n'(a \to 0)(a) = 0$ and $n(a \to 1)(a) = 1$ as well.

We can take a 'b'-change of an 'a'-change as follows: $n(a \to 1)(b \to 0)$ is the naming function which first differs from n in assigning 1 to 'a' and then differs from that resultant naming function in assigning 0 to 'b'. Order is important here. For example, $n(a \to 1)(a \to 0)(a) = 0$ whereas $n(a \to 0)(a \to 1)(a) = 1$. We read from left to right. Notice that $n = n(a \to 0)$ in the limiting case where $n(a)$ is already 0.

The formal definition of a **β-change** of a naming function n is a naming function n' such that for any name δ different from β, $n'(\delta) = n(\delta)$. Notice that according to this definition, for any name β, every naming function is a β-change of itself.

Exercise:

Prove that the notion of a β-change forms an equivalence
 relation. That is, prove: (1) n is a β-change of itself;
 (2) if n' is a β-change of n, then n is a β-change of n';
 and (3) if n_2 is a β-change of n_1 and n_3 is a β-change of
 n_2, then n_3 is a β-change of n_1.

In order to picture all values for 'a' and 'b' for the naming function $n(b \to e)(b \to e')$, it is helpful to draw an **assignment tree** for 'a' then 'b'. (An **assignment tree** is a diagram of all possible naming assignments for a given sequence of logical names. We use them only as a pictorial aid; no aspect of the formal development depends on them.) Here is the assignment tree for 'a' then 'b' for the naming function n that we have been discussing:

162   CHAPTER 6

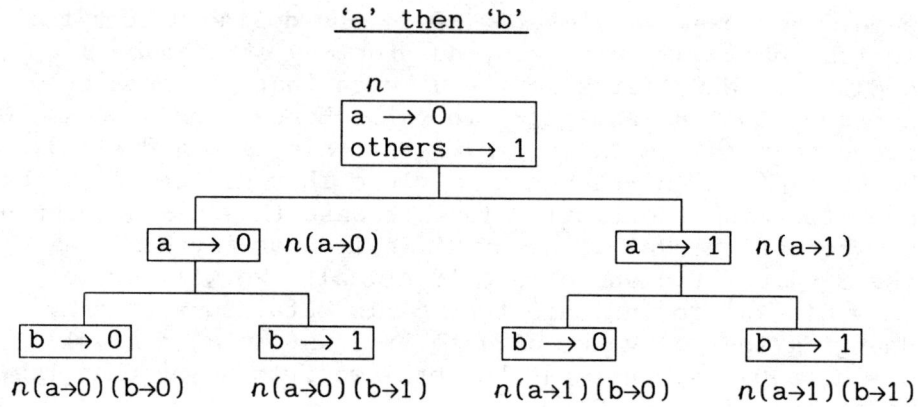

Exercises:

Let $\mathcal{D} = \{0,1\}$. Let n be such that $n(a) = 0$ and $n(\beta) = 1$ for any logical name β different from 'a'. Then, let n' be such that for any logical name β, $n'(\beta) = 1$.
1. For the naming function n' defined above, draw an assignment tree for 'a' then 'b'.
2. For the two naming functions n, n' answer whether the following statements are true or false:
   (a) $n = n'$ (b) $n(a\to 1) = n'$ (c) $n'(a\to 1) = n$
   (d) $n(a\to 1)(b) = n'(a\to 0)(b)$
   (e) $n'(a\to 0)(a) = n(b\to 0)(b)$
   (f) $n'(a\to 0)(b) = n(b\to 0)(a)$
   (g) $n'(a\to 0) = n$

There are some facts about naming functions and β-changes that we may as well mention now (n, n_1, n_2, n_2' are naming functions, while δ, β are names and e, e_0, e_1 are values of the domain):

(1) If n_1, n_2 are both naming functions for $\mathcal{D}$, then if $n_1(\beta) = e$, there is a β-change of n_2, say n_2', such that $n_2'(\beta) = e$. Clearly, n_2' is just $n_2(\beta\to e)$.

(2) $n(\beta\to e)(\beta) = n(\delta\to e)(\delta) = e$.

(3) If $\delta \neq \beta$, $n(\delta\to e_0)(\beta\to e_1) = n(\beta\to e_1)(\delta\to e_0)$.

Exercises:

Prove the above three facts about naming functions.

We now give "meanings" to every k-ary relation symbol R^k in L_{PL}.

6.2 INTERPRETATIONS

Another way of saying what the "meaning" of something is is to "interpret" it, so the function that gives "meanings" to relation letters is called an "interpretation". An **interpretation** $\mathcal{I}$ is a function that assigns to every 0-ary relation letter of L_{PL} a truth value **T** or **F**, assigns to every 1-ary relation letter of L_{PL} a subset of $\mathcal{D}$, assigns to every 2-ary relation letter a binary relation among elements of $\mathcal{D}$, and in general assigns to every k-ary relation letter R^k, where $k \geq 1$, a k-ary relation among elements of $\mathcal{D}$.

Here is a picture of a specific interpretation:

$\mathcal{I}$

```
𝒟 = {0,1}
all sentence letters → F
F¹ → {1}
F² → {<0,1>, <1,1>}
Others → ∅
```

(Recall that the empty set, ∅, is a subset of any set. So, since a k-ary relation is a set — a set of k-tuples — the empty set is a k-ary relation too because it is a subset of any set of k-tuples.)

An interpretation and a naming function give "meanings" to all atomic sentences of L_{PL}. To the above interpretation, let us add a naming function to provide denotations for all logical names:

n

```
a → 0
b → 1
others → 1
```

Now, given $\mathcal{I}$ and n, the sentence 'Fb' "says" that 1 is an element of {1}, which is **True**. 'Fa', however, is **False**.

For the naming function n, consider the following assignment tree for 'a':

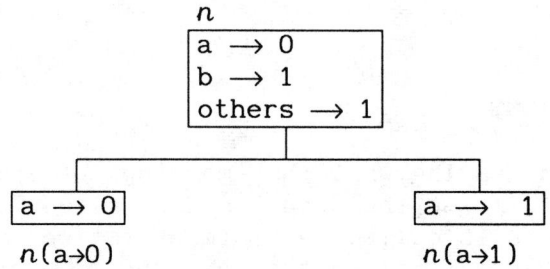

We can now say a few more things. Given $\mathcal{I}$ and n, we said above that 'Fa' was **False**. But, if we look at $\mathcal{I}$ together with $n(a \to 1)$, 'Fa' then becomes **True**. 'Fa' is **True** for $\mathcal{I}$ together with $n(a \to e_1)$ for <u>some</u> element e_1, namely for the element 1. 'Fa' is <u>not</u> **True** for $\mathcal{I}$ together with $n(a \to e_1)$ for <u>all</u> elements e_1, however, since 'Fa' is **False** when e_1 is the element 0.

Exercises:

Use the interpretation $\mathcal{I}$ and naming function n pictured above for the following questions:

1. **True** or **False**? a) Fb b) Fc c) F^2aa d) F^2ba e) F^2bb
2. Draw one assignment tree for 'a' then 'b' and another for 'b' then 'a'.
3. Use the assignment trees from **2** to determine whether the following are **True** or **False**:
    (a) 'F^2ab' for $\mathcal{I}$ together with $n(a \to e_1)$ for <u>all</u> e_1
    (b) 'F^2ab' for $\mathcal{I}$ together with $n(b \to e_1)$ for <u>all</u> e_1
    (c) 'F^2ab' for $\mathcal{I}$ together with $n(a \to e_1)(b \to e_2)$
        for <u>all</u> e_1 for <u>some</u> e_2
    (d) 'F^2ab' for $\mathcal{I}$ together with $n(b \to e_1)(a \to e_2)$
        for <u>all</u> e_1 for <u>some</u> e_2

We repeat a definition given in the preceding chapter: for a given sentence φ, a logical name β **is new to** φ iff β does not occur in φ. A second definition: β **is the first new name to** φ iff β is new to φ and β is the first such new one in the alphabetic listing of logical names.

We will from now on use the phrase 'naming interpretation' to mean an interpretation together with a naming function. So, $\langle \mathcal{I}, n \rangle$ is a **naming interpretation**, whereas $\mathcal{I}$ alone is an interpretation and n is a naming function.

6.3 TRUTH FOR A NAMING INTERPRETATION

For the following, φ is a sentence, α is an individual variable, β is the first name that is new to φ, $<\mathcal{I},n>$ is a naming interpretation, and e is an element of $\mathcal{D}$, the domain of $\mathcal{I}$. This is an inductive definition, following the inductive definition of a sentence of L_{PL}.

φ is true for $<\mathcal{I},n>$ (i.e., $<\mathcal{I},n>(\varphi) = T$) iff one of the following holds:

(i)a) φ is a 0-ary relation letter R^0 and $\mathcal{I}(R^0) = T$ (i.e., $\mathcal{I}$ assigns T to relation letter R^0);

b) φ is a k-ary relation letter followed by k individual constants, $R^k\delta_1\delta_2\ldots\delta_k$, where $k \geq 1$, and $<n(\delta_1), n(\delta_2),\ldots,n(\delta_k)> \in \mathcal{I}(R^k)$;

(ii) φ is a negation, $\neg\psi$, and $<\mathcal{I},n>(\psi) = F$;

(iii) φ is a conditional, $(\psi \rightarrow \chi)$ and either $<\mathcal{I},n>(\psi) = F$ or $<\mathcal{I},n>(\chi) = T$ (or both);

(iv) φ is a conjunction, $(\psi \wedge \chi)$, and $<\mathcal{I},n>(\psi) = <\mathcal{I},n>(\chi) = T$;

(v) φ is a disjunction, $(\psi \vee \chi)$, and either $<\mathcal{I},n>(\psi) = T$ or $<\mathcal{I},n>(\chi) = T$ (or both);

(vi) φ is a biconditional, $(\psi \leftrightarrow \chi)$, and $<\mathcal{I},n>(\psi) = <\mathcal{I},n>(\chi)$ (both T or both F);

(vii) φ is a universal generalization, $\forall\alpha\psi$, and $<\mathcal{I},n(\beta \rightarrow e)>(\psi\alpha/\beta) = T$ for every e;

(viii) φ is an existential generalization, $\exists\alpha\psi$, and $<\mathcal{I},n(\beta \rightarrow e)>(\psi\alpha/\beta) = T$ for some e.

Otherwise (if none of i-iv hold), φ is false for $<\mathcal{I},n>$ (i.e., $<\mathcal{I},n>(\varphi) = F$).

As before, we assert (but do not prove) that the Unique Truth Value Principle (UTV) holds of all sentences of L_{PL} (see UTV in sentential logic). This means that any sentence φ of L_{PL} is **True** or **False** for a given naming interpretation $<\mathcal{I},n>$ and not both.

Clause (vii) says that a universal sentence is true for a naming interpretation if, when the universal quantifier is removed and the first new name is substituted for those occurrences of the free variable, that resulting sentence is true no matter what element the name denotes.

An existential sentence is true for a naming interpretation, according to (viii), if, when the existential quantifier is removed and the first new name is substituted for those occurrences of the free variable, the resulting sentence is true for some denotation of the name.

6.4 BRACKET NOTATION

Suppose we have an interpretation $\mathcal{I}$, such that $\mathcal{I}(F^2) = \{<0,0>, <1,1>\}$, where $\mathcal{D} = \{0,1\}$. Suppose, additionally, that the naming function n is such that $n(a) = 0$ and $n(b) = 1$. Instead of writing 'Fab', we may write in certain contexts 'F[0,1]' to show that for the naming function under consideration 'a' denotes 0 and 'b' denotes 1. So, by simple inspection we can see that 'F[0,1]' is **False** since $<0,1> \notin \mathcal{I}(F^2)$. More generally, if we wish to show that e is assigned to some name, we write '[e]' in place of that name in a sentence. So 'F[e]b' means that the name in the first position denotes (according to the naming function in that context) the element e of the domain of discourse. Then 'F[e,e']' means that the first name denotes e and the second denotes e'.

We will also write 'ψ[e]' to indicate that the name in that context denotes e (for a given naming function). For example, in one of the rewritten truth clauses below, '$\exists\alpha\psi$' is an existential sentence, and '$\psi\alpha/\beta$' is that sentence with the existential stripped and with all free occurrences of α in ψ replaced by occurrences of β (which is the first new name to ψ). We write 'ψ[e]' to show that the β value (for that naming function) is e.

Using brackets as described above, we can rewrite our two truth clauses for sentences that begin with quantifiers, this way:

(Remember: φ is a sentence, α is an individual variable, β is the first name that is new to ψ, $\langle \mathcal{I}, n \rangle$ is a naming interpretation, and e is an element of $\mathcal{D}$, the domain of $\mathcal{I}$.)

φ is true for $\langle \mathcal{I}, n \rangle$ (i.e., $\langle \mathcal{I}, n \rangle(\varphi) = \mathbf{T}$) iff one of the following holds:

(vii) φ is a universal generalization, $\forall \alpha \psi$, and $\langle \mathcal{I}, n \rangle(\psi[e]) = \mathbf{T}$ for every e.

(viii) φ is an existential generalization, $\exists \alpha \psi$, and $\langle \mathcal{I}, n \rangle(\psi[e]) = \mathbf{T}$ for some e.

The bracket notation is sometimes clearer than the earlier notation. We will switch between them, using the one that seems preferable for the given context.

It should now be clearer that our formal definition of truth for a naming interpretation gives us intuitively what we want for quantified sentences. Informally, (vii) says that "for all α, ψ is true" just in case "$\psi[e]$ is true" for all elements e of our domain of discourse, and (viii) says that "for some α, ψ is true" just in case "$\psi[e]$ is true" for some element e of that domain. Our definition tells us that a "for all" sentence is true if it is true for all elements of the domain we are considering, and a "for some" sentence is true if it is true for some element of that domain.

We will now show that the sentence '$\forall y \exists x Fxy$' can be true for a naming interpretation while '$\exists x \forall y Fxy$' is false for that same interpretation. We actually show this twice, first using our original notation and second using our bracket notation (which may be easier to follow).

Let $\mathcal{I}$ be:

$\mathcal{I}$

| $\mathcal{D} = \{0,1\}$ |
| $F^2 = \{\langle 1,0 \rangle, \langle 0,1 \rangle\}$ |
| Others = $\varnothing$ |

and let n be:

| all $\rightarrow$ 0 |

168   CHAPTER 6

We now draw an assignment tree for 'a' then 'b':

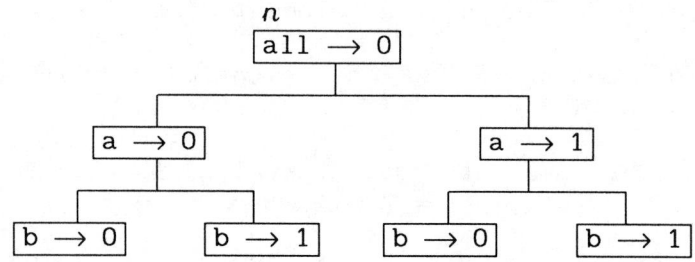

'∀y∃xFxy' is **True** for <𝓙,𝑛> iff (by vii)

'∃xFxa' is **True** for <𝓙,𝑛(a→e₁)> for <u>all</u> e₁ iff (by viii)

'Fba' is **True** for <𝓙,𝑛(a→e₁)(b→e₂)> for <u>all</u> e₁ for <u>some</u> e₂.

So, after applying the two truth clauses (vii) and (viii), we can see that the original sentence, '∀y∃xFxy' is **T** for <𝓙,𝑛> iff 'Fba' is **T** for <𝓙,𝑛(a→e₁)(b→e₂)> for <u>all</u> e₁ for <u>some</u> e₂. We look back at our assignment tree and first check <u>all</u> elements 'a' can name. 'a' can name 0 or 1. For <u>each</u> of them we look for <u>some</u> element 'b' can name such that 'Fba' comes out true. So, when 'a' names 0 and 'b' names 1 'Fba,' comes out **T** (because <1,0> ∈ {<1,0>, <0,1>}). <u>And</u>, when 'a' names 1 and 'b' names 0, 'Fba' comes out **T** (because <0,1> ∈ {<1,0>, <0,1>}). So, for <u>everything</u> 'a' can name there is <u>something</u> 'b' can name such that 'Fba' comes out **True**. Thus, '∀y∃xFxy' is **True** for <𝓙,𝑛>.

Now, we look at '∃x∀yFxy':

'∃x∀yFxy' is **T** for <𝓙,𝑛> iff (by viii)

'∀yFay' is **T** for <𝓙,𝑛(a→e₁)> for <u>some</u> e₁ iff (by vii)

'Fab' is **T** for <𝓙,𝑛(a→e₁)(b→e₂)> for <u>some</u> e₁ for <u>all</u> e₂.

For '∃x∀yFxy' to be **True** for <𝓙,𝑛>, 'Fab' must come out **T** for <u>something</u> 'a' can name, for <u>everything</u> 'b' can name. Suppose 'a' names 0. Then for <u>both</u> cases when 'b' names 0 <u>and</u> when 'b' names 1, 'Fab' must come out **T**. But, when 'a' names 0 and 'b' names 0, 'Fab' comes out **False**. So, let us try the case when 'a' names 1. When 'a' names 1 and 'b' names 1 also, 'Fab' comes out **False** again. So it is not the case that 'Fab' comes out **T** for something 'a' can name, for everything 'b' can name. Thus, '∃x∀yFxy' is **False** for $\langle \mathcal{I}, n \rangle$.

Using Bracket Notation

We can also show '∀y∃xFxy' is **True** for $\langle \mathcal{I}, n \rangle$ and that '∃x∀yFxy' is **False** for $\langle \mathcal{I}, n \rangle$ using our bracket notation, which may be easier to follow:

'∀y∃xFxy' is **T** for $\langle \mathcal{I}, n \rangle$ iff (by vii)

$\langle \mathcal{I}, n \rangle(\exists x Fx[e_1]) = $ **T** for all e_1 iff (by viii)

$\langle \mathcal{I}, n \rangle(F[e_2, e_1]) = $ **T** for all e_1 for some e_2.

As before, this means that when $e_1 = 0$ and when $e_1 = 1$ we need some e_2 such that $F[e_2, e_1]$ becomes **T**. When $e_1 = 0$, $e_2 = 1$ works (i.e., $\mathcal{I}(F[1,0]) = $ **T**), and when $e_1 = 1$, $e_2 = 0$ works (i.e., $\mathcal{I}(F[0,0]) = $ **T**). So, '∀y∃xFxy' is **T** for $\langle \mathcal{I}, n \rangle$.

Now,

'∃x∀yFxy' = **T** for $\langle \mathcal{I}, n \rangle$ iff (by viii)

$\langle \mathcal{I}, n \rangle(\forall y F[e_1]y) = $ **T** for some e_1 iff (by vii)

$\langle \mathcal{I}, n \rangle(F[e_1, e_2]) = $ **T** for some e_1 for all e_2.

This means that when $e_1 = 0$ or $e_1 = 1$, it must work when $e_2 = 0$ and when $e_2 = 1$. Let us try $e_1 = 0$ first. $\mathcal{I}(F[0,1]) = $ **T** but $\mathcal{I}(F[0,0]) = $ **F**. So, let us try $e_1 = 1$. Then, $\mathcal{I}(F[1,0]) = $ **T**, but $\mathcal{I}(F[1,1]) = $ **F**. So, it does not work when $e_1 = 0$ or when $e_1 = 1$. Thus, '∃x∀yFxy' = **F** for $\langle \mathcal{I}, n \rangle$.

Exercise:

1. Consider the sentences '∀y∃xFxy' and '∃x∀yFxy' again, this time using interpretation $\mathcal{I}'$, where $\mathcal{D} = \{0,1\}$ as before, but this time let $\mathcal{I}'(F^2) = \{\langle 0,0 \rangle, \langle 1,1 \rangle\}$. Take any naming function, n, for this domain and demonstrate whether either or both of '∀y∃xFxy' and '∃x∀yFxy' are **T** for $\langle \mathcal{I}, n \rangle$.

DIGRESSION ==> :

In defining "truth" for a universal sentence, it may seem more "natural" to say something like the following: $\forall \alpha \varphi$ is true for $\langle \mathcal{I}, n \rangle$ iff $\varphi \alpha / \beta$ is true for $\langle \mathcal{I}, n \rangle$ <u>for</u> <u>every</u> <u>name</u> β. The problem is that n <u>may</u> <u>not</u> <u>be</u> <u>thorough</u>. Thus, if $\mathcal{D} = \{0,1,2\}$ and $n(\delta) = 1$ for all names δ, then $\varphi \alpha / \beta$ can be true for $\langle \mathcal{I}, n \rangle$ <u>for</u> <u>every</u> β, but $\forall \alpha \varphi$ can be false. To show this, suppose $\varphi = $ 'Fx', let $\mathcal{I}(F) = \{1\}$, and let $n(\beta) = 1$ for any name β. Then <u>all</u> the sentences in this set are true: $\{Fa, Fb, Fc, Fd, Fa_1, \ldots\}$. But, since $\mathcal{I}(F) \neq \mathcal{D}$, '$\forall x Fx$' is <u>not</u> <u>true</u> for $\langle \mathcal{I}, n \rangle$. One way to eliminate this difficulty is to use only <u>thorough</u> naming functions, those that are onto. Then it <u>would</u> be the case that $\forall \alpha \varphi$ is true for $\langle \mathcal{I}, n \rangle$ iff $\varphi \alpha / \beta$ is true for $\langle \mathcal{I}, n \rangle$ for every β. One problem with that suggestion is that it <u>either</u> restricts the domains (of our interpretations) to those that are countable, i.e., either finite or denumerably infinite, <u>or</u> it requires us to add non-denumerably many names to our language in order to denote each element of a non-denumerable domain (e.g., the domain of real numbers).

Each of the above alternatives would necessitate certain changes in our presentation. The first choice, having thorough naming functions and only countable domains, would curtail our logic somewhat and limit our discussion of models to countable ones, some of which would be very "unnatural". The second choice is followed by Joseph Shoenfield in his <u>Mathematical</u> <u>Logic</u>. In that book, whenever an interpretation is considered, there are essentially <u>two</u> languages involved. One language is the "proper one," and a second language is the same as the first one except that it is augmented by having a name for every element in the domain under consideration.

Shoenfield's method does <u>not</u> curtail the resulting logical principles in any way, <u>nor</u> does it limit the discussion of models to countable ones. It does complicate some matters, however, and makes the resulting language non-denumerably infinite when interpretations are considered that have non-denumerable domains.

The first alternative, which can be identified as "the substitutional interpretation of the quantifiers" (about which the literature abounds), reflects a desire to always have a linguistic entity as a stand-in for any object of discourse. Though, as mentioned above, the resultant logic is somewhat curtailed and non-denumerable models are lost,

all countable models are retained. There is a metatheorem to
the effect that whenever a set of sentences has an infinite
model, it has a denumerable model as well — though this
denumerable model may not be a natural one (for example, a
denumerable model of the real numbers is not "natural", since
there are non-denumerably many reals.)
      We continue with the present alternative.
DIGRESSION <== :

Exercises:

Give a naming interpretation for which each sentence is
  true, or explain why such an interpretation cannot be
  given:
  1.   Fa ∧ ¬ Fb
  2.   ((Fa ∧ Ga) ∧ (Fb ∧ ¬Gb)) ∧ ((¬Fc ∧ Gc) ∧ (¬Fd ∧ ¬Gd))
       (At least how many elements must be in the domain?)
  3.   Fa → ∀xFx   4.   ¬(Fa → ∀xFx)
  5.   Fa → ¬∀xFx   6.   ¬(Fa → ¬∀xFx)
  7.   Fa → ¬∃xFx   8.   ¬(Fa → ¬∃xFx)
  9.   ∀xFx → Fa   10.   ¬(∀xFx → Fa)
 11.   ∀xFx → ¬Fa   12.   ¬(∀xFx → ¬Fa)
 13.   ∃xFx → Fa   14.   ¬(∃xFx → Fa)
 15.   ∃xFx → ¬Fa   16.   ¬(∃xFx → ¬Fa)

      It should be clear that in determining the truth of a
sentence for a given interpretation, only the interpretation
of relation symbols that occur in the sentence under
consideration matter to its truth-value. For example,
suppose we wish to provide a naming interpretation for which
'F^3abc' is true. Here is a very simple "partial
interpretation":

$$\begin{array}{|l|} \hline \mathcal{J} \\ \hline 4 \in \mathcal{D} \\ F^3 \to \{<4,4,4>\} \\ \hline \end{array} \qquad \begin{array}{|l|} \hline n \\ \hline a,b,c \to 4 \\ \hline \end{array}$$

      It is clear that no matter how we fill in the rest of $\mathcal{J}$
and n, 'F^3abc' will remain true. (Notice that in this case
we do not even need to specify $\mathcal{D}$ completely; all we need to
know is that 4 is an element of $\mathcal{D}$.) The fact that after a
domain has been assigned only the non-logical symbols and the
names in φ need to be given values to determine the truth of
φ allows us to use "partial interpretations". A **partial**

(naming) interpretation is a naming interpretation with some assignments to relation letters and/or logical names in L_{PL} left unspecified (but the domain must always be specified). Sometimes to emphasize that a given naming interpretation assigns values to <u>all</u> relation letters and logical names, and to distinguish it from a partial interpretation, we will call it a **total (naming) interpretation**. A partial interpretation for L_{PL} can be looked upon as a total interpretation for some language that is a proper subset of L_{PL}. The following metatheorem is a technical rendition of the fact that, after a domain has been chosen, the truth of a given sentence is fully determined by specifying values for only the names and relation symbols occurring in that sentence. Let's call it the Part. Interp. Theorem:

PART. INTERP. THEOREM: If for every k-ary relation letter R^k and for every logical name β in φ, two naming interpretations $<\mathcal{I},n>$ and $<\mathcal{I}',n'>$ are such that $\mathcal{I}(R^k) = \mathcal{I}'(R^k)$ and $n(\beta) = n'(\beta)$, then, provided their domains are equal:

$$<\mathcal{I},n>(\varphi) = <\mathcal{I}',n'>(\varphi) \text{ for every sentence } \varphi \text{ of } L_{PL}.$$

We won't rigorously prove the PART. INTERP. THEOREM at this point (see next chapter), but it should seem obvious. Notice that one special case of the PART. INTERP. THEOREM pertains to a sentence with <u>no</u> names. When φ has no names, the assignments of names to things named has no bearing on the truth of φ. Another way to look at this is that a given naming function can influence the truth only of sentences with names in them. If a sentence φ has no names in it, the assignment of values to names can have no effect on its truth. An interpretation alone (i.e., without a naming function) suffices.

The case when a sentence φ contains no occurrence of names is one motivation for our defining truth for an interpretation without a naming function. But there's another motivation. Sometimes a sentence φ has names, but they can be eliminated by "universalizing" φ. Let's take a simple example. A "universalization" of 'Fa' is '∀xFx'. To universalize 'Fa', we replace all occurrences of 'a' by 'x' and then put '∀x' in front of the resulting formula 'Fx', finally arriving at '∀xFx'. When the original sentence 'Fa' is true for $<\mathcal{I},n>$ for <u>all</u> n, then '∀xFx' is true for $<\mathcal{I},n>$, and vice versa. Since '∀xFx' has <u>no</u> names in it, we know that the choice of naming function n has no bearing on the fact that '∀xFx' is true for $\langle \mathcal{I}, n \rangle$. So, we want to say '∀xFx' is true for $\mathcal{I}$ alone, and drop any reference to a particular naming function n. Then, derivatively, we wish to say our original sentence 'Fa' is true for $\mathcal{I}$ alone, as well.

COROLLARY OF PART. INTERP.: For any interpretation, $\mathcal{I}$, and any two naming functions n_1, n_2, for $\mathcal{D}$, if φ is a sentence of L_{PL} containing no occurrence of names, then:

$$\langle \mathcal{I}, n_1 \rangle(\varphi) = \langle \mathcal{I}, n_2 \rangle(\varphi)$$

6.5 TRUTH FOR AN INTERPRETATION

We now define what it means for any sentence φ of L_{PL} to be **True for an interpretation** $\mathcal{I}$, where n is a naming function for $\mathcal{D}$, the domain of $\mathcal{I}$:

$$\mathcal{I}(\varphi) = T \quad \text{iff} \quad \langle \mathcal{I}, n \rangle(\varphi) = T \text{ for } \underline{\text{all}} \ n.$$

We now prove that for sentence 'Fa' and any interpretation $\mathcal{I}$:

$$\mathcal{I}(Fa) = T \quad \text{iff} \quad \mathcal{I}(\forall xFx) = T.$$

PF: Suppose $\mathcal{I}(Fa) = T$. Then $\langle \mathcal{I}, n \rangle(Fa) = T$ for all n (by definition). So, in particular, $\langle \mathcal{I}, n(a \rightarrow e) \rangle(Fa) = T$ for all n, for every e. Hence $\langle \mathcal{I}, n \rangle(\forall xFx) = T$ for all n. So, $\mathcal{I}(\forall xFx) = T$.

To prove the other direction, we just reverse our steps: Suppose $\mathcal{I}(\forall xFx) = T$. Then $\langle \mathcal{I}, n \rangle(\forall xFx) = T$ for all n. Then, for all n, $\langle \mathcal{I}, n(a \rightarrow e) \rangle(Fa) = T$ for every e. But then $\langle \mathcal{I}, n \rangle(Fa) = T$ for all n. Hence $\mathcal{I}(Fa) = T$. ∎

More generally, we can prove the following:

METATHEOREM: If δ is new to φ, then for any $\mathcal{I}$,

$$\mathcal{I}(\varphi \alpha / \delta) = T \quad \text{iff} \quad \mathcal{I}(\forall \alpha \varphi) = T.$$

The above METATHEOREM is a slight generalization of the instance we proved above. We prove it in the next chapter.

Actually, by a straightforward induction argument, we can prove the following stronger metatheorem for any number of distinct names $\delta_1, \delta_2, \ldots, \delta_k$ new to φ:

THEOREM: $\mathcal{I}(\varphi \alpha_1/\delta_1 \; \alpha_2/\delta_2 \; \ldots \; \alpha_k/\delta_k) = T$ iff
$$\mathcal{I}(\forall \alpha_1 \forall \alpha_2 \; \ldots \; \forall \alpha_k \varphi) = T.$$

Let us call $\forall \alpha_1 \forall \alpha_2 \ldots \forall \alpha_k \varphi$ a **universalization** of $\varphi \alpha_1/\delta_1 \; \alpha_2/\delta_2 \; \ldots \; \alpha_k/\delta_k$, where $\delta_1, \delta_2, \ldots, \delta_k$ are all the distinct names new to φ occurring in the sentence $\varphi \alpha_1/\delta_1 \; \alpha_2/\delta_2 \; \ldots \; \alpha_k/\delta_k$ in any order. As a limiting case, a sentence with no names is the universalization of itself. The above (meta-)theorem tells us, for any interpretation $\mathcal{I}$ and any sentence φ:

UNIVERSALIZATION THEOREM: $\mathcal{I}(\varphi) = T$ iff
$$\mathcal{I} \text{ (ALL UNIVERSALIZATIONS OF } \varphi) = T.$$

This UNIVERSALIZATION THEOREM will come in handy later when we examine mathematical theories. For example, a mathematical theory may "say" that the sentence 'a ≤ a' is true. Intuitively, this will mean that *any* element is less than or equal to itself, not just the particular element "a". The UNIVERSALIZATION THEOREM will make it possible for us to write 'a ≤ a' as shorthand for '∀x x ≤ x'.

It is important to distinguish carefully between a naming interpretation and an interpretation. For example, a few paragraphs back we proved (for any $\mathcal{I}$):

$$\mathcal{I}(Fa) = T \quad \text{iff} \quad \mathcal{I}(\forall xFx) = T,$$

which is the same as proving:

$$\mathcal{I}(Fa) = \mathcal{I}(\forall xFx).$$

But, we can<u>not</u> prove the following (for any n):

X $\qquad\qquad \langle \mathcal{I}, n \rangle(Fa) = \langle \mathcal{I}, n \rangle(\forall xFx) \qquad\qquad$ X

Here's a counter-example:

$$\begin{array}{cc} \mathcal{I} & n \\ \boxed{\begin{array}{l} \mathcal{D} = \{0,1\} \\ F \rightarrow \{0\} \end{array}} & \boxed{a \rightarrow 0} \end{array}$$

$\langle \mathcal{I},n \rangle(Fa) = T$, but $\langle \mathcal{I},n \rangle(\forall xFx) = F$.

Let's go over this carefully, since it is easy to get confused at this point. We <u>proved</u>:

$$\mathcal{I}(Fa) = T \quad \text{iff} \quad \mathcal{I}(\forall xFx) = T$$

which, by definition of truth for an interpretation, is:

* $\langle \mathcal{I},n \rangle(Fa) = T$ for all n iff $\langle \mathcal{I},n \rangle(\forall xFx) = T$ for all n.

What we <u>cannot</u> prove is:

**   X   For all n: $\langle \mathcal{I},n \rangle(Fa) = T$ iff $\langle \mathcal{I},n \rangle(\forall xFx) = T$.   X

To assure yourself that ** is false, look back at the counter-example in the previous paragraph.

The important point to notice here is the placement of 'for all n' in * and in **. The first statement, *, says that the left side holds for all n just in case the right side holds for all n as well. The second statement, **, makes a much stronger claim, that for <u>any</u> <u>particular</u> n the left side holds just in case the right side holds <u>for</u> <u>that</u> n. To see that ** is much stronger than *, notice that ** is like saying (or is an anolog of) 'for every person x, x is male just in case x is female'. Obviously, that statement is false. Yet the somewhat similar-sounding statement (which is like *) 'every person is male just in case every person is female' is true. Why? Because it is false that every person is male, and it is also false that every person is female. So, the two statements <u>are</u> equivalent in truth value. Yet (going back to **), it is false that for any <u>particular</u> person, <u>that</u> person is male just in case <u>that</u> person is female.

Similarly, it is <u>true</u> that $\langle \mathcal{I},n \rangle(Fa) = T$ for all n just in case $\langle \mathcal{I},n \rangle(\forall xFx) = T$ for all n. But it is <u>false</u> that for any n, that $\langle \mathcal{I},n \rangle(Fa) = T$ just in case $\langle \mathcal{I},n \rangle(\forall xFx) = T$ <u>for</u> <u>that</u> n.

Exercise:

Suppose 'G' stands for 'is male' and 'H' stands for 'is female'. then the first-order sentence '$\forall xGx \leftrightarrow \forall xHx$' "says" 'everyone is male just in case everyone is female' which is an analog of *. Similarly, $\forall x(Gx \leftrightarrow Hx)$ "says" 'anyone is male just in case he is female', which is an analog of **. Give an interpretation for which '$\forall xGx \leftrightarrow \forall xHx$' is **True**, yet '$\forall x(Gx \leftrightarrow Hx)$' is **False**.

6.6 SATISFIABILITY

The Model-Theoretic Definitions pertaining to the language of Sentential Logic, L_{SL}, carry over to the language of Predicate Logic, L_{PL}, with little change:

A naming interpretation $<\mathcal{I},n>$ **models** Γ, where Γ is a set of sentences of L_{PL}, if $<\mathcal{I},n>(\varphi) = \mathbf{T}$ for all sentences φ in Γ (we also say: $<\mathcal{I},n>$ **is a model of** Γ). As before, we say that $<\mathcal{I},n>$ **models** φ, where φ is a sentence of L_{PL}, as short for $<\mathcal{I},n>$ models $\{\varphi\}$. We also write '$<\mathcal{I},n>(\Gamma) = \mathbf{T}$' for '$<\mathcal{I},n>$ models Γ'. A set of sentences Γ **is satisfiable** iff Γ has a model.

Example: We show $\{\forall xFx \rightarrow P, \neg\forall x(Fx \rightarrow P)\}$ is satisfiable. (This shows, that it is *not* the case that '$\forall x(Fx \rightarrow P)$' is a "logical consequence" of '$\forall xFx \rightarrow P$' and it also shows that '$(\forall xFx \rightarrow P) \rightarrow \forall x(Fx \rightarrow P)$' is *not* valid. See the paragraphs below the *exercises* for definitions of 'logical consequence' and 'valid'.)

$\mathcal{I}$
$\mathcal{D} = \{0,1\}$
F: $\{0\}$
P: **F**

$\mathcal{I}(\forall xFx \rightarrow P) = \mathbf{T}$ because $\mathcal{I}(\forall xFx) = \mathbf{F}$. Yet, $\mathcal{I}(\forall x(Fx \rightarrow P)) = \mathbf{F}$ because $\mathcal{I}(F[0] \rightarrow P) = \mathbf{F}$. Thus, $\mathcal{I}(\neg\forall x(Fx \rightarrow P)) = \mathbf{T}$.

Exercises:

Provide a model for each set of sentences below:
1. $\{\forall xFx, Fa, Fb, Fc, Fd, Fa_1, \ldots\}$
2. $\{Fa, Fb, Fc, Fd, Fa_1, \ldots, \exists x\neg Fx\}$
3. $\{\forall x(Fx \rightarrow P), \neg P, \forall x\neg Fx\}$
4. $\{\forall xFx \rightarrow P, \neg P, \exists x\neg Fx, \exists xFx\}$

5. $\{\forall x \exists y Fxy, \neg \exists x \forall y Fyx\}$
6. $\{\forall x(Fx \rightarrow \neg Gx), \forall x Gx, \exists x \neg Fx\}$
7. $\{\exists x(Fx \rightarrow P), \neg(\exists x Fx \rightarrow P)\}$
8. $\{\exists y \forall x(Fx \rightarrow Gxy), \exists y(Fy \rightarrow Gyy)\}$
9. $\{\forall x \neg Fxx, \forall x \forall y(Fxy \rightarrow \neg Fyx), \forall x \forall y \forall z(Fxy \wedge Fyz \rightarrow Fxz),$
   $\forall x \exists y Fxy\}$

6.7 VALIDITY

If a sentence φ of L_{PL} is true for every first-order naming interpretation $\langle \mathcal{I}, n \rangle$, then φ **is valid**. We write '$\models \varphi$' for φ is valid.

Exercises:

Show by a purely semantic argument (like the one above) that the following sentences are valid:
1. $\forall x Fx \vee \exists x \neg Fx$
2. $\forall x Fx \rightarrow Fa$
3. $Fa \rightarrow \exists x Fx$
4. $\forall x(Fx \wedge Gx) \rightarrow \forall x Fx \wedge \forall x Gx$
5. $(\forall x Fx \wedge \forall x Gx) \rightarrow \forall x(Fx \wedge Gx)$
6. $\exists x Fx \vee \exists x Gx \rightarrow \exists x(Fx \vee Gx)$

It is difficult to prove sentences valid using purely semantic arguments. It is much easier to provide a derivation, which is a purely syntactic argument. For example, compare our semantic proof above that '$\forall x(Fx \rightarrow Gx) \rightarrow (\forall x Fx \rightarrow \forall x Gx)$' is valid with the following derivation:

1.	$\forall x(Fx \rightarrow Gx)$	1,A
2.	$\forall x Fx$	2,A
3.	Fa	2,UI
4.	$Fa \rightarrow Ga$	1,UI
5.	Ga	1,2,SL
6.	$\forall x Gx$	1,2,UG

Clearly, the above derivation is easier, both to follow and to do, than the semantic argument. But, the above <u>derivation</u> establishes only that '$\forall x(Fx \rightarrow Gx) \rightarrow (\forall x Fx \rightarrow \forall x Gx)$' is a <u>theorem</u>, whereas the semantic argument establishes the sentence's <u>validity</u>. Soon we will establish that all theorems are valid (i.e., that our system of

178   CHAPTER 6

derivational rules is sound). But, until this has been proved, we cannot rely on a theorem to be valid.

The soundness of a derivational system is not always an easy matter. At times systems have been asserted to be sound and later found to be unsound. So, we should not suppose our system really is sound until it has actually been proved.

6.8 INVALIDITY

Exercises:

Show that the following sentences are not valid by exhibiting an interpretation for which their negations are true. (The worked-out example below these *exercises* should be helpful.)
1. ∀xFx ∨ ∀x¬Fx
2. Fa → ∀xFx
3. ∃xFx → Fb
4. ∀x(Fx ∨ Gx) → (∀xFx ∨ ∀xGx)
5. ∃xFx ∧ ∃xGx → ∃x(Fx ∧ Gx)
6. ∀x∃yFxy → ∃y∀xFxy

Example: We show that '∀x∀y(Fx → Fy)' is invalid by exhibiting an interpretation for which it is false. First, we draw an assignment tree for 'a' then 'b' of a naming function n for {0,1}:

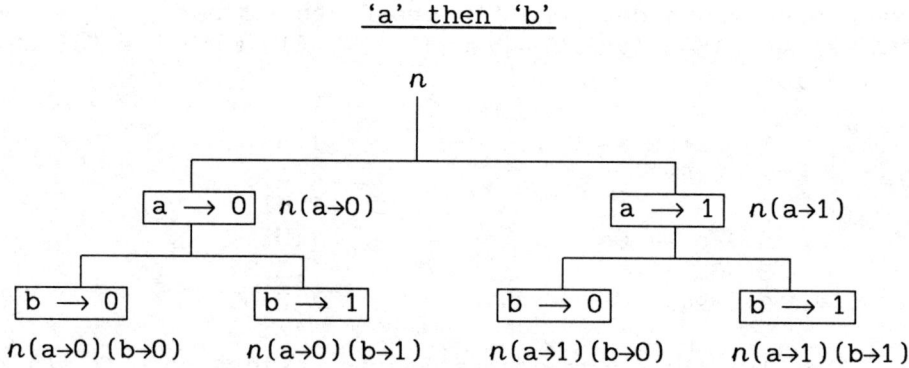

The (partial) interpretation, $\mathcal{I}$, is :

$\mathcal{I}$
$$\boxed{\begin{array}{l}\mathcal{D} = \{0,1\} \\ F = \{1\}\end{array}}$$

$\mathcal{I}(\forall x \forall y(Fx \rightarrow Fy)) = \mathbf{T}$ iff
$\mathcal{I}(\forall y(F[e] \rightarrow Fy)) = \mathbf{T}$ for <u>all</u> e iff
$\mathcal{I}(F[e] \rightarrow F[e_1]) = \mathbf{T}$ for <u>all</u> e for <u>all</u> e_1.

Now, look at the assignment tree. When e = 0, $\mathcal{I}(F[0] \rightarrow F[e_1]) = \mathbf{T}$ for <u>all</u> (i.e., both) e_1. When e = 1, <u>and</u> e_1 = 1, $\mathcal{I}(F[1] \rightarrow F[1]) = \mathbf{T}$. <u>But</u>, when e = 1 and e_1 = 0, then $\mathcal{I}(F[1] \rightarrow F[0]) = \mathbf{F}$, showing that it is <u>not</u> the case that $\mathcal{I}(F[e] \rightarrow F[e_1])$ for <u>all</u> e, for <u>all</u> e_1. This shows, in turn, that $\mathcal{I}(\forall x \forall y(Fx \rightarrow Fy)) = \mathbf{F}$.

Even though '$\forall x \forall y(Fx \rightarrow Fy)$' is <u>invalid</u>, '$\exists x \exists y(Fx \rightarrow Fy)$' <u>is</u> valid, which this short derivation shows (provided you believe the rules to be sound):

1. Fa $\rightarrow$ Fa   SL
2. $\exists y$(Fa $\rightarrow$ Fy)   EG
3. $\exists x \exists y$(Fx $\rightarrow$ Fy)   EG

<u>Or</u> you can take its negation, '$\neg \exists x \exists y(Fx \rightarrow Fy)$,' drive the negation in by QE, then make an internal conjunction, by noting that '$\neg(P \rightarrow Q) \leftrightarrow (P \wedge \neg Q)$' is a theorem, to form '$\forall x \forall y(Fx \wedge \neg Fy)$'. Then instantiate both 'x' and 'y' to 'a' to get 'Fa $\wedge \neg$Fa,' a contradiction. That <u>also</u> shows '$\exists x \exists y(Fx \rightarrow Fy)$' to be valid (by getting a contradiction from its negation). (This is not yet "a derivation," but can easily be turned into one.)

Exercise:

Try obtaining a contradiction from '$\forall x \forall y(Fx \rightarrow Fy)$' as we did for '$\exists x \exists y(Fx \rightarrow Fy)$' in the paragraph above. Compare your result with the interpretation above that shows '$\forall x \forall y(Fx \rightarrow Fy)$' to be <u>invalid</u>.

It is <u>easier</u> to show that a sentence is <u>invalid</u> using interpretations than to show that a sentence is <u>valid</u> using interpretations. To show that a sentence φ is <u>invalid</u> just requires us to find a single interpretation for which $\neg\varphi$ is true. We did this for several sentences in previous *exercises*.

To show that a given sentence φ is <u>valid</u> using interpretations requires us to show that φ <u>is true for all interpretations</u>. Equivalently, we must show that there is <u>no interpretation</u> for which φ <u>is false</u>. This requires an extended argument.

For example, let us prove that '$\forall x(Fx \rightarrow Gx) \rightarrow (\forall xFx \rightarrow \forall xGx)$' is valid using interpretations. To prove this, we will show that '$\forall x(Fx \rightarrow Gx) \rightarrow (\forall xFx \rightarrow \forall xGx)$' is true for <u>any interpretation</u> $\mathcal{I}$. (We will not need a naming function n, since no name occurs in that sentence. According to a corollary asserted (but not yet proved) several pages back, the naming function does not matter.)

PF: Prove for any interpretation $\mathcal{I}$, that either $\mathcal{I}(\forall x(Fx \rightarrow Gx)) = F$ or $\mathcal{I}(\forall xFx \rightarrow \forall xGx) = T$. We will suppose $\mathcal{I}(\forall x(Fx \rightarrow Gx)) = T$ and prove $\mathcal{I}(\forall xFx \rightarrow \forall xGx) = T$. To prove $\mathcal{I}(\forall xFx \rightarrow \forall xGx) = T$, we assume $\mathcal{I}(\forall xFx) = T$ as well. Thus, by our assumptions:

1. $\mathcal{I}(\forall x(Fx \rightarrow Gx)) = T$, and
2. $\mathcal{I}(\forall xFx) = T$.

                                   TO PROVE: $\mathcal{I}(\forall xGx) = T$.

3. Suppose $\mathcal{I}(\forall xGx) = F$.
4. Then for some $e \in \mathcal{D}$, $\mathcal{I}(G[e]) = F$, meaning $e \notin \mathcal{I}(G)$.
5. Since $\mathcal{I}(\forall xFx) = T$, $e \in \mathcal{I}(F)$. Another way to say that is: $\mathcal{I}(F[e]) = T$. Yet another way is: $<\mathcal{I},n(a \rightarrow e)>(Fa) = T$, for any naming function n on $\mathcal{D}$.
6. Since $\mathcal{I}(\forall x(Fx \rightarrow Gx)) = T$, $\mathcal{I}(F[e] \rightarrow G[e]) = T$.
7. This means either $\mathcal{I}(F[e]) = F$ or $\mathcal{I}(G[e]) = T$ (by truth definition).
8. Since $\mathcal{I}(F[e]) = T$, $\mathcal{I}(F[e]) \neq F$ and thus $\mathcal{I}(G[e]) = T$, contradicting our assumption, that $\mathcal{I}(\forall xGx) = F$.
9. Hence, $\mathcal{I}(\forall xGx) = T$ and '$\forall x(Fx \rightarrow Gx) \rightarrow (\forall xFx \rightarrow \forall xGx)$' is valid. ∎

6.9 PARTICULARIZATIONS

For some sentences of first-order logic (especially <u>monadic sentences</u>) there is a fairly straightforward technique for demonstrating invalidity. First, treat '$\exists$' as a big OR and treat '$\forall$' as a big AND. Then assume a two-element domain, say {0,1}, and drive the AND's and OR's (from the inside out) into the sentence to make respectively, ∧'s and ∨'s, while replacing the variable of quantification by each element of the domain. Let's take as an example the sentence '∀x∀y(Fx → Fy)' that was already shown to be invalid above, and let's "particularize" it by following the steps:

1. ∀x∀y(Fx → Fy)
2. ANDxANDy(Fx → Fy)
3. ANDx[(Fx → F[0]) ∧ (Fx → F[1])]
4. [(F[0] → F[0]) ∧ (F[0] → F[1])] ∧ [(F[1] → F[0]) ∧ (F[1] → F[1])]

We call the sentence on line 4 the **particularization** of '∀x∀y(Fx → Fy)' with respect to {0,1}. We now replace 'F[0]' by 'T' and 'F[1]' by 'F' (or the other way around), thus arriving at:

[(T → T) ∧ (T → F)] ∧ [(F → T) ∧ (F → F)],

which is obviously false because '(T → F)' is false. 'F[0]' and 'F[1]' are called **atomic components** of the particularization on line 4.

To clarify particularizations somewhat let us extend our substitution notation as follows. Suppose ANDα[...α...] is an innermost occurrence of an AND or an OR together with the scope of its variable. These are the steps to follow to **particularize** ANDα[...α...] with respect to {0,1}:

k   ANDα[...α...]
k+1 [...α...]α/∧{0,1}
k+2 [(...[0]...) ∧ (...[1]...)]

We similarly particularize an innermost ORα[...α...] with respect to {0,1} this way:

k   ORα[...α...]
k+1 [...α...]α/∨{0,1}
k+2 [(...[0]...) ∨ (...[1]...)]

Now, we fill in the intermediate steps in the particularization of '∀x∀y(Fx → Fy)' as follows:

1. ∀x∀y(Fx → Fy)
2. ANDxANDy(Fx → Fy)

3. ANDx[Fx $\rightarrow$ Fy]y/∧{0,1}
4. ANDx[(Fx $\rightarrow$ F[0]) ∧ (Fx $\rightarrow$ F[1])]
5. [(Fx $\rightarrow$ F[0]) ∧ (Fx $\rightarrow$ F[1])]x/∧{0,1}
6. [(F[0] $\rightarrow$ F[0]) ∧ (F[0] $\rightarrow$ F[1])] ∧ [(F[1] $\rightarrow$ F[0]) ∧ (F[1] $\rightarrow$ F[1])]

Suppose we wish to evaluate the validity of the sentence '(∃xFx $\rightarrow$ P) $\rightarrow$ (∀xFx $\rightarrow$ P)'. One way to do this is to particularize that sentence. To particularize a sentence letter, replace it with '**T**' or with '**F**', whichever seems indicated for what you're trying to do. Since we are attempting to make '(∃xFx $\rightarrow$ P) $\rightarrow$ (∀xFx $\rightarrow$ P)' **False**, '**F**' is the better choice here:

1. (∃xFx $\rightarrow$ P) $\rightarrow$ (∀xFx $\rightarrow$ P)
2. (∃xFx $\rightarrow$ **F**) $\rightarrow$ (∀xFx $\rightarrow$ **F**)
3. (ORxFx $\rightarrow$ **F**) $\rightarrow$ (ANDxFx $\rightarrow$ **F**)
4. ([Fx]x/∨{0,1} $\rightarrow$ **F**) $\rightarrow$ (ANDxFx $\rightarrow$ **F**)
5. ([F[0] ∨ F[1]] $\rightarrow$ **F**) $\rightarrow$ (ANDxFx $\rightarrow$ **F**)
6. ([F[0] ∨ F[1]] $\rightarrow$ **F**) $\rightarrow$ ([Fx]x/∧{0,1} $\rightarrow$ **F**)
7. ([F[0] ∨ F[1]] $\rightarrow$ **F**) $\rightarrow$ ([F[0] ∧ F[1]] $\rightarrow$ **F**)

In order to make 7. **False**, '[F[0] ∨ F[1]]' must be **False** and '[F[0] ∧ F[1]]', must be **True**. Can that happen? No. Can replacing 'P' by '**T**' instead of '**F**' help? No. Does it appear as though particularizing '(∃xFx $\rightarrow$ P) $\rightarrow$ (∀xFx $\rightarrow$ P)' with respect to a larger domain may make the resultant particularization **False**? Let's rephrase the question: is it possible to make '[F[0] ∨ F[1] ∨ F[2] ∨ ...]' **False** and '[F[0] ∧ F[1] ∧ F[2] ∧ ...]' **True**. Clearly, no! Hence, ... what? '(∃xFx $\rightarrow$ P) $\rightarrow$ (∀xFx $\rightarrow$ P)' must be valid! In attempting — and failing — to make it false we see why it must be true. In this way particularizations can help us to identify valid sentences, though they don't constitute a "proof" of validity. (Incidently, the sentence '(∃xFx $\rightarrow$ P) $\rightarrow$ (∀xFx $\rightarrow$ P)' "says" "if 'P' is true simply because <u>something</u> has **F**, then surely 'P' is true when <u>everything</u> has F". Phrasing it this way also helps to clarify its validity.)

Exercises:

Some of the sentences below are valid and some are invalid. Use particularizations to establish invalidity and to support (but not "establish") validity (**18** and **19** are done below):

1. $\forall x Fx \to \exists x Fx$
2. $\exists x Fx \to \forall x Fx$
3. $(\exists x Fx \to \exists x Gx) \to \exists x (Fx \to Gx)$
4. $\exists x (Fx \to Gx) \to (\exists x Fx \to \exists x Gx)$
5. $(\exists x Fx \land \exists x Gx) \to \exists x (Fx \land Gx)$
6. $(\exists x Fx \land \exists x Gx) \to \exists x (Fx \to Gx)$
7. $(\exists x Fx \land \exists x Gx) \to \exists x (Fx \to \exists x Gx)$
8. $(\exists x Fx \land \exists x Gx) \to \forall x (Fx \to \exists x Gx)$
9. $\forall x (Fx \to \exists x Gx) \to (\exists x Fx \land \exists x Gx)$
10. $\forall x (Fx \to P) \to (\forall x Fx \to P)$
11. $(\forall x Fx \to P) \to \forall x (Fx \to P)$
12. $\exists x (Fx \to P) \to (\exists x Fx \to P)$
13. $(\exists x Fx \to P) \to \exists x (Fx \to P)$
14. $\exists x (Fx \to P) \to \forall x (Fx \to P)$
15. $\forall x (Fx \to P) \to \exists x (Fx \to P)$
16. $\exists x (Fx \to Gx) \to (\forall x Fx \to \forall x Gx)$
17. $(\forall x Fx \to \forall x Gx) \to \exists x (Fx \to Gx)$
18. $(\exists x Fx \to \exists x Gx) \to \forall x (Fx \to Gx)$
19. $\forall x (Fx \to Gx) \to (\exists x Fx \to \exists x Gx)$

18) $[(F[0] \lor F[1]) \to (G[0] \lor G[1])] \to$
          T   T   F
$[(F[0] \to G[0]) \land (F[1] \to G[1])]$
       F   T   F   F   <u>in</u>valid

19) $[(F[0] \to G[0]) \land (F[1] \to G[1])] \to$
       F   T   F   F   T
$[(F[0] \lor F[1]) \to (G[0] \lor G[1])]$
       T   F   F   F   valid

Before continuing, let's look at one more example of particularization. Consider the sentence '$\exists y \forall x [Fxy \leftrightarrow \neg Fxx]$' (which is a first-order rendition of Russell's Paradox (see T61 in Chapter 5)). Let's particularize it with respect to $\{0,1\}$:

1. $\exists y \forall x [Fxy \leftrightarrow \neg Fxx]$
2. $OR y AND x [Fxy \leftrightarrow \neg Fxx]$
3. $OR y [Fxy \leftrightarrow \neg Fxx] x / \wedge \{0,1\}$
4. $OR y [(F[0]y \leftrightarrow \neg F[0,0]) \wedge (F[1]y \leftrightarrow \neg F[1,1])]$
5. $[(F[0]y \leftrightarrow \neg F[0,0]) \wedge (F[1]y \leftrightarrow \neg F[1,1])] y / \vee \{0,1\}$
6. $[(F[0,0] \leftrightarrow \neg F[0,0] \wedge (F[1,0] \leftrightarrow \neg F[1,1])] \vee$
   $[(F[0,1] \leftrightarrow \neg F[0,0]) \wedge (F[1,1] \leftrightarrow \neg F[1,1])]$

As long as there are at least two elements in the domain, we can see from line 6 that the particularization of '$\exists y \forall x [Fxy \leftrightarrow Fxx]$' will be a disjunction of conjunctions. For each disjunct the number of conjuncts will equal the cardinality of the domain, and for each disjunct one of the conjuncts will be of the form $F[e,e] \leftrightarrow \neg F[e,e]$, for an element of the domain. Since that conjunct can never be true, the conjunction of them can't be true either. Then, of course, the resultant disjunction must be false. (If the domain contains only one element e, then $F[e,e] \leftrightarrow \neg F[e,e]$ is itself the particularization of the sentence). Noticing that '$\exists y \forall x (Fxy \leftrightarrow \neg Fxx)$' will be false in a domain of any size, and thus, that its negation is valid, leads quite naturally to a derivation of its negation, '$\neg \exists y \forall x (Fxy \leftrightarrow \neg Fxx)$', which is T61 in Chapter 5.

Remember, every sentence is one of the three types: (1) valid, (2) unsatisfiable, or (3) neither valid nor unsatisfiable. Particularizations can help us to determine which type a given sentence is. Since the particularization of '$\exists y \forall x (Fxy \leftrightarrow \neg Fxx)$' with respect to $\{0,1\}$ revealed clearly that the sentence is unsatisfiable, we know its negation is valid, which we may wish to prove either by a purely semantic argument or by a syntactic argument (i.e., a derivation) together with the fact that our syntactic (derivational) rules are sound.

If a particularization of a given sentence yields both a false and a true interpretation, we know that the original sentence is of type (3) — neither valid nor unsatisfiable. The first sentence we particularized, '$\forall x \forall y (Fx \rightarrow Fy)$', was of type (3), since

$[(F[0] \rightarrow F[0]) \wedge (F[0] \rightarrow F[1])] \vee [(F[1] \rightarrow F[0]) \wedge (F[1] \rightarrow F[1])]$

easily yields true interpretations (let $\mathcal{I}(F) = \emptyset$, for one) and false ones (the one we found, for instance).

A few final words about particularizations.
Particularizing with respect to {0,1} will not always yield
an interpretation for which the given sentence is false. It
may be necessary to particularize with respect to {0,1,2} or
{0,1,2,...,k} for some larger (finite) k. But, even that
will not always work. There are invalid sentences that are
true in all interpretations of particularizations with
respect to {0,1,2,...,k} for <u>all</u> finite k (called **finite
particularizations**). We can demonstrate the invalidity of
these sentences using "infinite particularizations" with
respect to {0,1,2,...} (i.e., with respect to the infinite
set of natural numbers) if we first define an "infinite
particularization".

Given an innermost occurrence of ANDα, the
infinite particularization of ANDα[...α...] is
gotten this way:

```
k      ANDα[...α...]
k+1    [...α...]α/∧{0,1,2,...}
k+2    [(...[0]...) ∧ (...[1]...) ∧ . . .]
```

(The infinite particularization of ORα[...α...] is done
analogously.)

We can now say that there are invalid sentences that are
true for all interpretations of finite particularizations,
but false for some interpretations of infinite
particularizations.

Exercises:

1. Particularize the following three sentences with respect
   to {0,1} to demonstrate their invalidity:
   (a) $\exists x Fx \rightarrow \forall y Fy$
   (b) $\forall x(Fx \rightarrow \forall y Fy)$
   (c) $\forall y(\exists x Fx \rightarrow Fy)$
2. Consider the following sentence χ:
     (χ) [$\forall x \neg Fxx \land \forall x \forall y \forall z(Fxy \land Fyz \rightarrow Fxz) \land \forall x \exists y Fxy$].
   Argue informally that $\neg \chi$ is true in all interpretations of
   finite particularizations of $\neg \chi$, yet $\neg \chi$ is false in some
   interpretation of the infinite particularization of $\neg \chi$.
   [Hint: it may be easier to consider all partial
    interpretations whose domain is {0,1} (interpretating
    only 'F') and all of them with domain {0,1,2}. This
    should suggest that χ will be false for any such
    interpretations — and any larger finite one as well.

Hence ¬χ will be true for all these finite interpretations. But, then exhibit an interpretation whose domain is all of the natural numbers such that χ is true for this interpretation. Thus ¬χ is false for this interpretation.]

6.10 LOGICAL CONSEQUENCE

For a sentence φ of L_{PL} and a set of sentences Γ of L_{PL}, if any first-order naming interpretation that models Γ also models φ, then φ **is a logical consequence of** Γ. Another way to put this is that φ **is a logical consequence of** Γ if there is no naming interpretation that models Γ and does not model φ. We write 'Γ ⊨ φ' for 'φ is a logical consequence of Γ'. Sometimes we write 'consequence' for 'logical consequence'. We also write 'φ follows from Γ' for 'Γ ⊨ φ'.

Exercises:

Show by a purely semantic argument in each case below either that the sentence φ follows from the set of sentences Γ or that it does not.

Γ	φ
1. {Fa, Fb, Fc, ...}	∀xFx
2. {∀xFx, ∀x(Fx → Gx)}	∀xGx
3. {∀xFx, ∃x(Fx → Gx)}	∀xGx
4. {Fa, ∀x(Fx → Gx)}	∀xGx
5. {Fa, ∀x(Fx → Gx)}	Ga
6. {Fab → Fba, (Fab ∧ Fbc)→Fac, Fba, ¬Fac}	¬Fbc

CHAPTER 7

CONNECTIONS BETWEEN SYNTAX & SEMANTICS FOR FIRST-ORDER LOGIC

TOPICS: *Syntactic Metatheorems, Semantic Metatheorems, Strong Soundness, Consistency, Strong Completeness, Compactness.*

In this chapter we do essentially six things: (1) derive the rules EG and ES from our original set of rules for first-order logic; (2) prove an odd-sounding (and quite technical) theorem, the φ-epimorphism theorem, and infer several useful corollaries from it; (3) prove that each basic first-order rule is sound; (4) prove Strong Soundness for our system of derivations for first-order logic (using (3)); (5) prove Strong Completeness for first-order logic; and (6) prove Compactness.

Everything we prove in this chapter pertains either to the syntax of first-order logic alone (involving notions like 'sentence', 'formula', 'derivation', etc.), or to the semantics alone (involving notions like 'interpretation', 'truth', 'validity', etc.) or to relationships that hold between the syntax and the semantics of L_{PL}. We call concepts relating the syntax to the semantics "mixed concepts".

7.1 SYNTACTIC METATHEOREMS

First, we prove that our new rules EG and ES add no power to our system of basic rules consisting of: all sentential rules, UI, UG, and QE.

As in Chapter 2, we derive a new rule from basic ones by exhibiting a derivation form, which is similar to a fragment of a derivation, but which contains variables ranging over the elements of a derivation. The derivation form below provides a recipe for replacing any use of EG in a derivation by A, QE, UI, and CONTRA. Following the recipe is an example of how it can be used in a derivation.

Eliminating EG

n.	$\varphi\alpha/\beta$	$a_1, a_2, \ldots, a_p$ ← assumption numbers
n + 1.	$\neg\exists\alpha\varphi$	n + 1, A
n + 2.	$\forall\alpha\neg\varphi$	n + 1, QE
n + 3.	$\neg\varphi\alpha/\beta$	n + 1, UI
n + 4.	$\exists\alpha\varphi$	$a_1, a_2, \ldots, a_p$, CONTRA

Now, let's consider EG in the following derivation:

1.	$\forall xFx$	1, A
2.	$\forall x(Fx \rightarrow Gx)$	2, A
3.	$Fa \rightarrow Ga$	2, UI
4.	Fa	1, UI
5.	Ga	1, 2, MP
6.	$\exists xGx$	1, 2, EG

Let us use our recipe to replace EG in the above derivation by lines using only basic rules (the vertical bar indicates the extra lines, which correspond to n+1, n+2, and n+3 in the derivation form above):

1.	$\forall xFx$	1, A
2.	$\forall x(Fx \rightarrow Gx)$	2, A
3.	$Fa \rightarrow Ga$	2, UI
4.	Fa	1, UI
5.	Ga	1, 2, MP
6.	$\neg\exists xGx$	6, A
7.	$\forall x\neg Gx$	6, QE
8.	$\neg Ga$	6, UI
9.	$\exists xGx$	1, 2, CONTRA(5,6,8)

The above derivation form for the EG Rule is a proof of the following syntactic metatheorem about our basic rules:

EG THEOREM: If $\Gamma \vdash \varphi\alpha/\beta$, then $\Gamma \vdash \exists\alpha\varphi$

In words, the metatheorem above "says" that for our basic rules (<u>without</u> EG), whenever the sentence $\varphi\alpha/\beta$ can be derived, so can the sentence $\exists\alpha\varphi$ from the same set of assumptions. So, EG is theoretically dispensable.

We now want to show that our use of the ES rule is also theoretically dispensable — anything we can get using ES we can get without it. So we show below <u>without</u> using Existential Switch that if a sentence ψ can be obtained from an Existential Assumption (together with other assumptions), then ψ can be obtained from the existential sentence itself. So, this metatheorem justifies using ES as a short-cut rule:

ES THEOREM: If $\Gamma \cup \{\varphi\alpha/\beta\} \vdash \psi$, and β is new to φ and ψ, and any sentence of Γ used in the derivation, then $\Gamma \cup \{\exists\alpha\varphi\} \vdash \psi$.

As in the case of EG, we prove this metatheorem by exhibiting a recipe for replacing any use of ES by basic rules:

Eliminating ES

n.	$\exists\alpha\varphi$	$a_1, a_2, \ldots, a_p$
		$\checkmark$SHOW ψ
n + 1.	$\varphi\alpha/\beta$	n+1, EA (β is new to φ, ψ, and to assumptions $a_1, a_2, \ldots, a_p$)
$\vdots$		
n + k.	ψ	$b_1, b_2, \ldots, b_q$, n+1, (where β does not occur in $b_1, b_2, \ldots, b_q$)
n+k+1.	$\varphi\alpha/\beta \rightarrow \psi$	$b_1, b_2, \ldots, b_q$, CD (n+1, n+k)
		SHOW $\exists\alpha\varphi \rightarrow \psi$
		$\checkmark$SHOW $\neg\psi \rightarrow \neg\exists\alpha\varphi$
n+k+2.	$\neg\psi$	n+k+2, A
n+k+3.	$\neg\varphi\alpha/\beta$	$b_1, b_2, \ldots, b_q$, n+k+2, MT
n+k+4.	$\forall\alpha\neg\varphi$	$b_1, b_2, \ldots, b_q$, n+k+2, UG (where β does not occur in any assumption nor in φ)
n+k+5.	$\neg\exists\alpha\varphi$	$b_1, b_2, \ldots, b_q$, n+k+2, QE
n+k+6.	ψ	$a_1, a_2, \ldots, a_p, b_1, b_2, \ldots, b_q$, CONTRA (n, n+k+5)

Exercise:

1. Use the recipe provided above to eliminate ES from the following derivation (while adding the lines corresponding to n+k+1 through n+k+5 in the derivation form for ES):

1. $\exists x \forall y Fxy$ 1, A
2. $\forall y Fay$ 2, EA ('a' is new)

3.	Fab	2,UI
4.	∃xFxb	2,EG
5.	∀y∃xFxy	2,UG
6.	∀y∃xFxy	1,ES

Now that it is clear that only our basic rules count, we can prove Soundness and Completeness for the basic rules alone. To pave the way for those important theorems, we establish some needed semantic metatheorems in the next section, Section 7.2. Unfortunately, this entire section is quite technical, and it may be preferable to skip ahead to Section 7.3 where Soundness and Completeness are proved, before reading Section 7.2.

7.2 SEMANTIC METATHEOREMS [* Very Technical *]

As indicated in brackets above, this is a very technical section that should perhaps be skipped the first time through. Some of the theorems in this section are used in Section 7.3 in the proofs of Soundness and Completeness. You may prefer to look first at those more important results in Section 7.3 before reading this section.

At this point we want to establish a very useful and somewhat intuitive (with qualifications) theorem, called "The φ-epimorphism Theorem", whose statement and proof unfortunately require a multitude of technical detail. You should probably first try to grasp the main ideas embodied in the theorem, and then look at its many corollaries, and finally (if desired) return to the technical details of the proof itself.

We begin with a definition of a "counterpart" of a formula φ. Intuitively, a counterpart φ' of φ is structurally identical to φ, but φ' may contain different relation letters and names than φ has. What we want to do is to first define "counterpart" precisely and then to correlate naming interpretations for which φ' is true with interpretations for which φ is true.

For any formula φ of L_{PL} a formula φ' is a **counterpart of** φ if φ' differs from φ only in the following respects: having some number of distinct, new relation letters that replace all occurrences of whatever relation letters they replace, and having some number of distinct, new logical names replacing some or all of the occurrences of the logical names they replace. φ' may, however, have <u>no</u> relation letters not in φ. And, it is possible φ' has <u>no</u> logical names not in φ. In the limiting case when both are true (i.e., when φ' is <u>identical</u> to φ) φ' is still a counterpart of φ.

A formula φ' is a **counterpart of** φ, if φ' differs from φ only in the following respects:

1) For some number (possibly 0) of relation letters R_1^k, R_2^k, ..., R_m^k appearing in φ, the letters $R_1^{k'}$, $R_2^{k'}$, ..., $R_m^{k'}$ appear in φ' such that:

   a) All $R_i^{k'}$ are new to φ,
   b) If $R_i^k \neq R_j^k$, then $R_i^{k'} \neq R_j^{k'}$,
   c) $R_i^{k'}$ occurs in φ' <u>everywhere</u> R_j^k occurs in φ

2) For some number (possibly 0) of logical names δ_1, δ_2, ..., δ_m appearing in φ, the names δ_1', δ_2', ..., δ_m' appear in φ' such that:

   a) All δ_i' are new to φ,
   b) If $\delta_i \neq \delta_j$, then $\delta_i' \neq \delta_j'$
   c) δ_i' occurs in <u>some</u> places in φ' where δ_i occurs in φ.

Let us look at a few cases when φ' is a counterpart of φ:

φ	φ'
F^3axb	F^3axb
F^0	G^0
F^2ab	G^2cd
F^3axa	G^3axc
$\forall x F^3 axa$	$\forall x G^3 cxa$
$F^2aa \wedge G^2bb$	$H^2da \wedge G^2cc$

Here are some cases when φ' is <u>not</u> a counterpart of φ:

φ	φ'
$F^0 \vee \neg F^0$	$G^0 \vee \neg F^0$
$F^0 \rightarrow G^0$	$G^0 \rightarrow F^0$
$\neg(F^0 \rightarrow G^0)$	$\neg(H^0 \rightarrow H^0)$
F^3axb	F^3bxb
F^3axb	F^3cxc

CHAPTER 7

Exercises:

1. For the first set of cases above explain why each φ' <u>is</u> a counterpart of φ.
2. For the second set of cases above, locate the clause that is violated, showing why φ' is <u>not</u> a counterpart of φ.
3. Is φ always a counterpart of itself? Explain.
4. If φ' is a counterpart of φ, is φ also a counterpart of φ'? Explain why it is or give a counterexample.
5. If φ' is a counterpart of φ and φ'' is a counterpart of φ', is φ'' a counterpart of φ? Explain why it is or give a counterexample.

In obtaining a counterpart φ' of φ, two kinds of replacements were made — for relation letters and for logical names. We wish to unify the development somewhat by calling the replacement symbols **stand-ins**, subject to the following conditions. First, an occurrence of a relation letter or a name in a certain position in φ that also occurs in that position in φ' is a **stand-in for itself**. Let's use the following example:

$$\varphi: \quad \forall x(F^3axa \rightarrow G^3ayc)$$
$$\varphi': \quad \forall x(F^3bxb \rightarrow H^3ayc)$$

φ' in the example above is a counterpart of φ. The 10th symbol from the left in φ is 'a'. The 10th symbol from the left in φ' is also 'a'. That occurrence of 'a' in φ' is a stand-in for itself. In the above example, however, the 5th and 7th symbols in φ are both 'a'. In φ' the 5th and 7th symbols are both 'b'. Since 'b' in φ' occurs in those positions where 'a' occurs in φ, 'b' is also a stand-in for 'a'. Thus, a logical name occurring in φ may have <u>two</u> stand-ins, itself and the new name that replaces it (see clause (2)c)). Notice that 'c' occurs only in the next-to-last position in both φ and φ'. So, 'c' is also a stand-in for itself.

In φ' above, 'H³' occurs in the sole position where 'G³' occurs in φ. So, 'H³' is a stand-in for 'G³'. Since clause (1)c) does <u>not</u> permit some occurrences of a relation letter R in φ to be replaced by R' in φ', while other occurrences of R remain in φ', there can be only one stand-in in φ' for a relation letter occurring in φ. However, 'F³' in φ is also in φ'. So, 'F³' is a stand-in for itself.

We can sum up the above discussion of stand-ins by saying the following. A relation letter or logical name occurring in a given position in counterpart φ' is a **stand-in for the symbol in that position in** φ. Thus, a logical name occurring in φ can have two stand-ins, itself and its replacement symbol; whereas a relation symbol can have only one stand-in, either itself if it also occurs in φ' or the symbol appearing in its stead in all of its occurrences.[1]

Now, let's suppose that $<\mathcal{I},n>$ is appropriate for sentence φ, and $<\mathcal{I}',n'>$ is appropriate for φ' (where φ' is a counterpart of φ). $<\mathcal{I},n>(\varphi)$ may be **T** or $<\mathcal{I},n>(\varphi)$ may be **F**. What we wish to do is to formulate conditions under which $<\mathcal{I},n>(\varphi) = <\mathcal{I}',n'>(\varphi')$. Let's look at our second example above, where $\varphi = F^0$ and $\varphi' = G^0$. An appropriaate interpretation $\mathcal{I}$ for 'F^0' assigns either **T** or **F** to F^0. Let's say $\mathcal{I}(F^0) = $ **F**. Obviously $\mathcal{I}'(G^0)$ must be **F** as well. Then, $\mathcal{I}(F^0) = \mathcal{I}'(G^0)$.

Let's look at a more, complicated example, when $\varphi = $ 'F^{2}ab' and $\varphi' = $ 'G^{2}cd'. Let's suppose that $<\mathcal{I},n>$ is the following:

$\mathcal{I}$

$\mathcal{D} = \{0,1,2\}$
F^2 = $\{<0,2>,<1,0>,<2,1>\}$

n

a $\rightarrow$ 1
b $\rightarrow$ 0

Then $<\mathcal{I},n>(F^2ab) = $ **T**. How would we construct $<\mathcal{I}',n'>$ so that $<\mathcal{I}',n'>(G^2cd) = $ **T** as well? More generally, how would we transform $<\mathcal{I},n>$ into $<\mathcal{I}',n'>$ so that <u>any</u> atomic sentence φ that uses only symbols 'F', 'a', and 'b' is true for $<\mathcal{I},n>$ just in case φ' that uses only symbols 'G', 'c', and 'd' is true for $<\mathcal{I}',n'>$? Let's suppose there is a mapping h from $<\mathcal{I},n>$ to $<\mathcal{I}',n'>$, and let's try to figure out what constraints must be on h.

Since we are interested at this point <u>only</u> in the atomic sentences formed using 'F', 'a', and 'b', we can easily list them:

Faa, <u>Fab</u>, Fba, Fbb

---

[1] We <u>could</u> have permitted <u>some</u> occurrences of a relation letter in φ to also appear in φ', but there do not seem to be any natural benefits to this.

Among them, only 'Fab' is true for $\langle \mathcal{I}, n \rangle$. (That's why it's underlined.) Then, it is clear that for the following list of sentences, only the underlined one must be true for $\langle \mathcal{I}', n' \rangle$:

    Gcc, <u>Gcd</u>, Gdc, Gdd

From this, we can see that the element denoted by 'a' for naming function n must be mapped to the same element 'c' denotes for n'. That is, $h \circ n(a) = n'(c)$. And, similarly, whatever 'b' denotes (for n) must be mapped to the element 'd' denotes (for n'). That is, $h \circ n(b) = n'(d)$. What else do we need?

We need to say that 'F' holds of its two elements for $\langle \mathcal{I}, n \rangle$ whenever 'G' holds of those stand-ins for $\langle \mathcal{I}', n' \rangle$. In symbols (where δ_1, δ_2 range over 'a' and 'b' and δ_1' and δ_2' stand in for δ_1, δ_2 respectively): $\langle n(\delta_1), n(\delta_2) \rangle \in \mathcal{I}(F)$ iff $\langle n'(\delta_1'), n'(\delta_2') \rangle \in \mathcal{I}'(G)$. To rewrite these conditions, (where δ' is a stand-in for δ):

    (1) $h \circ n(\delta) = n'(\delta')$
    (2) $\langle n(\delta_1), n(\delta_2) \rangle \in \mathcal{I}(F)$ iff $\langle n'(\delta_1'), n'(\delta_2') \rangle \in \mathcal{I}'(G)$

Do the above two conditions ensure that for the symbols 'a', 'b' and 'F' and naming interpretation $\langle \mathcal{I}, n \rangle$ that every <u>atomic</u> sentence will be true just in case a counterpart of that sentence using only 'c', 'd' and 'G' is true for $\langle \mathcal{I}', n' \rangle$? Sure. We listed the counterparts of all the original atomic sentences, where 'c' is a stand-in for 'a' and 'd' is one for 'b', and we can see that (1) and (2) guarantee that conclusion. The only other case is when, instead of the above counterparts for 'Faa', 'Fab', 'Fba', and 'Fbb', we have 'Gdd', 'Gdc', 'Gcd' and 'Gcc', respectively, which can be handled in exactly the same way.

Can (1) and (2) be rewritten to apply to <u>any</u> relation letters and <u>any</u> names for <u>any</u> <u>atomic</u> <u>sentence</u> φ and its counterpart φ'? Sure. Why not? Will (1) and (2) then make <u>any</u> <u>molecular</u> <u>sentence</u> φ hold for $\langle \mathcal{I}, n \rangle$ just in case φ' holds for $\langle \mathcal{I}', n' \rangle$? Again, the answer is: sure, why not? Can we go even further? can we say that (1) and (2) guarantee that for <u>any</u> <u>sentence</u> φ, $\langle \mathcal{I}, n \rangle(\varphi) = \langle \mathcal{I}', n' \rangle(\varphi')$? No. Why not?

Let's back up a bit. First, let's state a theorem for the molecular case mentioned above. Since φ is not always a counterpart of φ', when φ' is a counterpart of φ, we must be careful.

Suppose h is a mapping from $\mathcal{D}$ to $\mathcal{D}'$ (the domain of $\mathcal{I}'$). And, suppose that for every logical name δ_i occurring in φ, where φ has no quantifiers, that n, the naming function for $\mathcal{I}$, is such that $n(\delta_i) = e_i$. Then, let δ_i' be a stand-in for δ_i in φ'. Then, <u>if</u> the following conditions are satisfied:[2]

(1) $h \circ n(\delta_i) = n'(\delta_i')$, and for any R^k in φ
(2) $\langle e_1, e_2, \ldots, e_k \rangle \in \mathcal{I}(R^k)$ iff
$\qquad \langle h(e_1), h(e_2), \ldots, h(e_k) \rangle \in \mathcal{I}'(R^{k'})$

<u>then</u>

$$\langle \mathcal{I}, n \rangle(\varphi) = \langle \mathcal{I}', n' \rangle(\varphi').$$

<u>PF</u>: We outline the proof and leave the details as an *exercise*. The proof is by Complete Induction on the height of φ. The height of φ is the number of quantifiers and connectives in φ. Since φ is quantifier-free, there are only the connectives '$\neg$', '$\rightarrow$', '$\wedge$', '$\vee$', and '$\leftrightarrow$' possibly in φ.

First, for Height(φ) = 0 (meaning that there are <u>no</u> connectives in φ) we need to prove:

$$\langle \mathcal{I}, n \rangle(R^k \delta_1 \delta_2 \ldots \delta_k) = \langle \mathcal{I}', n' \rangle(R^{k'} \delta_1' \delta_2' \ldots \delta_k').$$

This is not hard, given conditions (1) and (2).

Then, once we have proved our base case (when Height(φ) = 0), we <u>assume</u> that for any subsentence ψ of φ such that Height(ψ) < Height(φ), the theorem holds. This is our Inductive Hypothesis (IH), namely that:

$$\langle \mathcal{I}, n \rangle(\psi) = \langle \mathcal{I}', n' \rangle(\psi') \quad \text{(where } \psi \text{ is a (proper)}$$
$$\text{subsentence of } \varphi\text{).}$$

Then, one by one, we suppose (1) φ is of the form $\neg \psi$ (i.e., φ is a negation); (2) φ is $\psi_1 \rightarrow \psi_2$ (i.e., φ is a conditional); (3) φ is $\psi_1 \wedge \psi_2$ (i.e., a conjunction); (4) φ is $\psi_1 \vee \psi_2$ (a disjunction); or (5) φ is $\psi_1 \leftrightarrow \psi_2$ (a biconditional). In <u>each</u> of these 5 cases our Inductive Hypothesis will be sufficient to allow us to prove φ itself has the desired property from the assumption that every (smaller) subsentence of φ has the property. This concludes the proof. ∎

---

[2] This theorem could be called 'The φ-Homomorphism Theorem' on a par with The φ-Epimorphism Theorem that appears a few pages later.

Exercises:

1. Provide all the details of the proof sketched above.
2. For the mapping h in the theorem above, which of the following hold; <u>and</u> if so, give a proof, <u>and</u> if not, give a counter-example:
   (a) h is 1-1
   (b) h is onto
   (c) h restricted to the range of n is 1-1 (meaning: if $n(\delta_1) \neq n(\delta_2)$, then $h \circ n(\delta_1) \neq h \circ n(\delta_2)$. Since $h \circ n(\delta_i) = n'(\delta_i')$, we can also write: if $n(\delta_1) \neq n(\delta_2)$, then $n'(\delta_1') \neq n'(\delta_2')$).

Assuming you have done the above *exercises* or at least are convinced that 2a) is false, 2b) is false, and 2c) is true, we wish to enlarge the previous theorem to include sentences with quantifiers. In case you are not convinced of the falsity of 2a) and 2b), consider unnamed elements. The idea is that the unnamed elements of $\mathcal{D}$ (the domain of $\mathcal{I}$) can be mapped any which way to elements of $\mathcal{D}'$ without affecting the truth conditions of sentences containing names. (Remember: φ and φ' in the previous theorem contain no quantifiers.) Also, there may be elements of $\mathcal{D}'$ that are not images of elements of $\mathcal{D}$ under h.

The picture is rather like this:

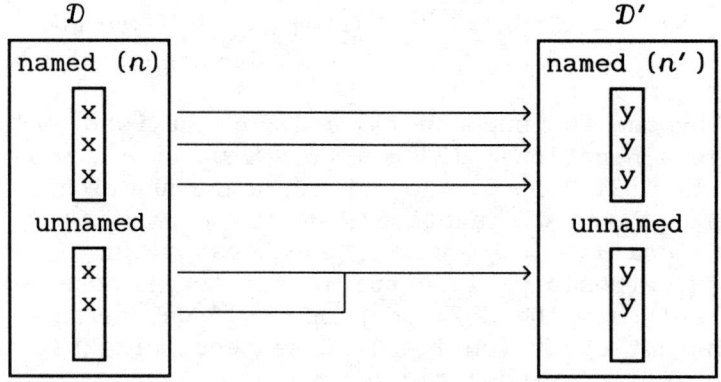

Now, if we wish to make $\forall \alpha \varphi'$ true whenever $\forall \alpha \varphi$ is true, then whenever $\varphi[e]$ is true for <u>all</u> $e \in \mathcal{D}$, $\varphi'[e']$ must be true for all e' ∈ $\mathcal{D}'$. Thus, the mapping h must be onto, though not necessarily 1-1. So, the picture needs to look something like this in order to accommodate quantifiers:

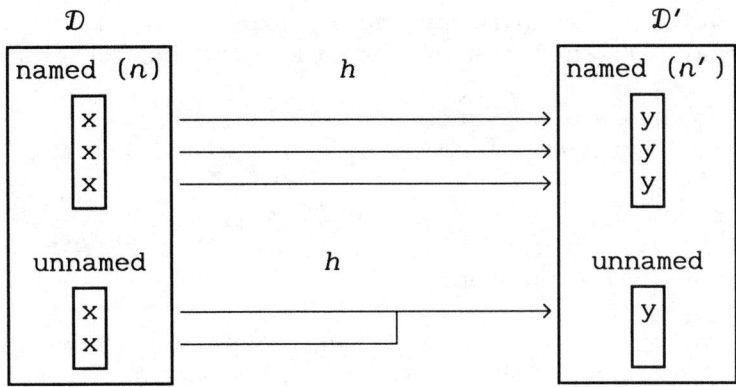

We still refrain from making h 1-1 (for the unnamed elements) unless something forces us to. You may think that with sufficient resourcefulness you could require h to be 1-1, but this is not so. One way to see this is to imagine $<\mathcal{I}',n'>$ to be $<\mathcal{I},n>$ itself. That is let h be the identity map and let φ and φ' be identical as well. Now, suppose $<\mathcal{I},n>(\varphi) = \mathbf{T}$. So, obviously $<\mathcal{I}',n'>(\varphi') = \mathbf{T}$ too (since $<\mathcal{I},n> = <\mathcal{I}',n'>$ and $\varphi = \varphi'$). Now, supose we throw an additional element into $\mathcal{D}'$ that behaves exactly as an element does that's already in $\mathcal{D}$ (and thus in $\mathcal{D}'$). That is, if '∃xFx' is true of the original element e ∈ $\mathcal{D}$, then '∃xFx' is true of the new element e' ∈ $\mathcal{D}'$ also. To distinguish e from e', we need a sentence that says so. Supposing e' is unnnamed, there's no way to do this. If we could say that <u>only</u> e has property 'F', we could distinguish e' from e, but we have no way to say of an unnamed object that <u>it</u> differs from any other. If we had equality in our language (which we will have later), we could write 'Fa ∧ ∀x(x ≠ a → ¬Fx)' which says that only a has F. Since we do <u>not</u> have equality in L_{PL}, the best we can do is to say that <u>at least</u> some number of elements have a given property. Thus, we can always create another naming interpretation with an additional element that behaves exactly like one already present, yet differs from it. Thus, we do not need to make h 1-1.

We will need, however, to specify the conditions for h so that $<\mathcal{I},n>(\varphi) = <\mathcal{I}',n'>(\varphi')$ when sentence φ can contain quantifiers. As before, φ' is a counterpart of φ, and $<\mathcal{I},n>$ is appropriate for φ, while $<\mathcal{I}',n'>$ is appropriate for φ'.

We call the map h a "φ-epimorphism" from $\langle \mathcal{I}, n \rangle$ to $\langle \mathcal{I}', n' \rangle$ because it is an onto map that is a homomorphism (see Chapter 10, *Models For First-Order Theories*, for an explanation of 'homomorphism' and related terminology).

We now define a φ-**epimorphism**, h, from $\langle \mathcal{I}, n \rangle$ to $\langle \mathcal{I}', n' \rangle$ for counterpart φ' in terms of the following conditions:

(1) h is an onto map such that $h: \mathcal{D} \to \mathcal{D}'$,
(2) $n'(\delta_i') = h \circ n(\delta_i)$ where δ_i' is a stand-in in φ' for δ_i in φ.
(3) For any $e_1, e_2, \ldots, e_k$ in $\mathcal{D}$: $\langle e_1, e_2, \ldots, e_k \rangle \in \mathcal{I}(R^k)$ iff $\langle h(e_1), h(e_2), \ldots, h(e_k) \rangle \in \mathcal{I}'(R'^k)$.

Now, here is the theorem:

φ-EPIMORPHISM THEOREM (7.1): For <u>all</u> φ-epimorphisms from $\langle \mathcal{I}_1, n_1 \rangle$ to $\langle \mathcal{I}_2, n_2 \rangle$ for φ', $\langle \mathcal{I}_1, n_1 \rangle(\varphi) = \langle \mathcal{I}_2, n_2 \rangle(\varphi')$.

PF: The proof is by Complete Induction on the height of φ, the number of connectives and quantifiers in φ.

Case 1: Height(φ) = 0. That is, $\varphi = R^k \delta_1 \delta_2 \ldots \delta_k$. (We assume below that $n(\delta_i) = e_i$, where each $e_i \in \mathcal{D}$.)

$\langle \mathcal{I}, n \rangle (R^k \delta_1 \delta_2 \ldots \delta_k) = T$ ⟺
$\langle n(\delta_1), n(\delta_2), \ldots, n(\delta_k) \rangle \in \mathcal{I}(R^k)$ (by truth definition) ⟺
$\langle e_1, e_2, \ldots, e_k \rangle \in \mathcal{I}(R^k)$ (since $n(\delta_i) = e_i$) ⟺
$\langle h(e_1), h(e_2), \ldots, h(e_k) \rangle \in \mathcal{I}'(R'^k)$ (by clause (3)) ⟺
$\langle h \circ n(\delta_1), h \circ n(\delta_2), \ldots, h \circ n(\delta_k) \rangle \in \mathcal{I}'(R'^k)$ ($n(\delta_i) = e_i$) ⟺
$\langle n'(\delta_1'), n'(\delta_2'), \ldots, n'(\delta_k') \rangle \in \mathcal{I}'(R'^k)$ (Either <u>every</u> occurrence of δ_i in φ is replaced by δ_i' in φ', or some are not. In the latter case, δ_i' just is δ_i) ⟺
$\langle \mathcal{I}', n' \rangle (R'^k \delta_1' \delta_2' \ldots \delta_k') = T$ (by truth definition).

This concludes the case that φ is atomic.

Case 2: φ is not atomic. Assume Height(φ) = m, where $m > 0$.

Induction Hypothesis (IH): For <u>all</u> ψ-epimorphisms from $\langle \mathcal{I}_1, n_1 \rangle$ to $\langle \mathcal{I}_2, n_2 \rangle$ for ψ', where the height of ψ is less than the height of φ, $\langle \mathcal{I}_1, n_1 \rangle(\psi) = \langle \mathcal{I}_2, n_2 \rangle(\psi')$.

Case 2a: We suppose φ is of the form of $\forall \alpha \psi$. We then assume $\langle \mathcal{I}, n \rangle (\forall \alpha \psi) = T$. then we try TO PROVE: $\langle \mathcal{I}', n' \rangle ((\forall \alpha \psi)') = T$, where h is a $\forall \alpha \psi$-epimorphism from $\langle \mathcal{I}, n \rangle$ to $\langle \mathcal{I}', n' \rangle$ for $(\forall \alpha \psi)'$. First, we use the truth definition of $\langle \mathcal{I}, n \rangle (\forall \alpha \psi) = T$ to get that $\langle \mathcal{I}, n(\beta \to e) \rangle (\psi \alpha / \beta) = T$ for <u>all</u> e (where β is the <u>first</u> name <u>new</u> to ψ). Then, for each $e \in \mathcal{D}$, $\langle \mathcal{I}, n(\beta \to e) \rangle (\psi \alpha / \beta) = T$. Since $\psi \alpha / \beta$ is of lesser height than $\forall\alpha\psi$, our Inductive Hypothesis tells us that for each e, $<\mathcal{I}',n'(\delta{\to}h(e))>(\psi'\alpha/\delta) = T$, where δ is new to $\psi\alpha/\beta$ and to $\forall\alpha\psi'$ (which is identical to $(\forall\alpha\psi)'$). To be specific, we make δ the <u>first</u> such new name.

At this point, we "almost" have what we wish <u>except</u> that we <u>cannot</u> claim that $<\mathcal{I}',n'>(\forall\alpha\psi)' = T$ simply on the basis that $<\mathcal{I}',n'(\delta{\to}h(e))>(\psi'\alpha/\delta) = T$ for each $e \in \mathcal{D}$. Why? One reason is that δ is not necessarily the <u>first</u> name new to $\forall\alpha\psi'$ (or $(\forall\alpha\psi)'$, which is the same). δ was taken to be the first name new to <u>both</u> $\psi\alpha/\beta$ <u>and</u> to $\forall\alpha\psi'$. So, our first task will be to prove that this result holds for the <u>first</u> name new to $\forall\alpha\psi'$. Suppose γ is the first name that is new to $\forall\alpha\psi'$ (γ may occur in $\psi\alpha/\beta$).

We want TO PROVE:

$<\mathcal{I}',n'(\gamma{\to}h(e))>(\psi'\alpha/\gamma) = T$ for all $e \in \mathcal{D}$.

This result follows from our Inductive Hypothesis, because (1) $\psi'\alpha/\gamma$ is a counterpart of $\psi'\alpha/\delta$ and (2) for each e, there is a $\psi'\alpha/\delta$-epimorphism from $<\mathcal{I}',n'(\delta{\to}h(e))>$ to $<\mathcal{I}',n'(\gamma{\to}h(e))>$ for $\psi'\alpha/\gamma$; and furthermore, the height of $\psi'\alpha/\delta$ is less than the height of $\forall\alpha\psi$ (on which we are performing the induction). What is that $\psi'\alpha/\delta$-epimorphism? It is the identity map Id: $\mathcal{D}' \to \mathcal{D}'$, such that $n'(\delta{\to}h(e))(\gamma) = n'(\delta{\to}h(e))(\delta) = h(e)$.

So, we now have:

$<\mathcal{I}',n'(\gamma{\to}h(e))>(\psi'\alpha/\gamma) = T$ for all $e \in \mathcal{D}$.

But, since h is onto, we can now arrive at:

$<\mathcal{I}',n'(\gamma{\to}e')>(\psi'\alpha/\gamma) = T$ for all $e' \in \mathcal{D}'$.

Hence:

$<\mathcal{I}',n'>(\forall\alpha\psi') = T$.

This concludes the left to right direction of this case. ∎

To proceed in the other direction, assume:

$<\mathcal{I}',n'>(\forall\alpha\psi') = T$,

and then continue in the same way we did for the left-to-right direction. The only important difference to note is the fact that h is onto is <u>not</u> needed in this direction. To see that, notice that when all elements in the range of a function (in this case all $e' \in \mathcal{D}'$ such that $e' = h(e)$) have a given property, every element of the domain of the function must be mapped to an element with that property. (In the left-to-right direction, though, we needed the fact that h is onto because the fact that the property held of all $e \in \mathcal{D}$ would otherwise not guarantee that it held of all $e' \in \mathcal{D}'$. Without the onto condition we could say only that it held of all $h(e) \in \mathcal{D}'$.) ∎

Exercise:

1. Finish the missing cases in the induction proof of the φ-EPIMORPHISM THEOREM, putting in the details of the right-to-left direction for the case when $\varphi = \forall\alpha\psi$, as explained above. [The sentential cases are all straightforward. The case when $\varphi = \exists\alpha\psi$ is much like the case when $\varphi = \forall\alpha\psi$. One difference to note is that in this case it is the right-to-left direction of the proof that requires the onto condition.]

Now that the φ-EPIMORPHISM THEOREM has been established, there are several helpful corollaries we can obtain as consequences of this theorem.

For one branch of corollaries we consider only one sentence rather than two:

COROLLARY 7.2 (PART. INTERP. THEOREM): If for every k-ary relation letter $\mathcal{R}^k$ and for every logical name β in φ, two naming interpretations $\langle \mathcal{I}, n \rangle$ and $\langle \mathcal{I}', n' \rangle$ are such that $\mathcal{I}(\mathcal{R}^k) = \mathcal{I}'(\mathcal{R}^k)$ and $n(\beta) = n'(\beta)$, then, provided their domains are equal:

$\langle \mathcal{I}, n \rangle(\varphi) = \langle \mathcal{I}', n' \rangle(\varphi)$ for every sentence φ of L_{PL}.

Exercises:

1. Explain why a sentence φ is a counterpart of itself, and
2. specify h, the φ-EPIMORPHISM. In particular, explain why conditions (1) - (3) are satisfied for h.

COROLLARY 7.2 rigorously justifies our use of partial interpretations, though it was intuitively clear that they were OK in the previous chapter.

Following from COROLLARY 7.2 is 7.2a, and then 7.2b is just a special case of 7.2a.

COROLLARY 7.2a: For any interpretation $\mathcal{I}$ and any logical name β occurring in a sentence φ, if two naming functions n_1, n_2 for $\mathcal{D}$ are such that $n_1(\beta) = n_2(\beta)$ then:
$$\langle \mathcal{I}, n_1 \rangle(\varphi) = \langle \mathcal{I}, n_2 \rangle(\varphi)$$

Exercises:

1. Why is COROLLARY 7.2a a special case of COROLLARY 7.2?
2. How do you know that the "two" interpretations interpret each R^k the same?

COROLLARY 7.2b: For any interpretation $\mathcal{I}$ and any two naming functions n_1, n_2, for $\mathcal{D}$, if φ is a sentence of L_{PL} containing <u>no</u> occurrence of names, then:

$$\langle \mathcal{I}, n_1 \rangle(\varphi) = \langle \mathcal{I}, n_2 \rangle(\varphi).$$

Exercise:

1. Is it true in COROLLARY 7.2b that $n_1(\beta) = n_2(\beta)$ for every name β occurring in φ (thus satisfying the condition for COROLLARY 7.2a)?

Now, let's start looking at corollaries of the φ-EPIMORPHISM THEOREM that involve two different sentences.

COROLLARY 7.3: Suppose φ and φ' are identical except that φ' has new occurrences of distinct logical names $\lambda_1', \lambda_2', \ldots, \lambda_k'$ where (but not necessarily everywhere) φ has occurrences of $\lambda_1, \lambda_2, \ldots, \lambda_k$. If $\mathcal{I}$ and $\mathcal{I}'$ are interpretations such that $\mathcal{D} = \mathcal{D}'$ and for each R^k in φ, $\mathcal{I}(R^k) = \mathcal{I}'(R^k)$, and if n is a naming function for $\mathcal{D}$ such that $n(\lambda_i') = n(\lambda_i)$ for each i from 1 to k, <u>then</u>:

$$\langle \mathcal{I}, n \rangle(\varphi) = \langle \mathcal{I}', n \rangle(\varphi').$$

202   CHAPTER 7

Exercises:

1. What is the relationship between $\mathcal{D}$, the domain of $\mathcal{I}$, and $\mathcal{D}'$, the domain of $\mathcal{I}'$?
2. What is the relationship between $\mathcal{I}(R^k)$ and $\mathcal{I}'(R^k)$ for each relation letter R^k in both φ and φ'?
3. Are there any relation letters in either φ or φ' that are not in the other?
4. What is h, the φ-EPIMORPHISM here?
5. Given your answer to (4), why are $\mathcal{I}$ and $\mathcal{I}'$ different interpretations?
6. What is the difference between COROLLARY 7.3 and COROLLARY 7.2?
7. Which of the two corollaries just named is a consequence of the other, and why?
8. State COROLLARY 7.2 and COROLLARY 7.3 in less technical terms so that their relationship to each other is clear.

A special case of COROLLARY 7.3 involves the replacement in φ of just one new name. In order to refer to this succinctly, we introduce another notational device.

$\{\alpha/\beta\}$ means **some or all of α are replaced by β**, where α can be a variable or a name, and β is a name. The difference between α/β and $\{\alpha/\beta\}$ is that $\{\alpha/\beta\}$ does not imply that all occurrences of α are replaced by β. The notation $\alpha/\beta\{\beta/\delta\}$, where α is either a variable or a name and β and δ are both names, means that all occurrences of α are first replaced by β and then some occurrences of β are replaced by δ. $\alpha/\beta\ \beta/\delta$ is, of course, just α/δ, since all occurrences of α are first replaced by β and then all occurrences of β are replaced by δ.

COROLLARY 7.4: If $\varphi' = \varphi\{\delta/\gamma\}$, where γ is new to φ, and $n(\delta) = e$, then $<\mathcal{I},n>(\varphi) = <\mathcal{I},n(\gamma\to e)>(\varphi')$.[3]

Incidentally, the proof of COROLLARY 7.4 alone, without the help of the φ-EPIMORPHISM THEOREM, involves almost as much detail as the proof of the φ-EPIMORPHISM THEOREM itself,

---

[3] When γ is not new, the conclusion does not follow. Let φ = 'Fab', φ' = 'Fbb', δ = 'a', γ = 'b', $\mathcal{D}$ = {0,1}, $\mathcal{I}(F)$ = {<0,1>}, $n(a)$ = 0, and $n(1)$ = 1. Then, $<\mathcal{I},n>$(Fab) = **T** (because <0,1> $\in \mathcal{I}(F)$), but $<\mathcal{I},n(b\to 0)>$(Fbb) = **F** (because <0,0> $\notin \mathcal{I}(F)$). When 'b' is substituted for 'a' in 'Fab' it "clashes" with the 'b' already there.

though it is easier if COROLLARY 7.2a is first independently established.

Exercise:

1. Prove COROLLARY 7.4 <u>without</u> the φ-EPIMORPHISM THEOREM, though you may use COROLLARY 7.2a [follow the steps in the proof of the φ-EPIMORPHISM THEOREM until COROLLARY 7.2a is needed (twice).]

Now, we wish to exchange relation letters in φ.

COROLLARY 7.5: Suppose φ and φ' are identical except that φ' has new occurrences of relation letters, $\mathcal{R}'^k$, where φ has occurrences of $\mathcal{R}^k$, for all R^k in φ. Then <u>if</u> (1) $\mathcal{D} = \mathcal{D}'$, and (2) $\mathcal{I}(\mathcal{R}^k) = \mathcal{I}'(\mathcal{R}'^k)$ for all $\mathcal{R}^k$ in φ,
<u>then</u>: $<\mathcal{I},n>(\varphi) = <\mathcal{I}',n>(\varphi')$.

Exercise:

1. What is the difference between COROLLARY 7.5 and COROLLARY 7.3?

And, of course we can combine COROLLARIES 7.3 and 7.5 to allow for substitution of both relation letters and any number of names.

COROLLARY 7.6: Suppose φ' is a counterpart in L_{PL} of φ. Then, if (1) $\mathcal{D} = \mathcal{D}'$; (2) $n'(\delta_i') = n(\delta_i)$, and $n'(\delta_i)$ for the remaining δ_i in φ'; and (3) $\mathcal{I}'(\mathcal{R}^{k'}) = \mathcal{I}(\mathcal{R}^k)$; then
$<\mathcal{I},n>(\varphi) = <\mathcal{I}',n'>(\varphi')$.

Exercise:

1. In what respects is COROLLARY 7.6 weaker than the φ-EPIMORHPISM THEOREM itself?

Finally, we are finished with the technical details of the φ-EPIMORPHISM THEOREM and its relatives. Some of these results will come in handy in the proofs of the two more familiar theorems below.

We should mention that all the semantic metatheorems of Chapter 3, *The Semantics of Sentential Logic*, carry over to first-order logic once the obvious semantic changes have been made. For example, THEOREM 3.2 is:

THEOREM 3.2: If $\mathcal{B}(\Delta) = \mathbf{T}$ and $\Gamma \subseteq \Delta$, then $\mathcal{B}(\Gamma) = \mathbf{T}$.

The proof of THEOREM 3.2 is essentially the argument that if any boolean-valued interpretation $\mathcal{B}$ interprets all sentences of Δ as **True**, then $\mathcal{B}$ must interpret all sentences of Γ as **True** as well, since $\Gamma \subseteq \Delta$. The analogous theorem for first-order logic is:

THEOREM 3.2′: If $\langle \mathcal{I},n \rangle(\Delta) = \mathbf{T}$ and $\Gamma \subseteq \Delta$, then $\langle \mathcal{I},n \rangle(\Gamma) = \mathbf{T}$.

The proof of THEOREM 3.2′ is identical to the proof of THEOREM 3.2 except that $\langle \mathcal{I},n \rangle$ replaces $\mathcal{B}$ in the proof. The rest of the theorems are handled similarly, with all semantic notions (e.g., "validity", "logical consequence") modified to apply to the first-order language. We shall assume in what follows that the analogs of THEOREMS 3.0 - 3.18 have been established for the language L_{PL}.

Exercises:

1. Prove the first-order analogs of THEOREMS 3.3, 3.8, 3.15, and 3.18.

The next theorem establishes the soundness of our UI Rule. It says that whenever a universal sentence is a consequence of a set of sentences, any instance of the universal is a consequence of that set.

THEOREM 7.2: If $\Gamma \models \forall \alpha \varphi$, then $\Gamma \models \varphi \alpha / \delta$.

PF: Assume $\Gamma \models \forall \alpha \varphi$.
Let $\langle \mathcal{I},n \rangle(\Gamma) = \mathbf{T}$. Thus $\langle \mathcal{I},n \rangle(\forall \alpha \varphi) = \mathbf{T}$.
TO PROVE $\langle \mathcal{I},n \rangle(\varphi \alpha / \delta) = \mathbf{T}$.
Since $\langle \mathcal{I},n \rangle(\forall \alpha \varphi) = \mathbf{T}$, if β is the first name new to φ, then $\langle \mathcal{I},n(\beta \to e) \rangle(\varphi \alpha / \beta) = \mathbf{T}$ for all e. Suppose $n(\delta) = e_0$, then $\langle \mathcal{I},n(\beta \to e_0) \rangle(\varphi \alpha / \beta) = \mathbf{T}$. δ may already occur in φ, but since β is new to φ, we let $\psi = \varphi \alpha / \delta$ and $\psi \{\delta / \beta\} = \varphi \alpha / \beta$. Thus (rewriting above to make assignments explicit),

$$\langle \mathcal{I}, n(\delta \to e_0)(\beta \to e_0) \rangle(\psi\{\delta/\beta\}) = \mathbf{T}.$$

Then, by COROLLARY 7.4

SYNTAX & SEMANTICS FOR FIRST-ORDER

$$\langle \mathcal{I}, n(\delta \to e_0) \rangle(\psi) = T.$$

$n(\delta) = e$ and $\psi = \varphi\alpha/\delta$. So, $\langle \mathcal{I}, n \rangle(\varphi\alpha/\delta) = T.$ ∎

The next theorem formally justifies our UG Rule.

THEOREM 7.3: For δ new to $\Gamma \cup \{\varphi\}$, if $\Gamma \models \varphi\alpha/\delta$, then $\Gamma \models \forall\alpha\varphi$.

PF: Assume $\Gamma \models \varphi\alpha/\delta$.
 Take any $\langle \mathcal{I}, n \rangle$ such that $\langle \mathcal{I}, n \rangle(\Gamma) = T$.
 Then $\langle \mathcal{I}, n \rangle(\varphi\alpha/\delta) = T$.
TO PROVE: $\langle \mathcal{I}, n \rangle(\forall\alpha\varphi) = T$.
TO PROVE: $\langle \mathcal{I}, n(\beta \to e) \rangle(\varphi\alpha/\beta) = T$ for all e.
 Take any e. Since $\delta \notin \Gamma$, $\langle \mathcal{I}, n(\delta \to e) \rangle(\Gamma) = T$, (by COROLLARY 7.2a) for any sentence in Γ. Since $\Gamma \models \varphi\alpha/\delta$,
$$\langle \mathcal{I}, n(\delta \to e) \rangle(\varphi\alpha/\delta) = T.$$

Since β is new to $\varphi\alpha/\delta$,

$$\langle \mathcal{I}, n(\delta \to e)(\beta \to e) \rangle(\varphi\alpha/\beta) = T, \quad \text{by COROLLARY 7.4.}$$

And, since $\delta \notin \varphi\alpha/\beta$,

$$\langle \mathcal{I}, n(\beta \to e) \rangle(\varphi\alpha/\beta) = T, \quad \text{by COROLLARY 7.2a again.}$$
∎

Recall from the previous chapter what it means for a sentence φ of L_{PL} to be **True** for an interpretation $\mathcal{I}$:

$$\mathcal{I}(\varphi) = T \quad \text{iff} \quad \langle \mathcal{I}, n \rangle(\varphi) = T \text{ for } \underline{\text{all}} \ n.$$

Dropping all references to naming functions when they are not needed will become standard practice in dealing with first-order theories. To prepare for this, we prove the following result:

UNIVERSALIZATION THEOREM (7.4): $\mathcal{I}(\varphi\alpha_1/\delta_1 \ \alpha_2/\delta_2 \ldots \alpha_k/\delta_k) = T$ iff $\mathcal{I}(\forall\alpha_1\forall\alpha_2\ldots\forall\alpha_k\varphi) = T$.

Exercise:

1. Prove the UNIVERSALIZATION THEOREM. [HINT: Technically, the proof is by Complete Induction on the number of occurrences of '∀' used to prefix φ. Case 1 is: $\mathcal{I}(\varphi\alpha_1/\delta_1) = T$ iff $\mathcal{I}(\forall\alpha_1\varphi) = T$. First, translate the above (by definition of $\mathcal{I}$) into:
$\langle\mathcal{I},n\rangle(\varphi\alpha_1/\delta_1) = T$ for <u>all</u> n iff $\mathcal{I}(\forall\alpha_1\varphi) = T$ for <u>all</u> n.
The above two theorems should prove helpful here. Once Case 1 is finished and Cases 1, 2, . . . , k have been assumed to hold (by the Induction Hypothesis), Case k + 1 (with k + 1 occurrences of '∀' prefixing φ) is just a repetition of Case 1.]

THEOREM 7.5a: $\Gamma \models \neg\forall\alpha\varphi$ iff $\Gamma \models \exists\neg\alpha\varphi$.
THEOREM 7.5b: $\Gamma \models \neg\exists\alpha\varphi$ iff $\Gamma \models \forall\neg\alpha\varphi$.

Exercises:

1. Prove 7.5a and 7.5b.

7.3   RELATIONS BETWEEN SYNTAX AND SEMANTICS

Strong Soundness

<u>Strong Soundness Metatheorem</u>:   if $\Gamma \vdash \varphi$ then $\Gamma \models \varphi$

PF: We prove this result by Complete Induction on the number of lines of a derivation. More specifically, we prove that for any arbitrary derivation, any line of it is such that the sentence on that line is a consequence of the assumptions of that line.
   <u>Case 1</u>: We consider line 1 of an arbitrary derivation. The sentence on this line can only be entered by A. We have already proved this case for sentential logic. The proof here is the same.
   <u>Case k</u>: We assume that what we wish to prove is true through k lines of an arbitrary derivation. More specifically, we assume that for an arbitrary derivation, lines 1, 2, 3, ..., k are such that each sentence on each of these lines is a consequence of its assumptions. Then, we wish to show that the sentence on line k+1 must also be a consequence of its assumptions.
   <u>Sentential Subcases</u>: We have already proved that if a sentence is entered by any sentential rule, it is a tautological consequence of its assumptions. For, all of

the sentential rules are derived from our five basic sentential rules: A, MP, CD, CONTRA, and DEF. This includes the Theorem Rule, T, which abbreviates an entire derivation within another. And, there is no circularity, since we permit T to be used only to abbreviate the result of an earlier derivation. <u>First-Order Subcases:</u> We need to prove that any sentence on line k+1 by UI, UG, or QE will still be a consequence of its assumptions if prior sentences are.

(i) Suppose a sentence $\psi\alpha/\delta$ is entered on line k+1 by the UI rule. In this case the sentence $\forall\alpha\psi$ must have appeared on an earlier line. By our inductive hypothesis, we know that $\forall\alpha\psi$ is a consequence of <u>its</u> assumptions, Γ say. Then, by the UI rule, the assumption numbers for the sentences in Γ will be written to the right of $\psi\alpha/\delta$. What we need to prove is that $\psi\alpha/\delta$ is a consequence of Γ, given that $\forall\alpha\psi$ is a consequence of Γ. In symbols, we need to prove: if $\Gamma \models \forall\alpha\psi$ then $\Gamma \models \psi\alpha/\delta$ for any name δ. But, this is just THEOREM 7.2. So, this case is finished.

(ii) Suppose a sentence $\forall\alpha\psi$ is entered on line k+1 by the UG rule. Then $\psi\alpha/\beta$ must have appeared on an earlier line. By our Inductive Hypothesis we know that $\psi\alpha/\beta$ is a consequence of its assumptions, Γ. Then we must prove the following: if $\Gamma \models \psi\alpha/\beta$ then $\Gamma \models \forall\alpha\psi$, if β does not occur in any assumption in Γ and β does not occur in ψ. But, this is just THEOREM 7.3. So, this case is finished.

(iii) Suppose a sentence ψ is entered on line k+1 by the QE rule. Then ψ is a sentence of one of the four forms below:

$$\neg\forall\alpha\varphi \qquad \exists\alpha\neg\varphi$$
$$\neg\exists\alpha\varphi \qquad \forall\alpha\neg\varphi$$

Whichever of the four forms above ψ is, we proved (as exercises) that whenever the sentence on one side of one of the above pairs is a consequence of a set of sentences, the sentence on the other side is also a consequence of that set of sentences.

There is no other way a sentence can be entered on line k+1 of a derivation. Hence, any sentence on a line of a derivation is a consequence of the assumptions of that line. This does not quite give us <u>exactly</u> what we wanted in the statement of the metatheorem. What we have proven is that for a finite set of assumptions, Γ', whose assumption numbers appear on the right of a sentence φ on a line of a derivation, $\Gamma' \models \varphi$. We are not quite finished because Γ' is finite, and we wish to show for a possibly infinite set Γ, if $\Gamma \vdash \varphi$, then $\Gamma \models \varphi$.

Consider any possibly infinite set Γ such that $\Gamma \vdash \varphi$. That is, φ is deriv<u>able</u> from Γ. Thus, there is some finite subset of Γ, Γ' say, such that <u>there is a derivation</u> of φ from the sentences belonging to Γ' (by definition of "derivability"). Then our above proof guarantees that $\Gamma' \models \varphi$ (i.e., that φ is a consequence of Γ'). Then, by THEOREM 3.2, φ is a consequence of Γ. That is, $\Gamma \models \varphi$. <u>Now</u> we are finished. ∎

Consistency

A set of sentences, Γ, is consistent, you will recall, if and only if for any φ <u>either</u> $\Gamma \nvdash \varphi$ or $\Gamma \nvdash \neg\varphi$. That is, it is not possible to derive both φ and $\neg\varphi$ from Γ, for any sentence φ. Our <u>rules</u> are consistent, or first-order predicate logic (as we have defined it) is consistent if and only if we do not have, for any sentence φ, that both φ and $\neg\varphi$ are theorems. There are several equivalent ways to define consistency. One of them is: our rules are consistent iff there is at least one sentence that is not a theorem.

The consistency of our rules follows straightforwardly from soundness. From (Weak) Soundness, we have

$$\text{if } \vdash \varphi, \text{ then } \models \varphi.$$

Suppose our rules were <u>in</u>consistent. Then, for some sentence φ, we would have $\vdash \varphi$ and $\vdash \neg\varphi$. By soundness, we would get these two results:

$$\text{if } \vdash \varphi, \text{ then } \models \varphi, \text{ and}$$
$$\text{if } \vdash \neg\varphi, \text{ then } \models \neg\varphi.$$

Then, by <u>in</u>consistency, we would get this:

$$\models \varphi \text{ and } \models \neg\varphi \text{ for some sentence } \varphi.$$

But, then φ would be true for every naming interpretation and so would $\neg\varphi$. But, we know that for <u>any</u> naming interpretation that φ is true for it iff $\neg\varphi$ is false for it. So, this could never happen. ∎

Strong Completeness

Before discussing Strong Completeness, we prove the major result that makes Strong Completeness hold. But before that, we'll need more logical names in our first-order language, and we'll need to explain what "witnesses" are. After explaining how to get more logical names into the language and describing "witnesses", we prove Lindenbaum's Theorem, from which Strong Completeness readily follows.

It should seem as though having infinitely many logical names in L_{PL} would be enough. But still, we need more.

Recall that these are the logical names of L_{PL}:

$$a, b, c, d, a_1, b_1, c_1, d_1, a_2, \ldots$$

Adding new names to L_{PL} is technically called "extending the language", about which we say much more in *First-Order Theories* and in *Model Theory For First-Order Theories*. For now, all we do is add the following logical names to the ones we already have. The new ones are:

$$k, k_1, k_2, k_3, \ldots$$

We call the new language that has all the same symbols as L_{PL} plus the new names: $L_{PL}*$. The <u>reason</u> for creating $L_{PL}*$ is that we are soon going to have a set Δ_0 of sentences of L_{PL} that may be infinite and thus may contain <u>all</u> of the original names. We will want to be sure that there are infinitely many names that do not occur in any sentence of Δ_0, and the new names in $L_{PL}*$ will fill the bill.

The other thing that needs to be explained is "witnesses". A **witness** for an existential sentence $\exists\alpha\psi$ is an instance $\psi\alpha/\delta$. We say a set of sentences Γ **has witnesses** if for every existential sentence in Γ, there is a witness for that existential that is also in Γ. For example, the set $\{\exists xFx, Fa, \exists xGxb, Gcb\}$ has witnesses but the set $\{\exists xFx, Fa, \exists xGxb, Hcb\}$ is lacking a witness.

We are now ready to state Lindenbaum's Theorem:

LINDENBAUM'S THEOREM: For any consistent set of sentences Δ_0 of L_{PL}, there is a set of sentences Δ of L_{PL*} such that:
  (1) $\Delta_0 \subseteq \Delta$,
  (2) Δ is consistent,
  (3) for any sentence ψ of L_{PL}, $\psi \in \Delta$ iff $\neg\psi \notin \Delta$, and
  (4) Δ has witnesses.[4]

PF: The construction of Δ is almost the same as in the sentential proof. One difference is that here we list <u>all</u> the sentences of L_{PL*} (whereas in the sentential case we listed <u>only</u> the atomic sentences). The second difference is that we make the present list of sentences adhere to the following condition: whenever an existential sentence $\exists\alpha\psi$ appears on the list, the very next sentence on the list is $\psi\alpha/\delta$, where δ is a name that does not occur earlier in the list (this is where our new names are necessary).

Here is our list of sentences of L_{PL*}:

$$\theta_0, \theta_1, \theta_2, \theta_3, \ldots$$

The construction of Δ from Δ_0 is identical to the construction of Δ in the proof of Strong Completeness for sentential logic, so we leave the details for you to provide.

Once Δ has been obtained from Δ_0, we need
TO PROVE:
  (1) $\Delta_0 \subseteq \Delta$,
  (2) Δ is consistent,
  (3) for any sentence ψ of L_{PL}, $\psi \in$ iff $\neg\psi \notin \Delta$, and
  (4) Δ has witnesses.

The proofs of (1) - (4) are left as exercises. ∎

---

[4]This clause is really due to Henkin, enabling him to construct an ingenious interpretation, which we will soon see.

Exercises:

1. If you have not already done so, write out the construction of Δ from Δ_0 (look back at the same construction in the proof of Strong Completeness for sentential logic). Then prove (1) - (4) above. [Hints: (1) is clear by the construction of Δ; (2) is virtually identical to the proof of the same thing for sentential logic; for (3) look back at the similar proof for sentential logic, but notice that for the sentential proof we cared only about derivability. Here we want <u>exactly one</u> of ψ and $\neg\psi$ to actually <u>belong</u> to Δ. The argument <u>against both</u> ψ and $\neg\psi$ belonging to Δ is that this leads to the conclusion that for some Δ_k, both belong to Δ_k. The argument <u>against neither</u> one belonging to Δ leads by THEOREM 4.12 to the conclusion that for some Δ_k, $\Delta_k \vdash \psi$ and $\Delta_k \vdash \neg\psi$; (4) is fairly clear by our manner of listing sentences and by the method of constructing Δ.]

Lindenbaum's Theorem is the most important element needed in the proof of Strong Completeness. We state Strong Completeness (for any set of sentences Γ and any sentence φ of L_{PL}) this way:

STRONG COMPLETENESS THEOREM: If $\Gamma \models \varphi$, then $\Gamma \vdash \varphi$,

but the version we actually prove is this:

STRONG COMPLETENESS (CONTRAPOSITIVE): If $\Gamma \nvdash \varphi$, then $\Gamma \nvDash \varphi$.

That is, if φ is <u>not</u> derivable from Γ, then φ is <u>not</u> a logical consequence of Γ.

As in the sentential proof of Strong Completeness, we assume $\Gamma \nvdash \varphi$ and we use THEOREM 4.12, which is:

$\Gamma \cup \{\neg\varphi\}$ is consistent   iff $\Gamma \nvdash \varphi$,

to give us that $\Gamma \cup \{\neg\varphi\}$ is consistent. We then define $\Delta_0 = \Gamma \cup \{\neg\varphi\}$ and then use LINDENBAUM'S THEOREM to give us the set Δ. Armed with LINDENBAUM'S THEOREM, we now define interpretation $\langle \mathcal{I}_\delta, n_\delta \rangle$. Here's what we do. We take our domain of objects in $\mathcal{I}_\delta$ to be <u>all the logical names of</u> L_{PL^*} <u>themselves</u>. That is:

$$\mathcal{D} = \{a, b, c, d, k, a_1, b_1, c_1, d_1, k_1, a_2, \ldots\}.$$

And we let every name stand for ... <u>itself</u>. That is:

$$n_\delta(a) = a, \; n_\delta(d_1) = d_1, \; n_\delta(k_{33}) = k_{33}, \ldots.$$

This turns out to be very convenient and is, quite ingenious as well (due to Leon Henkin [1949] as is, essentially, the entire proof).

We now need to say how $\mathcal{I}_\delta$ interprets each k-ary relation letter R_j^k. For any k-ary relation letter R_j^k in L_{PL}, where k > 0:

$$\mathcal{I}_\delta(R_j^k) = \{\langle\delta_1,\delta_2,\ldots,\delta_k\rangle: \text{ the sentence } R_j^k\delta_1,\delta_2,\ldots,\delta_k \in \Delta\}.$$

That is, $\mathcal{I}_\delta$ assigns to any relation letter R_j^k the set of ordered k-tuples of logical names $\langle\delta_1,\delta_2,\ldots,\delta_k\rangle$ such that the sentence beginning with R_j^k followed by those names, $\delta_1, \delta_2,\ldots,\delta_k$, belongs to the set of sentences Δ. For any sentence letter R_j^0, $\mathcal{I}(R_j^0) = T$ iff $R_j^0 \in \Delta$.

Now, the properties of Δ (from LINDENBAUM'S THEOREM) in conjunction with the clever way of defining $\langle\mathcal{I}_\delta,n_\delta\rangle$ (due to Henkin) make it the case that for any sentence χ of L_{PL^*}, $\langle\mathcal{I}_\delta,n_\delta\rangle(\chi) = T$ iff $\chi \in \Delta$. Let's call this the Lindenbaum-Henkin Theorem:

LINDENBAUM-HENKIN THEOREM: If Δ is constructed à la Lindenbaum, and $\langle\mathcal{I}_\delta,n_\delta\rangle$ is constructed à la Henkin, then for any sentence χ of L_{PL^*}:

$$\langle\mathcal{I}_\delta,n_\delta\rangle(\chi) = T \quad \text{iff} \quad \chi \in \Delta.$$

PF: This theorem is proved by Complete Induction on the height of the sentence χ. The atomic case is handled by the definition of $\langle\mathcal{I},n\rangle$. The most important molecular case is when χ is a negation (i.e., when χ is of the form $\neg\psi$). This case is handled by clause (3) of LINDENBAUM'S THEOREM. The rest of the molecular cases are straighforward once the case when χ is a negation has been proved. The two quantifier cases depend on the fact that Δ has witnesses, as well as other facts from LINDENBAUM'S THEOREM. We leave the details as exercises. ■

Exercises:

1. Provide the details of the proof of the LINDENBAUM-HENKIN THEOREM sketched above.

So, what's left to prove for Strong Completeness to hold? Not much, for our naming interpretation $\langle \mathcal{I}_\delta, n_\delta \rangle$ interprets all sentences in Δ as **True**. Since $\Delta_0 \subseteq \Delta$, all sentences in Δ_0 are **True** for $\langle \mathcal{I}_\delta, n_\delta \rangle$ as well. But, $\Delta_0 = \Gamma \cup \{\neg\varphi\}$. So, $\langle \mathcal{I}_\delta, n_\delta \rangle(\Gamma) = \mathbf{T}$, and $\langle \mathcal{I}_\delta, n_\delta \rangle(\neg\varphi) = \mathbf{T}$. Thus, $\langle \mathcal{I}_\delta, n_\delta \rangle(\varphi) = \mathbf{F}$. But, this establishes that $\Gamma \not\models \varphi$ holds (since we have exhibited an interpretation for which Γ is **True** and φ is **False**). This is what we have been striving for ever since we assumed $\Gamma \not\vdash \varphi$. Hence (the contrapositive of) Strong Completeness holds, namely: if $\Gamma \models \varphi$, then $\Gamma \vdash \varphi$. (The only tiny fly in the ointment is that $\langle \mathcal{I}_\delta, n_\delta \rangle$ is defined not for L_{PL}, but for L_{PL*}, the first-order language with the extra logical names. We rectify this easily by restricting n_δ to only the original logical names in L_{PL}. (Notice that this does not eliminate any of the new names as elements of the <u>domain</u> of the interpretation. It just eliminates them from the first-order language, that is, from being names of themselves.) The resulting naming interpretation interprets all sentences of L_{PL} the same (technically, by COROLLARY 7.2a), and thus the theorem holds. ∎

Exercises:

1. Summarize the argument for Strong Completeness, beginning with $\Gamma \not\vdash \varphi$ and arriving at $\Gamma \not\models \varphi$. Mention the result that justifies each step in the reasoning, but don't stop to prove that result.
2. Prove each result mentioned in (1) that you have not already proved in earlier exercises.
3. Suppose instead of <u>adding</u> infinitely many new logical names to L_{PL}, we decided to eliminate infinitely many of them, say every second one from L_{PL}. Our original list of names is:

    a, b, c, d, a_1, b_1, c_1, d_1, a_2, ... ,

    And, our shortened list would be, then:

    a, c, a_1, c_1, a_2, ...

    Now, let's call the language using names from the shortened list: L_{PL-}. Suppose there is a mapping m from the set of all names to the smaller set such that the first element of the big set is mapped to the first element of the small set, the second element of the big set is mapped to the second element of the small set, and so on. Now, for each sentence ψ of L_{PL} there is a counterpart ψ' of L_{PL-}.

(a) Prove: if $\Gamma \not\vdash \varphi$, then $\Gamma' \not\vdash \varphi'$.
(b) Using LINDENBAUM'S THEOREM to arrive at Δ (in L_{PL}), define a naming interpretation $\langle \mathcal{I}_\delta, n_\delta \rangle$ for L_{PL} such that every sentence of Δ is **True**.
(c) Explain why the set of sentences $\Gamma' \cup \{\neg\varphi'\}$ is **True** for $\langle \mathcal{I}_\delta, n_\delta \rangle$.
(d) Modify the interpretation $\langle \mathcal{I}_\delta, n_\delta \rangle$ for the language L_{PL} appropriately to arrive at $\langle \mathcal{I}_{\delta'}, n_{\delta'} \rangle$ for L_{PL-}, and now prove that the set of sentences $\Gamma \cup \{\neg\varphi\}$ is **True** for $\langle \mathcal{I}_{\delta'}, n_{\delta'} \rangle$.
(e) Explain why (a) – (c) constitute an alternative proof of Strong Completeness.

Compactness

Suppose our system is strongly complete. Then if $\Gamma \models \varphi$, $\Gamma \vdash \varphi$. Let Γ be infinite. Since φ is a consequence of Γ, φ is derivable from Γ. Let us imagine what a derivation of φ from Γ would look like. Do you think that the last line of the derivation could have infinitely many numbers to the right of it? Let us see why not. Suppose the derivation was one billion lines long. And, let us suppose that half of the lines have sentences entered by the A rule. That means that we could only have one-half billion numbers to the right of φ. We could <u>never</u> have infinitely many numbers to the right because every derivation has only finitely many lines, and only <u>some</u> of them are assumptions.

Hence, if φ is a consequence of an infinite set of sentences, Γ, φ can be <u>derived</u> from Γ', where Γ' is some finite subset of Γ. But if whenever $\Gamma \models \varphi$ there is a <u>finite</u> set, Γ', such that $\Gamma' \subseteq \Gamma$, and $\Gamma' \vdash \varphi$, then by Strong Soundness φ is a consequence of this <u>finite</u> set Γ'. Let us restate this last fact: if φ is derivable from a <u>finite</u> set Γ', then $\Gamma' \models \varphi$. Strong Soundness tells us this is so.

The above chain of reasoning gives us the following facts:

(1) if $\Gamma \models \varphi$, then $\Gamma \vdash \varphi$
(2) if $\Gamma \vdash \varphi$, then there is a <u>finite</u> Γ' such that $\Gamma' \subseteq \Gamma$ and $\Gamma' \vdash \varphi$

(3) if $\Gamma' \vdash \varphi$ then $\Gamma' \models \varphi$
(4) if $\Gamma \models \varphi$ then $\Gamma' \models \varphi$

Line (4) IS COMPACTNESS! One paraphrase of it is that from an infinite set of sentences you cannot get any more consequences than you can get from some finite subset of them. Loosely speaking, then, an infinite set of sentences cannot "carry any more information" than some finite subset of them. This seems to say that in first-order logic, an infinitude of information tells us nothing more than a finite amount does.

Another description of Compactness is: if every finite subset of a set of sentences has a model, then the entire infinite set has a model.

<u>Compactness</u> 1): If $\Gamma \models \varphi$, then for some finite $\Gamma' \subseteq \Gamma$, $\Gamma' \models \varphi$.

<u>Compactness</u> 2): If every finite subset of Γ has a model, then Γ itself has one.

Exercises:

1. Prove (1) => (2).
2. Prove (2) => (1).

Notice that even though Compactness is a purely semantic concept, the proof we presented used Strong Completeness and Strong Soundness, which are both "mixed concepts", consisting of syntax and semantics. This raises the question whether Compactness can be proved by purely semantic means alone. The answer is yes, it can, and in fact the proof of it is quite beautiful. Unfortunately, we cannot do full justice to it within the scope of the present textbook, but we do outline it in Chapter 10, *Models for First-Order Theories*, and hopefully give some idea of its beauty.

CHAPTER 8

FIRST-ORDER THEORIES

TOPICS: *the first-order theory of equality EQ, axioms, theses, initial sentences, mathematical completeness, the theory of relations $\mathcal{R}$, operation symbols, terms, modified UG rule, group theory $\mathcal{G}$, theory of addition $\mathcal{A}$, theories of arithmetic: $\mathcal{P}, \mathcal{K}, \mathcal{Q}$, mechanical procedure, decidable set, axiomatizable theory, extension and conservative extension of a theory.*

8.1 INTRODUCTION TO FIRST-ORDER THEORIES

In ordinary conversation a "theory" can include many different kinds of things. Here, we concentrate on so-called "formal theories". A formal theory is so-called because everything about it is based on considerations of form, namely <u>syntactic considerations</u>. Its alphabet consists of non-logical symbols specified in advance, together with purely logical symbols. Then, its sentences are defined to be certain of these symbol strings. When sentences of a formal theory have been singled out to be "axioms", those axioms must be identifiable in terms of their formal symbol patterns. Then, from the axioms, other sentences can be derived, based on rules of derivation. Each rule of derivation specifies a syntactic relation that must hold between the earlier sentences of the derivation and a later sentence. Thus, every application of the rule can be checked for correctness. When sentences, axioms, and derivational rules have all been specified formally — that is, in terms of patterns of recognizable symbols — the resultant theory is considered to be "formal".

Formal theories can be traced back to Euclid's system of geometry, around 300 B.C.. His system was not completely formal in the present-day sense, since some of his proofs depended on associating the correct "meanings" to some of the terms. (There were other reasons as well.) For example, the statement in English 'any two distinct points determine a unique line' contains the terms 'point' and 'line', and requires us to understand that "two points determine a line" in this sense means that those points are both <u>on</u> the line in question. We could "formalize" that geometric statement as this sentence of first-order logic:

$$\forall x \forall y [(F^1x \land F^1y \land x \ne y) \longrightarrow$$
$$\exists z(G^1z \land H^3xyz \land \forall w(G^1w \land H^3xyw \longrightarrow w = z))],$$

supposing that "a is a point", "a is a line", and "a and b are on c" are represented by the non-logical symbols 'F^1', 'G^1', and 'H^3'. But, sentences like the above in a formal system can have other meanings as well. For example, suppose we interpret the above sentence to be "about" the positive integers, and we make 'F^1' mean "a is a prime number", 'G^1' mean "a is a composite number", and 'H^3' mean "c is a product of a and b". Then the sentence above says: "Every two distinct primes have a unique composite number that is their product". Thus, the sentence could be singled out as being "about" geometry or numbers. And then, any theorem derived from that sentence could also be interpreted as being "about" geometry or numbers, or any other realm for which the original sentence can be interpreted as true. The theorems of a formal theory, then, become the sentences that are derivable from the sentences singled out as axioms, regardless of how the non-logical symbols in the theory are interpreted.

An example of a formal theory expressed in the language of first-order logic — and, thus a "first-order theory" — is the Theory of Quasi-Equivalence. This theory has one non-logical symbol 'F_0^2', and here are its two (non-logical) axioms:

1) $\forall x \forall y (F_0^2 xy \longrightarrow F_0^2 yx)$
2) $\forall x \forall y \forall z (F_0^2 xy \land F_0^2 yz \longrightarrow F_0^2 xz)$

Note that any interpretation of the symbol 'F_0^2' for which both 1) and 2) are true must make the relation corresponding to 'F_0^2' symmetric and transitive. We can derive "quasi-reflexivity" from 1) and 2) as follows:

QUASI-REFLEXIVITY THEOREM: $\forall x (\exists y F_0^2 xy \longrightarrow F_0^2 xx)$

√SHOW $\forall x (\exists y F_0^2 xy \longrightarrow F_0^2 xx)$

                                √SHOW $\exists y F_0^2 ay \longrightarrow F_0^2 aa$
1. $\exists y F_0^2 ay$ 1,A
                                √SHOW $F_0^2 aa$
2. $F_0^2 ab$ 2,EA ('b')
3. $\forall x \forall y (F_0^2 xy \longrightarrow F_0^2 yx)$ Axiom 1
4. $\forall y (F_0^2 ay \longrightarrow F_0^2 ya)$ UI (3)
5. $F_0^2 ab \longrightarrow F_0^2 ba$ UI (4)

6. F_0^2ba 2, MP (2,5)
7. $\forall x \forall y \forall z (F_0^2 xy \land F_0^2 yz \longrightarrow F_0^2 xz)$ Axiom 2
8. $\forall y \forall z (F_0^2 ay \land F_0^2 yz \longrightarrow F_0^2 az)$ UI (7)
9. $\forall z (F_0^2 ab \land F_0^2 bz \longrightarrow F_0^2 az)$ UI (8)
10. $F_0^2 ab \land F_0^2 ba \longrightarrow F_0^2 aa$ UI (9)
11. $F_0^2 ab \land F_0^2 ba$ 2, Conj (2,6)
12. $F_0^2 aa$ 2, MP (2,11)
13. $F_0^2 aa$ 1, EI
14. $\exists y F_0^2 ay \longrightarrow F_0^2 aa$ CD (1,13)
15. $\forall x (\exists y F_0^2 xy \longrightarrow F_0^2 xx)$ UG (14)

Quasi-reflexivity "says": any element that bears **F** (the binary relation associated with F_0^2) to some, element bears **F** to itself. The above derivation establishes that for any first-order theory in which 1) and 2) are axioms the QUASI-REFLEXITY THEOREM holds.

Exercises:

Consider the Theory of Quasi-Equivalence to include the axioms 1) and 2) above. Then, add strong connectivity, by adding this sentence "saying" that for any two elements, one must bear **F** to the other:

$$3)\ \forall x \forall y (Fxy \lor Fyx)$$

1. Derive (genuine) reflexivity from 3). That is, derive:

$$\forall x Fxx$$

from '$\forall x \forall y (Fxy \lor Fyx)$' as the only assumption.

2. Derive (weak) connectivity from strong connectivity. That is, derive:

$$\forall x \forall y (x \neq y \longrightarrow Fxy \lor Fyx)$$

from 3).

The results of **1.** and **2.** lead to the question whether (weak) connectivity together with axioms 1) and 2) yield (genuine) reflexivity. To ensure a "yes" answer to this question, we add a sentence that says there are at least two elements in our domain of discourse. But, unfortunately, in order to continue with our reasoning, we must <u>assume</u> we also know all the familiar properties of "=", for example, that "substituting equals for equals" is permissible. Now, <u>provided</u> we can use any principle of "=" in a derivation, it

is possible to obtain reflexivity. So, for this exercise alone, any equality principle may be used on a line of a derivation with the citation 'equality' written to the right of that line. (In Section 8.2, The Theory of Equality, the desired equality principles are made fully explicit.) So:
3. Given:
   axioms 1), 2), and
   $\forall x \forall y (x \neq y \longrightarrow Fxy \vee Fyx)$, and
   $\exists x \exists y (x \neq y)$,
derive (with the help of any principles of equality):

$$\forall x Fxx.$$

Now that we have before us an example of a first-order theory (the Theory of Quasi-Equivalence), we now explain what a first-order theory is from a more general perspective.

The **language of a first-order theory** L_T is exactly like L_{PL}, the language of predicate logic, <u>except</u> for the following differences:

(1) L_T contains the logical relation symbol '=' for equality.
(2) L_T contains a special subset of the non-logical relation letters of L_{PL}.
(3) L_T may contain some non-logical operation letters (which L_{PL} doesn't have at all), and
(4) L_T contains no logical names.

According to (1), L_T contains a logical symbol not contained in L_{PL}. The new symbol, '=', is intended to represent the notion of "equality" in any first-order theory. Its formal semantics will be presented later.

For each first-order theory $\mathcal{T}$, there is a special subset of relation letters and operation letters that constitute the subject matter of that particular theory, according to (2) and (3). For example, in the Theory of Quasi-Equivalence presented above, the single non-logical relation letter was 'F_0^2'. That theory contained no operation letters. Operation letters, which do not occur in L_{PL} at all, will be explained when we encounter them in a first-order theory.

According to (4), the language of a theory, L_T, does not contain any of the infinitely many logical names of L_{PL}. But, we do permit logical names to occur in <u>derivations</u> of a theory. For example, the sentence '$\forall x (\exists y F_0^2 xy \longrightarrow F_0^2 xx)$', which we derived as the QUASI-REFLEXIVITY THEOREM, contains no logical names. But, some of the sentences in the derivation of that theorem do contain logical names. For any theory $\mathcal{T}$, the derivations are carried out in the language L_{T^+}, which includes all the symbols of L_T plus all the logical names of L_{PL}. Recall that the logical names are:

$$a, b, c, d, a_1, b_1, c_1, d_1, a_2, \ldots$$

8.2 THE THEORY OF EQUALITY $\mathcal{EQ}$

Let us now look at a theory with the simplest possible language: one with no non-logical symbols whatsoever. The theory we look at is $\mathcal{EQ}$, **The Theory of Equality**. The only relation symbol in L_{EQ} is the logical symbol '='. Its formulas, thus, are very restricted.

Now that L_{EQ} has been specified we need only specify the "axioms" of $\mathcal{EQ}$. **Axioms** for a given theory $\mathcal{T}$ are simply sentences of L_T that have been singled out for special recognition. We will say more about axioms later. The first axiom of $\mathcal{EQ}$ is (brackets are added to improve readability):

$\mathcal{EQ}$-A1: $\quad \forall x[x = x]$

$\mathcal{EQ}$-A1 says that everything is equal to itself, which is a fact about equality. Another fact about equality we wish to express is that whenever x and y are equal, whatever is true of x is true of y. But, we cannot express the idea of "whatever is true of x" with a single first-order sentence. The phrase "whatever is true of x" implicitly refers to all formulas that have only 'x' free.

We have previously written 'φx' to signify any formula having only 'x' free. At this point, though, we rewrite 'φx' as 'φxx' to emphasize that 'x' may occur several times in φ (though it is possible 'x' occurs only once). Then, we write 'φxy' to signify the same formula (whatever it is) with 'y' substituted freely for some occurrences of 'x'. Since 'φxx' and 'φxy' are expressions of the metalanguage — expressions we use to talk about the object language — they are the forms of formulas rather than actual formulas. The form 'φxx' signifies any formula having only 'x' free and 'φxy' signifies 'φxx' with some occurrences of 'x' freely replaced by 'y'. Thus, '$\varphi xx \rightarrow \varphi xy$' expresses the idea, using forms, "whatever is true of x is also true of y".

So, in order to capture the idea that if x equals y, then whatever is true of x is also true of y, we write the following axiom form:

$\mathcal{E}Q$-A2: $\forall x \forall y(x=y \rightarrow (\varphi xx \rightarrow \varphi xy))$

(where φxy is just like φxx except having 'y' substituted freely in some places where 'x' was free.)

To repeat, axiom form $\mathcal{E}Q$-A2 conveys the idea that whenever two things are equal (x and y), whatever is true of one is true of the other.

Theorems of a given theory $\mathcal{T}$ are those sentences of L_T derivable from the axioms of $\mathcal{T}$. We will now derive some theorems of $\mathcal{E}Q$. But, first, let us ask ourselves what we desire of a "respectable" theory of equality. What theorems should we expect of $\mathcal{E}Q$? One theorem should be the law of symmetry. We know that the equality relation is symmetric: whenever x = y, then y = x. So, if $\mathcal{E}Q$ is a "respectable" theory of equality, then symmetry should be derivable in $\mathcal{E}Q$:

$\mathcal{E}Q$-T1 $\forall x \forall y(x=y \rightarrow y=x)$

		SHOW $\forall x \forall y(x=y \rightarrow y=x)$
		SHOW $a=b \rightarrow b=a$
1.	a=b	1,A
		√SHOW b=a
2.	$\forall x \forall y(x=y \rightarrow (x=x \rightarrow y=x))$	A2
3.	$a=b \rightarrow (a=a \rightarrow b=a)$	UI (twice from 2)
4.	a=a	A1,UI
5.	b=a	1,SL (1,3,MP,4,MP)

We should also expect the transitivity of equality to be derivable in $\mathcal{E}Q$ if $\mathcal{E}Q$ is to be accepted as "respectable":

$\mathcal{E}Q$-T2 $\forall x \forall y(x=y \land y=z \rightarrow x=z)$

		SHOW $\forall x \forall y \forall z(x=y \land y=z \rightarrow x=z)$
		SHOW $(a=b \land b=c) \rightarrow a=c$
1.	$a=b \land b=c$	1,A
		√SHOW a=c
2.	$\forall x \forall y(x=y \rightarrow (x=z \rightarrow y=z))$	$\mathcal{E}Q$-Ax2
3.	$b=a \rightarrow (b=c \rightarrow a=c)$	UI(2)
4.	a=b	SL(1)
5.	$\forall x \forall y(x=y \rightarrow y=x)$	$\mathcal{E}Q$-T1
6.	$a=b \rightarrow b=a$	UI twice(5)
7.	b=a	SL(4,6)

8.	b=c → a=c	SL(3,7)
9.	b=c	SL(1)
10.	a=c	SL

Principle $\mathcal{E}Q$-A2 is sometimes called Leibniz's Law.[1] It expresses Leibniz's principle of "the indiscernibility of identicals", according to which if two things are equal, there is no way to "discern" one from the other. (Sometimes Leibniz's Law is taken to be the converse principle ("the identity of indiscernibles"): if two things are indiscernible, they are identical. And sometimes Leibniz's Law is thought to go in both directions: two things are identical if and only if they are indiscernible.)

We will often write **LL** (to abbreviate Leibniz's Law, in the sense of the indiscernibility of identicals) for $\mathcal{E}Q$-A2, and also write **Id** (as the Law of Identity) for $\mathcal{E}Q$-A1. Similarly, we write **Trans** for $\mathcal{E}Q$-T2 and **Symm** for $\mathcal{E}Q$-T1.

Id is a single sentence of L_{EQ}, but LL includes infinitely many sentences. The set of **initial sentences** Δ_{EQ} **of theory** $\mathcal{E}Q$ includes the single sentence Id and the sentences LL. We say that a sentence φ of L_{EQ} is a **thesis** (or a theorem) of the Theory of Equality, $\mathcal{E}Q$, iff $\Delta_{EQ} \vdash \varphi$. We usually write $\mathcal{E}Q \vdash \varphi$ for $\Delta_{EQ} \vdash \varphi$. Since any sentence φ of L_{EQ} is "recognizable" as Id or as being one of the sentences LL, the initial sentences are also called **axioms** of $\mathcal{E}Q$. ("Recognizable" will be explained more thoroughly later in this chapter.) The theory whose axioms are Id and LL is: **The (first-order) Theory of Equality**. (An axiom is trivially a thesis or theorem of a theory, since it can appear in a one-line derivation from the set of axioms of that theory.)

Of course, the above definitions apply more generally to any theory $\mathcal{T}$: (1) Δ_T is the set of **initial sentences** of $\mathcal{T}$, and (2) φ is a **thesis** of $\mathcal{T}$ iff $\Delta_T \vdash \varphi$, usually written $\mathcal{T} \vdash \varphi$. If φ is a thesis of $\mathcal{T}$, it is sometimes said that φ **belongs to** $\mathcal{T}$.

In deference to actual mathematical usage, we usually drop initial universal quantifiers from axioms and theorems. For example, instead of writing $\mathcal{E}Q$-T2 as:

$$\forall x \forall y \forall z ((x=y \land y=z) \rightarrow x=z),$$

---

[1] Named after the mathematician and philosopher, Baron Gottfried Wilhelm von Leibniz (1646-1716), who is probably best known for co-inventing the infinitesimal calculus independently of Isaac Newton.

we will sometimes write $\mathcal{EQ}$-T2 as:

$$(x=y \land y=z) \rightarrow x=z.$$

So, if a given thesis of theory $\mathcal{T}$ is of the form $\forall\alpha_1\forall\alpha_2\ldots\forall\alpha_n\varphi$, where $\alpha_1, \alpha_2, \ldots, \alpha_n$ are the free variables of φ in alphabetic order, we will often just write φ as an abbreviation of that sentence.

The Theory of Equality, $\mathcal{EQ}$, is a special case of **a first-order theory with** equality. **A first-order theory with equality** is any first-order theory that includes Id and all LL sentences among its theses. Note that $\mathcal{EQ}$ has the simplest language possible, one with <u>no</u> non-logical symbols. Other theories will normally have <u>several</u> non-logical symbols and thus many additional LL sentences. For any theory to qualify as a first-order theory with equality, it must be possible to derive the Law of Identity, $\mathcal{EQ}$-A1, and all sentences of the form of Leibniz's Law, $\mathcal{EQ}$-A2, for the language of that particular theory. From now on, we will take a **first-order theory** to be a first-order theory with equality.

Exercise:

In *exercise* 3. of the previous set of exercises, we had to leave equality principles on an intuitive level in obtaining the reflexivity of **F**. Now that those principles have been made fully explicit, we can redo that exercise:

1. Let the only non-logical symbol of the Theory of Quasi-equivalence be 'F', as presented above. Recall that its two axioms are:
    1) $\forall x\forall y(Fxy \longrightarrow Fyx)$
    2) $\forall x\forall y\forall z(Fxy \land Fyz \longrightarrow Fxz)$

   We now suppose that this theory is a theory with equality, and thus these principles hold as well:
    3) $\forall x[x = x]$
    4) $\forall x\forall y(x=y \rightarrow (\varphi xx \rightarrow \varphi xy))$

   We now add "connectivity":
    5) $\forall x\forall y(x \neq y \longrightarrow Fxy \lor Fyx)$

   and the principle that there are at least two elements:
    6) $\exists x\exists y(x \neq y)$.

   Now, using only 1) - 6), derive:

$$\forall x Fxx.$$

Given <u>any</u> first-order theory **with** equality that has at least two non-logical relation symbols 'F¹' and 'G²', the following theorems can be obtained (We begin the numbering with T3, where Symm and Trans are T1 and T2, respectively):

T3 $\forall x \forall y (x=y \rightarrow (Gxy \rightarrow Gxx))$
       [Use $\mathcal{EQ}$-A2. Let φxx be 'Gxx $\rightarrow$ Gxx', and let φxy be 'Gxy $\rightarrow$ Gxx'.]
T4 $\forall x \forall y (x=y \rightarrow (Gxy \rightarrow Gyx))$
T5 $\forall x \forall y (x=y \rightarrow (Gxx \leftrightarrow Gxy))$
T6 $\forall x (Fx \leftrightarrow \forall y (y = x \rightarrow Fy))$
T7 $\forall x (Fx \leftrightarrow \exists y (y = x \rightarrow Fy))$
T8 $\forall x \exists y (y = x)$
T9 $\forall x \exists y (y = x \land \forall z (z = x \rightarrow z = y))$
T10 $\forall x \exists y \forall z (z = x \leftrightarrow z = y)$

Exercises:

   Derive some of the above theorems, especially: one of T3 and T4; one of T6 and T7; and T9.

      The symbol '=' for "equality", as mentioned earlier, is treated as a logical symbol like '$\rightarrow$' and '$\neg$'. In the chapter *Models For First-Order Theories*, we explain the semantics of '=' by supplying a truth clause for it, similar to the truth clauses provided for the other logical symbols (in Chapter 6, *Semantics For First-Order Logic*). The intuitive idea is that '=' stands for "real equality" or "identity". That is, for any naming interpretation and names β and δ, the sentence '$\beta = \delta$' = **T** if and only if β and δ denote the identical object in that naming interpretation. (The reason '=' is defined as a logical symbol rather than letting it be any mathematical relation satisfying the equality axioms is that no first-order axioms could then constrain '=' to be the relation of identity that we intend. The best our axioms could do — without the intervention of a truth definition — would be to coerce '=' to behave like an equivalence relation. So, for example, given that the elements 0, 1, and 2 satisfy our equality axioms, nothing would prevent '=' from being interpreted as, say: {<0,0>,<0,1>,<1,0>,<1,1>,<2,2>}. The only way to ensure that '=' is interpreted as {<0,0>,<1,1>,<2,2>}, which is the identity relation for those three elements, is to make '=' a logical symbol and to <u>define it as identity</u>.)[2]

8.3 MATHEMATICAL COMPLETENESS

A theory $\mathcal{T}$ is **mathematically complete** if for any sentence φ of L_T, either $\mathcal{T} \vdash \varphi$ or $\mathcal{T} \vdash \neg\varphi$. Notice the difference between <u>mathematical completeness</u> and <u>logical completeness</u>. Our derivation rules are <u>logically</u> complete for L_{PL}, as we proved in Chapter 7, since every logical consequence of any set of sentences is derivable from that set. In symbols, Strong (Logical) Completeness is the following: if $\Gamma \models \varphi$, then $\Gamma \vdash \varphi$.

Strong Soundness for our rules of derivation is the converse of Strong Completeness. According to Strong Soundness: if $\Gamma \vdash \varphi$, then $\Gamma \models \varphi$. This means that every sentence derivable from a set of sentences <u>should</u> be derivable, for it is a logical consequence of that set.

The proofs of Strong (Logical) Completeness and Strong Soundness establish a correlation between the distinct concepts of "logical consequence" and "derivable" for the derivational rules we have chosen. Thus, for any theory $\mathcal{T}$ and any sentence $\varphi \in L_T$:

$$\mathcal{T} \vdash \varphi \quad \text{iff} \quad \mathcal{T} \models \varphi.$$

That is, the <u>theses</u> of any theory are identical to the logical consequences of (the initial sentences of) that theory. But, even though the above correspondence holds between the theses and the logical consequences of any theory $\mathcal{T}$, $\mathcal{T}$ may still be mathematically <u>in</u>complete. When $\mathcal{T}$ is mathematically complete, <u>every</u> sentence of the theory is either true for every interpretation that makes all the initial sentences of that theory true or false for every such interpretation. Another way to say this is that any sentence or its negation is a logical consequence of the initial sentences of that theory. Thus, since all consequences of a theory are derivable, every sentence or its negation is derivable in a mathematically complete theory. So, one way

---

[2]Some authors, for example Elliott Mendelson [1987], allow '=' to be a non-logical symbol. They then separate interpretations into two kinds, "normal" ones and "non-normal" ones. The "normal" ones are those that interpret '=' as "real identity", and the "non-normal" ones interpret '=' as some other equivalence relation.

to show that a sentence φ of a complete theory is true for every interpretation of that theory is to produce a derivation of that sentence. Hence, an important theoretical goal is to establish mathematically complete theories.

But first, before the creation of a theory, there must be a "realm of objects" of interest to mathematicians. For example, one such realm of interest is the arithmetic of natural numbers. One way to discover the truths of that mathematical realm is to establish a formal theory of arithmetic and then to investigate the properties of that theory.

We created a first-order theory of equality, $\mathcal{E}Q$, in order to capture all the truths of the "equality realm". Did we capture them all in $\mathcal{E}Q$, or did we miss some of them? Well, suppose we ask how many distinct objects there are in the "equality realm". Is there only one? Are there two? Three? Infinitely many? Notice that we don't have an intuitive answer to this question. It looks as though the realm we have in mind is not specific enough to allow us to determine how many objects there are. The theory $\mathcal{E}Q$ is <u>not</u> mathematically complete, but the "realm" itself seems incomplete. Our very conception of equality seems incomplete.

If we first arbitrarily <u>decide</u> how many objects there are in an "equality realm", we can create a mathematically complete theory for <u>that</u> realm. For example, we could create $\mathcal{E}Q_1$ as the theory of the "one-object-equality-realm" and $\mathcal{E}Q_2$ as the two-object-equality realm, and so forth for every finite n. We could also create $\mathcal{E}Q_\infty$, the theory that "says" there are infinitely many objects in the "equality realm".[3]

Exercises:

For each theory below, write one or more sentences that "say" there are that many objects in the realm under consideration. (See previous footnote.)
1. Prove $\mathcal{E}Q_1$ is complete.
2. Prove $\mathcal{E}Q_2$ is complete.
3. Prove that for any n ≥ 1, $\mathcal{E}Q_n$ is complete.

---

[3]When we write that a first-order sentence (or set of sentences) "says" something or other, we are implicitly referring to interpretations for which the given sentence is true. For example, the sentence '$\exists x \forall y (x \neq y \land \forall z (z=x \lor z=y))$' "says" there are exactly two elements, because any interpretation for which the sentence is true must have exactly two elements.

4. Prove $\mathcal{E}Q_\infty$ is complete.

8.4 IS THE REALM OF ARITHMETIC COMPLETE?

Often the intuitive mathematical realms we begin with are <u>not</u> <u>complete</u> <u>conceptions</u>, like equality, and to make them complete requires us to resort to an arbitrary decision, like choosing the number of objects that there are. Realms that are described by literary works are always intuitively <u>in</u>complete, since many details of those realms are not relevant to the work itself. For example, suppose we ask whether Sherlock Holmes, in the stories by Sir Arthur Conan Doyle, parted his hair from the left or from the right, or whether he even parted his hair at all. Since he is a fictional character, some things about him, such as whether he parted his hair, are not completely determined by the Sherlock Holmes stories. To capture this idea, we say that the "realm" of Sherlock Holmes is intuitively <u>in</u>complete. The realm in which we live — the real world — is intuitively <u>complete</u> with respect to large-scale properties.[4] For example, it is a fact of the real world whether some specific detective really parts his hair or not. Some mathematical realms are intuitively complete, as well. For example, it is a fact about numbers that when the constant 'C' is given some numeric value that the sentence 'C = 6' is either true or false. There seems to be no other alternative, and thus it seems obvious that the realm of ordinary arithmetic assertions is complete. Some philosophers question this claim, however, when it pertains to unproved (and possibly unprov<u>able</u>) assertions. For example, an unproved assertion about numbers is that any even number greater than 2 is equal to the sum of two (not necessarily distinct) primes. This hypothesis is called Goldbach's Conjecture, after Christian Goldbach who first posed it in 1742. To this day no one has been able to prove that it is true or that it is false. Nevertheless, it is thought by almost anyone who pauses to think about this conjecture that it is <u>either</u> true <u>or</u> false, regardless of whether a proof is <u>ever</u> found. The basis for this belief seems to be the intuitive conviction that the realm of ordinary arithmetic is <u>complete</u>, as mentioned earlier. Some philosophers, though, think that the truth of Goldbach's

---

[4] "large-scale", because quantum mechanics may provide possible counter-examples.

228   CHAPTER 8

Conjecture is indeterminate — neither true nor false — until a proof of it has been constructed or a counterexample to it has been found. In the present book, we pursue the consequences of the traditional view that arithmetic is a fully determinate reality.[5] Soon, we begin investigating first-order theories of arithmetic to gain a better understanding of that reality.

8.5 THE THEORY OF ORDERINGS[6]

For now, let us look at another first-order theory which we call: **The Theory of Orderings ($\mathcal{O}$)**. The mathematical (or non-logical) symbols of $L_\mathcal{O}$ are "officially" two binary relation symbols from our infinite stock of them (presented in Chapter 5). But, for perspicuity, we use '<' and '>' as surrogates for those symbols. (The "official" symbols actually selected will not matter to anything we do.) The first two axioms of $\mathcal{O}$ are those for the Theory of Equality, $\mathcal{EQ}$.

AXIOMS OF $\mathcal{O}$

($\mathcal{EQ}$-A1 & $\mathcal{EQ}$-A2)
$\mathcal{O}$-A3 $x=y \lor (x < y \lor x > y)$
$\mathcal{O}$-A4 $x < y \rightarrow \neg(y < x)$
$\mathcal{O}$-A5 $x > y \rightarrow \neg(y > x)$
$\mathcal{O}$-A6 $(x < y \land y < z) \rightarrow x < z$
$\mathcal{O}$-A7 $(x > y \land y > z) \rightarrow x > z$

$\mathcal{O}$-T1 and $\mathcal{O}$-T2 are the same as $\mathcal{EQ}$-T1 and $\mathcal{EQ}$-T2.

$\mathcal{O}$-T3 $\neg(x < x)$

```
                                    SHOW ∀x¬(x < x)
   1.   ¬∀x¬(x < x)                 1,A
                                    √SHOW A CONTRADICTION
   2.   ∃x¬¬(x < x)                 1,QE
   3.   ¬¬(a < a)                   3,A
   4.   a < a                       3,SL
   5.   (a < a) → ¬(a < a)          O-A4,UI
   6.   ¬(a < a)                    3,SL(4,5)
```

---

[5]Other viewpoints could be presented with equal rigor. We choose to investigate the traditional view in this book, while noting that other viewpoints are possible.
[6]Much of the material of this section is founded on Tarski [1965], particularly chapters VII - X.

Exercises:

From this point on, certain theorems will be specifically assigned as *exercises* because either the theorems themselves or their derivations feature some special properties. The theorems <u>not</u> explicitly assigned as *exercises* should be derived as well, in order to better understand the concepts involved and to become better at constructing derivations and proofs.

O-T4 ¬(x > x)

O-T5 (x > y) ↔ (y < x)

```
                                    SHOW ∀x∀y(x > y ↔ y < x)
                                    SHOW (a > b) → (b < a)
1.   a > b                          1,A
2.   b=a ∨ ((b < a) ∨ (b > a))      O-A3,UI
3.   (a > b) → ¬(b > a)             O-A5,UI
4.   ¬(b > a)                       1,SL(1,3)
5.   b=a ∨ b < a                    1,SL(3,4)
6.   b=a                            6,A
7.   b=a → ((a > b) → (a > a))      LL,UI
8.   a > a                          1,6,SL(1,6,7)
9.   ¬(a > a)                       O-T4,UI
10.  ¬(b=a)                         1,CONTRA(6,8,9)
11.  b < a                          1,SL(5,10)
```

Exercise:

Show the other direction, namely:

$$\text{SHOW} \quad (b < a) \to (a > b)$$

O-T6 ¬(x=y) → (x < y ∨ y < x)

O-T7 ¬(x=y) → (y > x ∨ y > x)

Theorem 8 to follow is in parts. What we wish to express is that for any two numbers x and y, exactly <u>one</u> of these three alternatives is true:

$$x=y, \qquad x < y, \qquad x > y.$$

230   CHAPTER 8

That is the Law of Trichotomy. We know by $\mathcal{O}$-A3 that <u>at least</u> one of the above three possibilities must obtain. But, we need to show that <u>no more than one</u> of these three alternatives is true. Thus, $\mathcal{O}$-T8 is in 3 parts.

$\mathcal{O}$-T8a $x=y \rightarrow \neg(x < y \vee x > y)$

$\mathcal{O}$-T8b $x < y \rightarrow \neg(x=y \vee x > y)$

$\mathcal{O}$-T8c $x > y \rightarrow \neg(x=y \vee x < y)$

Given that the Law of Trichotomy holds, namely that <u>exactly one</u> of these alternatives holds:

$$x=y \qquad x < y \qquad x > y,$$

we can easily derive the following:

$\mathcal{O}$-T9 $(x < y \vee x=y) \leftrightarrow \neg(x > y)$

$\mathcal{O}$-T10 $(x > y \vee x=y) \leftrightarrow \neg(x < y)$

We want to continue the development of the Theory of Orderings in order to have a first-order theory of arithmetic. We want to "say" in the theory that for any two numbers x, y, that x+y = y+x. But, intuitively, "+" is a binary operation on numbers — the operation that produces the sum of two numbers — and we do not have any operation symbols in our first-order language.

So, first we are going to permit operation symbols to be in the language of a first-order theory.

8.6 OPERATION (or FUNCTION) SYMBOLS

The language L_T may contain some number of the following operation (or function) letters:

$$A_i^k, \ B_i^k, \ C_i^k, \ D_i^k \qquad \text{for } k, i \geq 0.$$

Now we will explain how these letters may occur in formulas of L_{T+}. But, first, we need to define **terms**. A **term** of L_{T+} is:

(1) A variable or logical name, or
(2) A k-ary operation symbol followed by k terms within parentheses.

Here are a few examples of terms:

$$x, \quad b, \quad A_7^0, \quad A_6^1(x), \quad C_4^4(A_7^0 x A_6^1(x) b), \quad D_0^2(ab).$$

A term is intended to stand for a single object in a domain of discourse. Intuitively, an operation letter stands for an operation or a function. So, intuitively, '$D_0^2(ab)$' stands for the object resulting from the two-place operation (or function) D_0^2 on the two objects a and b (in that order). 0-ary operation letters like 'A_7^0' are to be thought of as standing for distinguished elements of the domain of discourse. Formulas of L_{T^+} can contain terms anywhere that formulas of L_{PL} can contain names or variables, except immediately after quantifiers.

Here are some examples of formulas of L_{T^+} containing operation letters. We use only the one-place relation letter 'F' here (compare these examples with the examples of terms above):

$$\exists x F x, \quad Fb, \quad FA_7^0, \quad \forall x F A_6^1(x), \quad \forall x \exists y F C_4^4(A_7^0 x A_6^1(x) b), \quad F D_0^2(ab).$$

Formulas of L_{T^+} containing operation letters can become extremely difficult to read, as one can see from the above examples. So, we immediately agree to use more natural symbols for operation letters. For example, in the Theory of Addition we will use '+' for the operation of "+", instead of using an official operation letter such as 'D_0^2'. (Note that '+' is slightly different from '+'.) Also, we will write 'x + y' instead of writing '+(xy)'.

8.7 MODIFIED UG RULE

One rule of derivation for L_{T^+} must be restricted somewhat because of the presence of operation letters in the language. Before operation symbols were introduced, the restrictions on the rules sufficed for variables and names. But, the introduction of operation letters creates an additional concern.

First we define a "free term". A **free term** is one that contains no occurrence of a bound variable. Now, we restrict the UG Rule (Universal Generalization). For the language of a first-order theory that includes logical names, the UG Rule can<u>not</u> be applied to arbitrary terms. It can only be used as before, for names. Since the Existential Switch Rule ES was derived from UG, ES is similarly restricted. UG and ES are now defined this way:

UG: k. $\varphi\alpha/\delta$ where δ must be a <u>name</u>.
     .
     .
     k+n. $\forall\alpha\varphi$ by UG, provided all previous UG restrictions have been observed.

ES: k. $\exists\alpha\varphi$
     k+1. $\varphi\alpha/\delta$ where δ must be a <u>name</u>.
     .
     .
     k+n. ψ using assumption $\varphi\alpha/\delta$ and possibly others.
     k+n+1. ψ by ES, provided all previous ES restrictions have been observed.

US and derived rule EG <u>can</u> be used on arbitrary terms, and so their use is <u>not</u> further restricted:

US : k. $\forall\alpha\varphi$
     .
     .
     k+n. $\varphi\alpha/\tau$ by US, where τ is any term.

UG : k. $\varphi\alpha/\tau$ where τ is any term.
     .
     .
     k+n. $\exists\alpha\psi$ by EG.

The reason US needs no restriction, whereas UG must be restricted, can be seen by this reasoning: if a property holds of all elements of a domain of discourse, it holds of <u>each</u> of them, <u>including</u> those resulting from a given operation. But, if a property holds for the <u>result</u> of an operation performed on <u>each</u> element of the domain, it does <u>not</u> follow that the property holds of <u>all</u> <u>elements</u> of the domain. As counter-example for this latter inference, suppose the property we have in mind is "being an even number" and suppose the operation is "multiplying by 2". Then, if we did not restrict UG, we could make the following <u>unsound</u> inference:

> Every result of multiplying by 2 is an even number
> ✗ Therefore: every number is even. **(FAULTY)** ✗

If 'A^1' is the one-place operation of multiplying by 2 and 'F' is the property of being an even number, the <u>faulty</u> conclusion can be obtained as follows, <u>assuming no restriction</u> on UG:

	1.	$\forall x FA^1(x)$	1, A	
	2.	$FA^1(a)$	1, UI	
✗	3.	$\forall x Fx$	1, UG	**(FAULTY)** ✗

Suppose, though, for any x, $A^1(x) = B^1(x)$. Then, for any x, if '$FA^1(x)$' holds, then '$FB^1(x)$' holds as well. This is an example of Leibniz's Law holding for operation symbols. In words: For any input, if two functions (operations) produce the same output, then if the output of one of the functions has a given property (for that input) so does the output of the other one. To derive this fact, do *exercise* 2. below:

Exercises:
1. Derive: $\forall x \forall y(x = y \rightarrow (A^1(x) = A^1(y)))$
   [Hint: take an axiom of the form $\mathcal{EQ}$-A2 such that φxy is '$A(x) \rightarrow A(y)$' and such that φxx is derivable when 'x' is replaced by a logical name.]
2. Derive: $\forall x(A^1(x) = B^1(x) \rightarrow (FA^1(x) \rightarrow FB^1(x)))$
3. Derive: $\forall x((A^1(x) = B^1(x)) \rightarrow \forall x(FA^1(x) \rightarrow FB^1(x)))$
   independently of **2**, from which it follows.
4. Derive either:
   $\forall x((A^1(x) = B^1(x)) \rightarrow (\forall x FA^1(x) \rightarrow \forall x FB^1(x)))$
   or $\forall x((A^1(x) = B^1(x)) \rightarrow (\exists x FA^1(x) \rightarrow \exists x FB^1(x)))$
   independently of **3** from which both follow.

8.8 CHARACTERIZATION OF "FIRST-ORDER THEORY (WITH EQUALITY)"

We can now explain a first-order theory (with equality) $\mathcal{T}$ more generally. $\mathcal{T}$ consists of three items: (1)

the language L_T; (2) an initial set of sentences of L_T, Δ_T, and (3) theorems with only sentences of Δ_T as assumptions. We can write this more succinctly as:

$$\mathcal{T} = \langle L_T, \Delta_T, \text{Thms}_T \rangle$$

The language L_T is fully determined by its non-logical vocabulary, that is, its special operation symbols and/or relation symbols. Next, let us recall that Id ("Identity") and LL ("Leibniz's Law") must be included in Thms_T.

Also, often a theory $\mathcal{T}$ is referred to as if it is <u>merely</u> its set of theorems. Hence, in this loose sense, to say $\varphi \in \mathcal{T}$ means that φ belongs to the set of theorems of $\mathcal{T}$ (i.e., $\varphi \in \text{Thms}_T$).

Truth Clause For '=' In L_{T^+}

We have previously explained what naming interpretations are for L_{PL}, the language of first-order logic. We now wish to explain naming interpretations for L_{T^+}, the language of a first-order theory with names. The difference between the two languages is that L_{T^+} contains '=' and may contain operation symbols. Another difference is that an interpretation for L_{T^+} must treat '=' as genuine equality. This will be explained shortly.

First, a naming interpretation $\langle \mathcal{I}, n \rangle$ for L_{T^+} must assign values to a variable-free term $O^k(\tau_1, \tau_2, \ldots, \tau_k)$ in the following manner:

$$\langle \mathcal{I}, n \rangle (O^k(\tau_1, \tau_2, \ldots, \tau_k)) = $$
$$\mathcal{I}(O^k)(\langle \mathcal{I}, n \rangle(\tau_1), \langle \mathcal{I}, n \rangle(\tau_2), \ldots, \langle \mathcal{I}, n \rangle(\tau_k)).$$

And, of course, if there are no names in the term:

$$\mathcal{I}(O^k(\tau_1, \tau_2, \ldots, \tau_k)) = \mathcal{I}(O^k)(\mathcal{I}(\tau_1), \mathcal{I}(\tau_2), \ldots, \mathcal{I}(\tau_k)).$$

Now, we can define what it means for an atomic sentence of L_{T^+} containing '=' to be **True** for a naming interpretation. Note that according to the following definition, '=' is treated as "genuine equality".

Given an interpretation $\mathcal{I}$ that is appropriate for L_T, a naming function n, and two variable free terms, τ_1, τ_2, of L_{T^+}:

$$\langle \mathcal{I}, n \rangle (\tau_1 = \tau_2) = \mathbf{T} \quad \text{iff} \quad \langle \mathcal{I}, n \rangle (\tau_1) = \langle \mathcal{I}, n \rangle (\tau_2).$$

If the two terms, τ_1 and τ_2, belong to the language of the theory <u>without</u> names, then the naming function drops out, and we have:

$$\mathcal{I}(\tau_1 = \tau_2) = T \quad \text{iff} \quad \mathcal{I}(\tau_1) = \mathcal{I}(\tau_2).$$

To formally justify dropping reference to the naming function requires a proof. But, it should be clear that when there are no names in the terms that the naming function is dispensable.

Exercise:

Let the definition of "truth for an interpretation" in Chapter 6 now range over sentences of L_{T^+}, so that for any sentence φ of L_{T^+}:

$$\mathcal{I}(\varphi) = T \quad \text{iff} \quad \langle \mathcal{I}, n \rangle(\varphi) = T \text{ for } \underline{\text{all}}\ n.$$

Then, formally prove:

$$\mathcal{I}(\tau_1 = \tau_2) = T \quad \text{iff} \quad \mathcal{I}(\tau_1) = \mathcal{I}(\tau_2).$$

All the rest of the truth clauses are unchanged. "Logical consequence" for L_{T^+} is now defined exactly as before (except that the language and interpretations are extended). For any sentence φ of L_{T^+} and any set Γ of such sentences, φ **is a logical consequence of** Γ if any naming interpretation that models Γ also models φ, written: $\Gamma \models \varphi$. Again, the 'naming' part drops out when all the sentences belong to L_T.

Exercise:

Suppose all sentences of Γ and φ have no names, then prove:

$$\Gamma \models \varphi \quad \text{iff} \quad \text{for all } \mathcal{I},\ \text{if } \mathcal{I}(\Gamma),\ \text{then } \mathcal{I}(\varphi) = T.$$

Often a theory $\mathcal{T}$ is defined in terms of its logical consequences rather than its theorems (an alternative we cover in Chapter 10). That is, if 'Cns_T' stands for the logical consequences of Δ_T, a sentence φ sometimes is said to belong to $\mathcal{T}$ if and only if $\varphi \in Cns_T$. Assuming Strong

Soundness and Strong (Logical) Completeness[7], it is easy to prove for any sentence φ:

$$\varphi \in \text{Thms}_T \quad \text{iff} \quad \varphi \in \text{Cns}_T.$$

Exercise:

Prove the above equivalence (assuming Strong (Logical) Completeness and Strong Soundness) for any first-order theory $\mathcal{T}$).

It is possible for the initial sentences Δ_T of a theory and its theorems Thms_T to be coextensive, though normally there are infinitely more theorems than initial sentences. For example, suppose for a given first-order language, we let Δ_T be the sentences true in all interpretations of the language. Then, Δ_T is the set of logical truths for that language. In this case the set of theorems, Thms_T, is the same as Δ_T by the completeness and soundness theorems (extended to include the equality symbol and operation letters).

8.9 GROUP THEORY $\mathcal{G}$

Before extending the Theory of Orderings (to a theory that more closely approximates arithmetic), we introduce the Theory of Groups, $\mathcal{G}$, which has three non-logical symbols: (1) a binary operation symbol 'D_1^2', (2) a unary operation symbol 'D_1^1', and (3) a 0-ary operation symbol 'D_1^0'. The axioms for $\mathcal{G}$ using the official symbols 'D_1^2', 'D_1^1', and 'D_1^0' are:

$\mathcal{G}$-A1: $\forall x \forall y \forall z [D_1^2(D_1^2(xy)z) = D_1^2(xD_1^2(yz))]$
$\mathcal{G}$-A2: $\forall x [D_1^2(xD_1^0) = x]$
$\mathcal{G}$-A3: $\forall x [D_1^2(xD_1^1(x)) = D_1^0]$

Now, we rewrite the three operation symbols 'D_1^2', 'D_1^1', and 'D_1^0', respectively as '∘', '*', and 'ı', and then we position '∘' *between* its two operands and position '*' *after* its single operand. We also drop the initial universal

---

[7]The proofs of Strong Soundness and Strong Completeness for any theory $\mathcal{T}$ require us to modify our earlier proofs to accommodate: (1) the semantics for '=' and operation symbols; and (2) the axioms of $\mathcal{EQ}$, and the modified UG Rule.

quantifiers. The above axioms are now easier to read and to comprehend:

$\mathscr{G}$-A1: $(x \circ y) \circ z = x \circ (y \circ z)$
$\mathscr{G}$-A2: $x \circ I = x$
$\mathscr{G}$-A3: $x \circ x^* = I$

Now, let us derive some theorems about groups. We wish to use LL in these derivations without being so deliberate about exactly how LL is being used. When operation letters were introduced and the UG Rule was modified accordingly, there was a brief discussion of Leibniz's Law, followed by a few *exercises* on deriving sentences containing operation symbols. At this point, though, we wish to use LL more informally, so we now provide an example of how the informal use of LL can be replaced by a derivation. For example, suppose on a line of a derivation we have:

$$a = b$$

and we wish to obtain on a later line,

$$a \circ c = b \circ c$$

Intuitively, we have "multiplied" both a and b by c. What guarantees that this can be done?

Consider the following:

1. $a = b$ \hfill 1, A
   √SHOW $a \circ c = b \circ c$
2. $\forall x \forall y (x = y \rightarrow (\forall z [x \circ z = x \circ z] \rightarrow \forall z [x \circ z = y \circ z]))$ \hfill LL
3. $a = b \rightarrow (\forall z [a \circ z = a \circ z] \rightarrow \forall z [a \circ z = b \circ z])$ \hfill 2, UI (twice)
4. $\forall z [a \circ z = a \circ z] \rightarrow \forall z [a \circ z = b \circ z]$ \hfill 1, 3, SL
   √SHOW $\forall z [a \circ z = a \circ z]$
   √SHOW $a \circ c = a \circ c$
5. $\forall x [x = x]$ \hfill Id
6. $a \circ c = a \circ c$ \hfill 5, UI
7. $\forall z [a \circ z = a \circ z]$ \hfill 6, UG
8. $\forall z [a \circ z = b \circ z]$ \hfill 4, 7, SL
9. $a \circ c = b \circ c$ \hfill 8, UI

ANALYSIS: The critical step in the above derivation is to notice that after we have:

$$a=b$$

we can get something[8] like the following sentence by using $\mathcal{EQ}$-A2:

$$a=b \rightarrow ([a \circ c = a \circ c] \rightarrow [a \circ c = b \circ c]).$$

Then, since we have 'a=b' and we know we can get 'a∘c = a∘c' by the Law of Identity, we can get our desired conclusion:

$$a \circ c = b \circ c$$

We will call the inference from 'a = b' to 'a∘c = b∘c' a principle of Substitution (Subst) for '=', since it relies on our substituting 'b' for 'a' in 'a ∘ c = a ∘ c'. All such substitutions can be turned into derivations involving Leibniz's Law.

$\mathcal{G}$-T1: $\forall x \forall y \forall z (x \circ z = y \circ z \rightarrow x = y)$

	SHOW a∘c = b∘c → a=b	
1.	a∘c = b∘c	1, A
	√SHOW a = b	
2.	(a∘c)∘c* = (b∘c)∘c*	1, Subst.
3.	(a∘c)∘c* = a∘(c∘c*)	$\mathcal{G}$-A1
4.	c∘c* = I	$\mathcal{G}$-A3
5.	(a∘c)∘c* = a∘I	3, 4, Subst.
6.	a∘I = a	$\mathcal{G}$-A2
7.	(a∘c)∘c* = a	5, 6, Subst.
8.	(b∘c)∘c* = b∘(c∘c*)	$\mathcal{G}$-A1
9.	(b∘c)∘c* = b∘I	4, 8, Subst.
10.	b∘I = b	$\mathcal{G}$-A2
11.	(b∘c)∘c* = b	9, 10, Subst.
12.	a = (b∘c)∘c*	2, 7, Subst.
13.	a = b	11, 12, Subst.

---

[8]The sentence actually used is the one on line 2 of the derivation. Recall that only sentences without names are permitted as axioms (in this case as an instance of $\mathcal{EQ}$-A2). The sentence on the displayed line can be easily obtained on a further line of the derivation, where names are permitted.

It may appear at a glance that $\mathcal{G}$-T1 is really a principle of equality, like '$\forall x \forall y \forall z(x=y \rightarrow x \circ z = y \circ z)$', since the two sentences superficially resemble each other. But $\mathcal{G}$-T1 goes beyond notions pertaining only to equality and says something specific about the operation '$\circ$'. Let's look at an example of $\mathcal{G}$-T1 <u>failing</u> to hold.

Suppose '$\circ$' is the operation of addition MOD 2, and we have in our domain of discourse 0, 1, 2, and 3. Then $(1 +_{MOD2} 1) = (3 +_{MOD2} 1)$ (since both sums = 0), yet $1 \neq 3$. If we look back at the axioms of $\mathcal{G}$ to see which one(s) fail when this counter-example holds, we can see that $\mathcal{G}$-A2 must fail. $(1 +_{MOD2} 0) = 1$ but $(3 +_{MOD2} 0) \neq 3$ (since $(3 +_{MOD2} 0) = 1$ and $3 \neq 1$). Thus, $\mathcal{G}$-T1 holds not because of facts about equality, but because of principles embodied in the Group Theory axioms.

$\mathcal{G}$-T2: $x \circ I = I \circ x$
SHOW $a \circ I = I \circ a$
  [HINT: First try to SHOW: $(a \circ I) \circ a^* = (I \circ a) \circ a^*$. Then use the right-cancellation law given to us by $\mathcal{G}$-T1 to SHOW: $a \circ I = I \circ a$.]

$\mathcal{G}$-T3: $\forall y[y \circ x = y] \rightarrow x = I$

		SHOW $\forall y[y \circ a = y] \rightarrow a=I$	
1.	$\forall y[y \circ a = y]$		1,A
		√SHOW $a=I$	
2.	$I \circ a = I$		1,UI
3.	$a \circ I = I \circ a$		$\mathcal{G}$-T2
4.	$a \circ I = I$		1,Subst(2,3)
5.	$a \circ I = a$		$\mathcal{G}$-A2
6.	$a = I$		1,Subst(4,5)

$\mathcal{G}$-T4: $\forall x[x \circ x = I] \rightarrow \forall x \forall y[x \circ y = y \circ x]$
1. $\forall x[x \circ x = I]$
     SHOW $a \circ b = b \circ a$
  [HINT: Prove $(a \circ b)(b \circ a) = I$. Then use line 1.]

$\mathcal{G}$-T5: $x \circ x^* = x^* \circ x$
     √SHOW $a \circ a^* = a^* \circ a$
1. $a^* \circ (a \circ a^*) = a^* \circ I$
2. $a^* \circ I = (a^* \circ a) \circ a^*$
3. $I \circ a^* = (a^* \circ a) \circ a^*$
4. $(a \circ a^*) \circ a^* = (a^* \circ a) \circ a^*$
5. $a \circ a^* = a^* \circ a$

𝒢-T6:   (x*)* = x
   [Straightforward]

𝒢-T7:   I = I*

𝒢-T8:   ∀y[y∘x = y] → x=x*

𝒢-T9:   z∘x = z∘y → x=y
   [HINT: Similar to the derivation of 𝒢-T1, except that an additional theorem is needed.]

𝒢-T10: x∘y = I → y=x*
[Fairly straightforward]

𝒢-T11:   (i) ∃z(x = y∘z)
         (ii) ∃z(x = z∘y)
   [HINT: In showing for (i) that a = b∘c for some c, figure out what c must equal for it to be true and then derive it. The same for (ii).]

𝒢-T12: (x∘y)* = y*∘x*

𝒢-T13: ∃z(x = y∘z ∧ ∀w(x = y∘w → w=z))

𝒢-T14: ∃z(x = z∘y ∧ ∀w(x = w∘y → w=z))

Exercises:

A formulation of Group Theory that does not include mathematical symbols for the identity element and the inverse function (operation) has the following axioms:

   i) (x∘y)∘z = x∘(y∘z)
   ii) ∃x(∀y(x∘y = y) ∧ ∀y∃z(z∘y = x))

1. Prove that 𝒢A1 - 𝒢A3 ⟹ i) & ii)
2. Prove that i) & ii) ⟹
      ∃x(∀y([x∘y = y] ∧ [∀z(z∘y = y) → z = x]))

   Group Theory can obviously be developed much further. But, we now wish to create a theory that looks more like ordinary arithmetic.

8.10 THEORY OF ADDITION $\mathcal{A}$

We now present the Theory of Addition $\mathcal{A}$, whose first seven axioms are those for $\mathcal{O}$ (whose first two axioms are those of $\mathcal{EQ}$). Axioms $\mathcal{A}$-A8 — $\mathcal{A}$-A12 introduce the new operation symbol '+':

$\mathcal{O}$-A1 — $\mathcal{O}$-A7
$\mathcal{A}$-A8 $x+y = y+x$
$\mathcal{A}$-A9 $x + (y+z) = (x+y) + z$
$\mathcal{A}$-A10 $\exists z(x = y+z)$
$\mathcal{A}$-A11 $x < y \rightarrow z+x < c+y$
$\mathcal{A}$-A12 $x > y \rightarrow z+x > z+y$

Theorems $\mathcal{A}$-T1 — $\mathcal{A}$-T10 are those for $\mathcal{O}$, obtained from $\mathcal{O}$-A1 — $\mathcal{O}$-A7. The following theorems involve our new addition symbol:

$\mathcal{A}$-T11 $x + (y+z) = (x+z) + y$
$\mathcal{A}$-T12 $x=y \rightarrow z+x = z+y$

$\mathcal{A}$-T13 $x+y = x+z \rightarrow y=z$

		SHOW $a+b = a+c \rightarrow b=c$
1.	$a+b = a+c$	1, A
		SHOW $b=c$
2.	$\neg(b=c)$	2, A
		√SHOW A CONTRADICTION
3.	$\neg((a+b < a+c) \lor (a+b > a+c))$	1, $\mathcal{O}$-T8a, UI, MP
4.	$\neg(a+b < a+c) \land \neg(a+b > a+c)$	SL(3)
5.	$(b < c) \lor (c < b)$	2, $\mathcal{O}$-T6, UI, MP
6.	$b < c$	6, A
7.	$a+b < a+c$	6, $\mathcal{A}$-A11
8.	$\neg(a+b < a+c)$	SL(4)
9.	$\neg(b < c)$	CONTRA(7,8)
10.	$c < b$	2, SL(5,9)
11.	$b > c$	2, $\mathcal{O}$-T5
12.	$a+b > a+c$	2, $\mathcal{A}$-A12(11)
13.	$\neg(a+b > a+c)$	SL(4)

$\mathcal{A}$-T14 $x+y < x+z \rightarrow y < z$
$\mathcal{A}$-T15 $x+y > x+z \rightarrow y > z$

𝒜-T14 and 𝒜-T15 are very much like 𝒜-T13 and can be derived similarly. Here, though, is a new idea:

𝒜-T16 $x+y < z+w \rightarrow (x < z \lor y < w)$

Informally, this is the thought behind the derivation: we want to SHOW that if a+b < c+d, then a < c or b < d. Suppose this conclusion is false. Then (a=c or (a > c)) <u>and</u> (b=d or (b > d)). This gives us four distinct counter-possibilities:

(1)	a = c	and	b = d
(2)	a = c	and	b > d
(3)	a > c	and	b = d
(4)	a > c	and	b > d

Let us consider case (1). From a=c we get that b+a = b+c by 𝒜-T12. By A8, b+a = a+b. By LL, a+b = b+c. Then, by 𝒜-A8 again, b+c = c+b, so a+b = c+b (by LL again). Then, since b=d, a+b = c+d by LL. This gives us that ¬(a+b < c+d) by 𝒪-T8a. We now have a contradiction, since our hypothesis was that a+b < c+d.

The <u>derivation</u> of 𝒜-T16 is then quite long, since the above <u>informal</u> <u>argument</u> represents only one case.

Exercises:

1. Turn the informal argument above into a derivation of that case.
2. Write an informal argument for one of the other three cases.

𝒜-T17 $\exists z(x=y+z \land \forall w(x=y+w \rightarrow w=z))$

𝒜-T17 says that for any two numbers x,y there is <u>exactly</u> <u>one</u> z such that x = y+z. Given 𝒜-T17, we can define subtraction as follows: x-y = z iff y+z = x. We know by 𝒜-T17 there is <u>one and only one</u> such z, so no ambiguity can occur.

However, instead of enriching this theory further to include subtraction, we will define a new Theory of Arithmetic that enables us to add and to multiply.

8.11 THE THEORIES OF ARITHMETIC

Since there are numerous first-order theor*ies* of arithmetic, it may be instructive to look at a few of them. One theory of arithmetic, Theory $\mathcal{A}r$, from Tarski, Mostowski, and Robinson [1953][9], consists of the true sentences "about the natural numbers". This means that the theses of $\mathcal{A}r$ are those sentences of the language of the theory that are true when the mathematical symbols in those sentences are interpreted in "the standard way" and the variables are considered to range over the natural numbers. The set of mathematical (non-logical) symbols for that language is $\{+, \circ, \underline{0}, \underline{1}, <\}$. Let us now consider a sentence of that language. Take: $\exists x \exists y (\neg(x=\underline{0}) \wedge \neg(y=\underline{0}) \rightarrow x+y = \underline{0})$. We will check to see whether it belongs to (or: is a thesis of) Theory $\mathcal{A}r$. First, we interpret the mathematical symbols in that sentence in the standard way, which means that '+' is interpreted as the "+" relation and '$\underline{0}$' is interpreted as the natural number 0. Further, we consider all the variables in '$\exists x \exists y (\neg(x=\underline{0}) \wedge \neg(y=\underline{0}) \rightarrow x+y = \underline{0})$' to range over the natural numbers. Once we do this, we can see that '$\exists x \exists y (\neg(x=\underline{0}) \wedge \neg(y=\underline{0}) \rightarrow x+y = \underline{0})$' is **False**, since it says (under this interpretation) that there are two natural numbers different from 0 that add up to 0. So, that sentence does not belong to $\mathcal{A}r$ (though it does belong to the corresponding theory of integers, since $(-1) + (+1) = 0$).

But we do not wish to check infinitely many sentences to determine whether they are true of the natural numbers in order to say whether they belong to $\mathcal{A}r$. We want to find some recognizable subset of those truths (the axioms of $\mathcal{A}r$) that we know belong to $\mathcal{A}r$, from which we can derive (using inference rules that are sound) all the rest of the truths (the theorems) of the natural numbers. Though $\mathcal{A}r$ contains all (and only) the truths about the natural numbers, we do not know which ones they are because the theory is not axiomatized. $\mathcal{A}r$ is in fact not a "formal theory" at all, since its theses are not identifiable in terms of syntactic patterns. This does not mean that there is no formal theory whose theses coincide with those of $\mathcal{A}r$, just that $\mathcal{A}r$ so-described is not a formal theory.

---

[9]They actually call it 'Theory N', but 'N' looks too much like our 'N' and possibly '$\mathfrak{N}$' as well.

Theory $\mathcal{P}$

We begin our quest for a "good theory" of arithmetic with the axiomatized theory $\mathcal{P}$, where '$\mathcal{P}$' stands for Peano Arithmetic. L_P contains the one-place operation symbol 'S', interpreted as the successor operation (or adding-one operation); the two-place operation symbol '+', interpreted as "+"; the two place operation symbol '∘', interpreted as multiplication; and the 0-place operation symbol '$\underline{0}$', interpreted as the natural number 0. And, of course, the variables are considered to range over the natural numbers.

$\mathcal{P}$-A1. $\forall x \forall y (S(x) = S(y) \rightarrow x=y)$
$\mathcal{P}$-A2. $\forall x \neg (\underline{0} = S(x))$
$\mathcal{P}$-A3. $\forall x (x+\underline{0} = x)$
$\mathcal{P}$-A4. $\forall x \forall y (x+S(y) = S(x+y))$
$\mathcal{P}$-A5. $\forall x (x \circ \underline{0} = x)$
$\mathcal{P}$-A6. $\forall x \forall y (x \circ S(y) = ((x \circ y) + x))$
$\mathcal{P}$-A7. $(\varphi(\underline{0}) \wedge \forall x [\varphi x \rightarrow \varphi S(x)]) \rightarrow \forall x \varphi x$

Again, we rewrite all of the above axioms more perspicuously, and we abbreviate '$\neg(\tau_1 = \tau_2)$' as '$\tau_1 \neq \tau_2$', where τ_1 and τ_2 are any terms :

$\mathcal{P}$-A1. $S(x) = S(y) \rightarrow x=y$
$\mathcal{P}$-A2. $\underline{0} \neq S(x)$
$\mathcal{P}$-A3. $x+\underline{0} = x$
$\mathcal{P}$-A4. $x+S(y) = S(x+y)$
$\mathcal{P}$-A5. $x \circ \underline{0} = x$
$\mathcal{P}$-A6. $x \circ S(y) = (x \circ y) + x$
$\mathcal{P}$-A7. $(\varphi(\underline{0}) \wedge \forall x [\varphi x \rightarrow \varphi S(x)]) \rightarrow \forall x \varphi x$

It should be fairly clear that $\mathcal{P}$-A1 — $\mathcal{P}$-A6 hold for the interpretation provided above. $\mathcal{P}$-A6 "says", for example, that the following principle holds for the natural numbers: $x(y + 1) = xy + x$. We know this to be true by elementary Algebra.

$\mathcal{P}$-A7 has a different status than the rest of the Peano axioms. 'φ' is not a symbol of the language of $\mathcal{P}$, it is a variable ranging over formulas in L_P having one free variable. Hence, '$\varphi(\underline{0})$' stands for the sentence of L_P that results from the uniform replacement of '$\underline{0}$' for that free variable in φ. '$\varphi(\underline{0})$' "says" that 0 has property φ. '$\forall x[\varphi x \rightarrow \varphi S(x)]$' "says" that whenever some number x has φ, the next number (x + 1) also has φ. Then '$\forall x \varphi x$' "says" that every natural number has property φ. So, $\mathcal{P}$-A7 expresses a principle of induction for the formulas of L_P having one free variable. Since we know induction holds of the natural numbers, and we know all the other axioms hold for the natural numbers, we know that $\mathcal{P}$-A7 must be true for the natural numbers as well.

But, there are two primary reasons for <u>not</u> using $\mathcal{P}$ as our preferred theory of arithmetic. One reason is that $\mathcal{P}$ has infinitely many axioms, because $\mathcal{P}$-A7 is an axiom form (or axiom <u>schema</u>), rather than a single axiom. Let us look more closely at one instance of $\mathcal{P}$-A7. Consider the formalization of the fact that any number different from 0 is the successor of some number:

$$\forall x(x \neq \underline{0} \rightarrow \exists z(x = S(z)))$$

First, we remove the outside quantifier in order to turn it into a formula with one free variable (in this case, the variable 'x'):

$\varphi x:$ $\qquad x \neq \underline{0} \rightarrow \exists z(x = S(z))$

Now, we plug <u>this formula</u> in for φ in $\mathcal{P}$-A7. That gives us:

$\mathcal{P}$-A7* $\quad [\ (\underline{0} \neq \underline{0} \rightarrow \exists z[\underline{0} = S(z)]) \land$
$\forall x([x \neq \underline{0} \rightarrow \exists z(x = S(z))] \rightarrow [S(x) \neq \underline{0} \rightarrow \exists z(S(x) = S(z))]) \ ]$
$\qquad\qquad\qquad\qquad \rightarrow \forall x(x \neq \underline{0} \rightarrow \exists y[x = S(y)]).$

And, of course there are infinitely many other formulas that can be plugged in for φ in $\mathcal{P}$-A7.

So that is one objection to $\mathcal{P}$, that it includes infinitely many axioms. A second objection to $\mathcal{P}$ is that it does <u>not</u> include axioms about the equality relation, "=".

A popular system that <u>does</u> include axioms for equality, but <u>still includes infinitely many axioms</u>, is gotten by adding two equality axioms to $\mathcal{P}$. This system (with one difference) seems to have originated with Kleene [1952]. After Kleene, we call this theory, Theory $\mathcal{K}$:

Theory $\mathcal{K}$

$\mathcal{P}$-A1 — $\mathcal{P}$-A7
$\mathcal{K}$-A8 $\quad x=y \rightarrow x=z \rightarrow y=z$
$\mathcal{K}$-A9 $\quad x=y \rightarrow S(x) = S(y)$

Theory $\mathcal{K}$ can be found in Mendelson [1987]. (The difference between this theory and Kleene's is that Kleene has axiom 2 reversed. That is, Kleene's axiom 2 is: $\forall x \neg (S(x) = \underline{0})$.)

At this point you may well wonder why we added axioms $\mathcal{K}$-A8 and $\mathcal{K}$-A9 when we know that our two axioms for equality Id and LL would suffice instead. One advantage that $\mathcal{K}$-A8 and $\mathcal{K}$-A9 have over the two equality axioms is that the second equality "axiom", LL, is really infinitely many axioms, as we explained above, while axioms $\mathcal{K}$-A8 and $\mathcal{K}$-A9 are <u>really</u> two axioms, not infinitely many.

But, you may ask the following question: if axioms $\mathcal{P}$-A1 — $\mathcal{P}$-A7 are <u>already</u> <u>infinite</u>, what difference does it make if "one" of the "two" equality axioms is really an axiom schema and <u>also</u> represents infinitely many axioms?

One response to the above question is to introduce a theory with only a finite set of number-theoretic axioms as well as a finite number of equality axioms. The most popular finitely axiomatized theory for the natural numbers <u>excluding</u> axioms for equality is Raphael M. Robinson's Theory Q:

Theory Q

$\mathcal{P}$-A1 — $\mathcal{P}$-A6
Q-A7 $x \neq \underline{0} \rightarrow \exists y(x = S(y))$

Since we can derive Q-A7 from $\mathcal{P}$-A7 plus $\mathcal{P}$-A1 — $\mathcal{P}$-A6 (given <u>any</u> suitable axioms for equality), Theory $\mathcal{P}$ is at least as strong as Theory Q (To perform this derivation, take $\mathcal{P}$-A7* above and derive its antecedent by deriving the first conjunct (whose antecedent is false, so there is not much to show here) and then deriving the second conjunct. Then, by MP, the conclusion follows. (The converse direction will not work, so $\mathcal{P}$ is strictly <u>stronger</u> than Q.))

But, can we now add finitely many equality axioms to the finite set of number-theoretic axioms of Q to get all the theorems we need? Yes, we can. The main advantage of this is the ease of then proving the "undecidability" of first-order logic, a concept explained a few paragraphs below.

Even though "undecidable" has yet to be explained, we can still follow the intuitive argument for the proof of "undecidability" when there are only finitely many axioms for equality and arithmetic. Here is how that argument goes: first, it is established that any theory that is "undecidable" remains "undecidable" when only <u>finitely many</u> of its axioms are removed. We then get some theory of arithmetic (including equality) that is strong enough to establish its undecidability <u>and</u> has only finitely many axioms. We then remove its (finitely many) axioms and voilà: first-order logic is undecidable. (This argument is due to Alonzo Church and is explained in greater detail in Chapter 11, *Gödel's Theorems*.).

On the other hand, if we begin with a theory of arithmetic that <u>already</u> has infinitely many axioms (in order to make derivations easier), then there is not much point in restricting our equality axioms just to make them finite. When a theory of arithmetic has only finitely many equality axioms, many of the derivations are long and tedious because of the added difficulty in deriving the basic properties of equality. We prefer to select a theory of arithmetic in which the facts about equality are transparent, so that we can focus attention on the purely number-theoretic properties of the theory.

The theory of arithmetic we choose to highlight, called Number Theory (N), results from combining the infinitely many axioms of P with the infinitely many axioms of EQ. N contains the strongest combination of axioms about equality and arithmetic that we have looked at so far. Thus, N is quite convenient for deriving many important truths about the arithmetic of the natural numbers. Rather than presenting Number Theory now as just another theory of arithmetic, we devote all of Chapter 9 to it.

In the remainder of the present chapter, we explain some properties of first-order theories (with equality) from a more general perspective.

8.12 MECHANICAL, DECIDABLE, & AXIOMATIZABLE

A "**mechanical**" procedure is one that fully specifies exactly what steps to follow and produces the same determinate result in a finite amount of time. Cooking recipes sometimes come close enough for us to accept them as being "mechanical" in a loose sense, when every step is fully spelled out and "the same dish" is always produced. Of course, in mathematics "the same result" is the <u>identical</u> result.

The word 'mechanical' was enclosed in scare-quotes in bold because the notion is intuitive rather than a precisely defined mathematical concept. We will use this informal and imprecise (but we hope somewhat intuitive) notion in what follows.

A subset Γ of the set of all sentences in the language of a theory, L_T, is **decidable** if, given a sentence of L_T, there is a mechanical procedure or method for deciding whether that sentence belongs to Γ or whether it does not. For example, the set of all sentences of L_T that are universal sentences (i.e., those beginning with a universal quantifier) is decidable, since we can mechanically decide whether a sentence begins with a universal quantifier or not.

An **axiomatizable** theory is one for which there is a decidable set of axioms (from which all theses can be generated). An **axiomatic** or **axiomatized** theory is one whose theses are all sentences derivable from a set of sentences that have been previously designated as axioms. For example, all the theories we have looked at so far except 𝒜𝓇 have been axiomatic theories. Theory 𝒜𝓇, you'll recall, is the theory of all truths of arithmetic (for the language containing '+' '∘', and '0'). So, the initial sentences Δ_{Ar} for 𝒜𝓇 are all true sentences about the natural numbers. Since we can't just look at the form of a sentence to tell whether or not it belongs to Δ_{Ar}, 𝒜𝓇 is not an axiomatized theory. Using earlier terminology, the sentences of Δ_{Ar} are not recognizable. The initial sentences of 𝒜𝓇 are the same as the theorems of 𝒜𝓇 in this special case, but Δ_{Ar} is not a set of axioms.

The axioms of an axiomatic theory are, by definition decidable, but what about its theorems? Are the theorems decidable if the axioms are? First, a definition: 𝒯 is a **decidable theory** if all the theses of 𝒯 are decidable. So, we rephrase our question: are axiomatic theories (those whose axioms are decidable) decidable (i.e., are their theses decidable)?

The answer is 'no', but it is tempting to think that we might somehow be able to use axioms together with our derivation rules to decide whether a sentence is or is not a thesis of an axiomatized theory 𝒯. We will follow this line of reasoning to show where such an attempt at a proof breaks down:

X TO PROVE: If 𝒯 is axiomatizable, 𝒯 is decidable. X **(FALSE)**

  X PF: Assume Δ is some decidable set of axioms for 𝒯. Let's consider any sentence φ of L_T and try to mechanically determine whether φ is a thesis of 𝒯. To do this we construct a mechanical derivation procedure to generate all theses of $\mathcal{T}$. You may not believe this can really be done, but let's accept it for now, because there is a far more serious problem. We're supposing that we can construct a "derivation machine" that spits out all theses of $\mathcal{T}$ one at a time. Given this supposition, the way to test whether φ is a thesis is just to wait until φ shows up. If, unbeknownst to us, φ <u>is</u> a thesis, φ will in a finite amount of time be spit out by our derivation machine. But — and this is a big BUT — if, unbeknownst to us, φ is <u>not</u> a thesis, we would have to wait forever (i.e., an infinite amount of time) until our machine finished spitting out the infinitely many theorems of $\mathcal{T}$. You may counter this by saying that you are going to construct a "speed-up derivation machine" that derives its first theorem in one second, its second theorem in one-half second, its third theorem in one-quarter second, and so forth. Then, you say that if φ is <u>not</u> a thesis, your machine will be through spitting out all the theses of $\mathcal{T}$ (and, let's say also comparing each one with φ) in only two seconds! Thus, you say that you have found a mechanical method that finishes the test of whether φ is a thesis or whether it is not in two seconds. The response to this interesting suggestion is to say that our notion of a "mechanical procedure" is intended to rule out "speed-up machines" of this type. We assume that any "machine" takes some fixed unit of time, however small the unit is, to carry out each step of its operations.

Thus, any "derivation machine" taking a fixed amount of time to carry out each step of its operations would take forever to finish spitting out each of the theses of $\mathcal{T}$. Then, if φ is <u>not</u> a thesis, we'd have to wait forever to find out φ isn't a theorem. You may try to think of ways to rectify this, but nothing will work.

You may suggest that the way to test whether φ is a thesis is to run our derivation machine on φ and $\neg\varphi$ simultaneously. Then, if $\neg\varphi$ is spit out, φ is <u>not</u> a thesis, or else $\mathcal{T}$ is inconsistent. If $\mathcal{T}$ is <u>inconsistent</u>, then <u>all</u> sentences of L_T are theorems and the sentences of L_T make up a decidable set. So, let's assume $\mathcal{T}$ is consistent. But, even if we assume $\mathcal{T}$ <u>is</u> consistent, this will not guarantee a result in a finite time. Why? Because it is possible <u>neither</u> φ <u>nor</u> $\neg\varphi$ is a thesis of $\mathcal{T}$. So, we'd still have to wait forever.

250   CHAPTER 8

You may say that your derivation machine is so constructed as to generate the shortest theses, those with the fewest symbols, first. Somehow your machine, which we'll call 'Short First', produces all theses of length n (if there are any of that length) before producing those of length n+1. As Short First finishes spitting out the shortest ones, it moves on to the very next shortest, spitting out those when they are theses, and so on. If <u>this</u> machine starts spitting out theses <u>longer</u> than φ without having spit out φ itself, then it must have passed over φ, and thus φ is <u>not</u> a thesis. <u>And</u>, <u>this</u> <u>machine</u> takes only a finite amount of time to produce this result.

Does Short First violate our notion of "mechanical procedure"? No, it doesn't. It just turns out that such a machine is <u>not</u> <u>possible</u>.

X ■ X

These are some of the ways in which such reasoning breaks down if we try to prove that an axiomatizable theory is decidable. Soon, we will exhibit an important axiomatic theory that is <u>not</u> decidable (in a very strong sense).

Exercises:

1. Prove that every theory has infinitely many theses.
2. Prove that the theses of a decidable theory are mechanically listable (i.e., there is a mechanical procedure for arriving at a "first" thesis, a "second", and so on).
3. Prove that a complete, axiomatizable theory is decidable. [HINT: consider the false statement 'if 𝒯 is axiomatizable, 𝒯 is decidable' that we tried — and failed — to prove. One line of reasoning above was blocked because 𝒯 could be incomplete. If 𝒯 is complete (which is part of the hypothesis of this exercise), the reasoning goes through.]
4. Prove: If Thms$_T$ are mechanically listable, 𝒯 is axiomatizable. [Hint: Suppose you consider the theorems, say $\varphi_1, \varphi_2, \ldots, \varphi_n, \ldots$, themselves to be the axioms of 𝒯. Then, suppose you consider a sentence φ that happens <u>not</u> to be an axiom and you try to find that out in a mechanical way. Note that there is <u>not</u> a finite stage (in the mechanical listing of theorems) at which you can be sure φ is <u>not</u> an axiom, so this attempt fails (though if φ <u>were</u> an axiom, there <u>would</u> <u>be</u> a finite stage at

which φ would show up). So, the task is to somehow create axioms from the mechanically generated theorems, $\varphi_1, \varphi_2, \ldots, \varphi_n, \ldots$, so that when a sentence φ is not an axiom, there is a finite stage at which that can be determined.]

Two of the above exercises are important enough to be theorems in their own right, so we will highlight them:

LISTABILITY THEOREM: The theses of a decidable theory are mechanically listable.

PF: By *exercise* 2. above.

DECIDABILITY THEOREM: A complete axiomatizable theory is decidable.

PF: By *exercise* 3. above.

DIGRESSION ==>:

Notice that a kind of anomaly exists in the proof of the DECIDABILITY THEOREM (*exercise* 3.). If $\mathcal{T}$ is inconsistent, all sentences of L_T are theorems. So, the mechanical procedure for testing whether a sentence φ of L_T is a theorem is to do nothing. φ is automatically a theorem. If $\mathcal{T}$ is consistent, there is some other mechanical test for theoremhood (based on the fact that either $\mathcal{T} \vdash \varphi$ or $\mathcal{T} \vdash \neg\varphi$). Let's call these two tests, the do-nothing test and the derivability test. The next question is: which test do we use? Since we don't know whether $\mathcal{T}$ is consistent or not, we don't know which test applies. But one does. And that's all that's needed for decidability.

The situation we're calling attention to raises a kind of *epistemic* concern, as opposed to a *metaphysical* one. We know there is a procedure, but we don't know which procedure to use. We know (by the Law of the Excluded Middle) that either $\mathcal{T}$ is consistent or $\mathcal{T}$ is inconsistent. If there is a mechanical procedure for deciding whether a sentence is a theorem for each of the two possibilities, then $\mathcal{T}$ is decidable whether consistent or not. But, it seems somewhat unsettling to know that one of two decision procedures works but not to know which one. Possibly this is because 'decidable' suggests that we can decide which procedure to use. But, of course, we can't decide which procedure to use if we don't know which possibility obtains. So, $\mathcal{T}$ is decidable based on our metaphysical acceptance of the Law of the Excluded Middle, not on our epistemic awareness of which possibility obtains. The Law of the Excluded Middle, which is an assumption used throughout this book, may seem more philosophically questionable in this particular context.

<==:END OF DIGRESSION

8.13 EXTENSION, CONSERVATIVE EXTENSION

To prove Strong (Logical) Completeness in Chapter 7, we extended the first-order language L_{PL} to $L_{PL}*$ by adding infinitely many new names. (Recall that L_{PL} has denumerably many relation letters, but no operation letters.) We can consider the theorems of L_{PL} that do not contain names as belonging to the theory of first-order logic *without equality*. That theory, $\mathcal{T}_{PL}$, has no non-logical axioms whatsoever. Its theses are all valid sentences of L_{PL} without names, some of which were derived in Chapter 5. In a similar way, we can consider $\mathcal{T}_{PL}*$ to be the first-order theory *without equality* obtained from $\mathcal{T}_{PL}$ by adding infinitely many mathematical constant symbols, but not furnishing any axioms. So, the theorems of $\mathcal{T}_{PL}*$ are all valid sentences of L_{PL} plus those containing the new constants. So, for example, if k is a new constant, then the sentence '(Fk ∨ ¬Fk)' ∈ $Thms_{PL}*$. Thus, $Thms_{PL} \subseteq Thms_{PL}*$, since every theorem of L_{PL} is certainly a theorem of $L_{PL}*$ also. This is an example of one theory being an "extension" of another theory (though now we focus on first-order theories with equality whose relation letters and operation letters are specified by the theory).

A first-order theory with equality $\mathcal{T}'$ is an **extension** of theory $\mathcal{T}$ if $Thms_T \subseteq Thms_{T'}$. If $\mathcal{T}'$ extends $\mathcal{T}$ (written: $\mathcal{T} \subseteq \mathcal{T}'$), then $L_{T'}$ must have the same or more non-logical symbols as L_T. If $\mathcal{T}'$ has more theorems than $\mathcal{T}$ and they have the same language, then $\Delta_{T'}$, the set of initial sentences of $\mathcal{T}'$, must include some that are not in Δ_T. When $\mathcal{T}$ and $\mathcal{T}'$ are axiomatic theories, it is sometimes the case that the axioms of $\mathcal{T}$ are a subset of the axioms of $\mathcal{T}'$. When that occurs, it is clear that $\mathcal{T} \subseteq \mathcal{T}'$ ($\mathcal{T}'$ extends $\mathcal{T}$). But, it is often the case that $\mathcal{T} \subseteq \mathcal{T}'$ and the axioms for the two theories are simply different from each other.

There is another situation that occurs. Suppose we create a new theory $\mathcal{T}'$ just by adding more mathematical constants to the language of a theory $\mathcal{T}$. Clearly, there will be more theorems in $L_{T'}$, because there are more sentences. To take a trivial example, if 'c' is a new constant and φ is a theorem of $\mathcal{T}$, we can obtain the sentence $\varphi \wedge c = c$ as a theorem of $\mathcal{T}'$. So $\mathcal{T}'$ has strictly more theorems than $\mathcal{T}$, but only because of the extra non-logical symbols. The theorems of $\mathcal{T}'$ expressed in the poorer language L_T are exactly the same as the theorems of $\mathcal{T}$. When this occurs, $\mathcal{T}'$ is a "conservative" extension of $\mathcal{T}$.

If for any sentence φ of L_T, if $\varphi \in \text{Thms}_{T'}$ then $\varphi \in \text{Thms}_T$, then $\mathcal{T}'$ is a **conservative extension** of $\mathcal{T}$. So, for example, $\mathcal{T}_{PL}*$ is a conservative extension of $\mathcal{T}_{PL}$, since introducing new constants without adding any axioms does not increase the number of theorems in the language containing no constants. But, it can also be the case that $\mathcal{T}'$ is a conservative extension of $\mathcal{T}$ when the axioms are not the same. All that's needed is that the <u>theorems</u> of the poorer language be identical.

NEW CONSTANTS THEOREM: Let $L_{T'} = L_T \cup \{\text{new constants}\}$ and let $\mathcal{T}'$ have the same initial sentences as $\mathcal{T}$. Then $\mathcal{T}'$ is a conservative extension of $\mathcal{T}$. That is, for any sentence $\varphi \in L_T$:

$$\varphi \in \text{Thms}_T \quad \text{iff} \quad \varphi \in \text{Thms}_{T'}$$

PF:

Exercises:

1. Prove the NEW CONSTANTS THEOREM.
   [HINT: $\Rightarrow$ is clear. To prove $\Leftarrow$, consider any derivation in $\mathcal{T}'$ that uses k new, distinct constants, say $m_1, m_2, \ldots, m_k$. Get the first k logical names <u>not</u> occurring in that derivation. Create a new derivation in $\mathcal{T}$ by substituting those logical names respectively for $m_1, m_2, \ldots, m_k$.] ∎
2. Suppose $\mathcal{T} \subseteq \mathcal{T}'$, $\text{Thms}_T \neq \text{Thms}_{T'}$, and $L_T = L_{T'}$.
   (a) Prove there is at least one sentence φ, such that $\varphi \in \Delta_{T'}$ and $\varphi \notin \Delta_{T'}$.
   (b) Provide a counter-example to this claim: $\Delta_T \subseteq \Delta_{T'}$.

Recall that a set of sentences Γ is consistent if and only if for <u>no</u> sentence φ of the given language is it the case that $\Gamma \vdash \varphi$ and $\Gamma \vdash \neg\varphi$. See Chapter 4, THEOREMS 4.10 and 4.11, for two equivalent formulations of consistency. A first-order **theory $\mathcal{T}$ is consistent** iff there is <u>no</u> sentence φ of L_T such that $\mathcal{T} \vdash \varphi$ and $\mathcal{T} \vdash \neg\varphi$.

EXTENSION THEOREM: $\mathcal{T}$ is (mathematically) complete iff $\mathcal{T}$ has <u>no</u> consistent extension $\mathcal{T}'$ in L_T such that $\mathcal{T}' \neq \mathcal{T}$.

COMPLETE EXTENSION THEOREM: Every consistent theory $\mathcal{T}$ has a complete, consistent extension $\mathcal{T}'$ in the language L_T.

COMPLETE, DECIDABLE EXTENSION THEOREM: Every consistent, decidable theory $\mathcal{T}$ has a complete, consistent, decidable extension in L_T.

Chapter Exercises:

1. Prove the EXTENSION THEOREM
2. Prove the COMPLETE EXTENSION THEOREM.
   [Use LINDENBAUM'S THEOREM in Chapter 7 without clause 4 requiring witnesses (which was added by Henkin).]
3. Prove the COMPLETE, DECIDABLE EXTENSION THEOREM.
   [Use Lindenbaum's construction for the COMPLETE EXTENSION THEOREM, and associate every sentence of L_T with a unique natural number n such that given n, φ_n can be mechanically determined; and, given a sentence φ, its associated natural number can be mechanically determined. Then argue that for a given sentence φ, it can be mechanically determined whether or not $\mathcal{T}' \vdash \varphi$, where $\mathcal{T}'$ is the complete theory found à la Lindenbaum.]
4. Prove that the non-theses of a decidable theory are mechanically listable.
5. Prove that if the theses of $\mathcal{T}$ are mechanically listable <u>and</u> the non-theses of $\mathcal{T}$ are also mechanically listable, then $\mathcal{T}$ is decidable.
6. Put *exercises* 4 and 5 together with the LISTABILITY THEOREM, to prove:

   A Theory is decidable iff
   its theses and non-theses are each mechanically listable.

CHAPTER 9

FIRST-ORDER NUMBER THEORY

Number Theory, N, consists of the axioms for P plus the axioms of EQ in Chapter 8, *First-Order Theories*. Let us list them again:

9.1 THEORY N ($EQ + P$)

N-A1: $x = x$
N-A2: $x = y \rightarrow (\varphi xx \rightarrow \varphi xy)$
N-A3: $S(x) \neq \underline{0}$
N-A4: $S(x) = S(y) \rightarrow x = y$
N-A5: $x + \underline{0} = x$
N-A6: $x + S(y) = S(x+y)$
N-A7: $x \circ \underline{0} = \underline{0}$
N-A8: $x \circ S(y) = (x \circ y) + x$
N-A9: $[\varphi(\underline{0}) \land \forall x(\varphi x \rightarrow \varphi S(x))] \rightarrow \forall x \varphi x$

Notice that in N-A3, we have written '$S(x) \neq \underline{0}$' as an abbreviation of '$\neg(S(x) = \underline{0})$'. We will continue to use such abbreviations. In Chapter 8, we referred to N-A1 as Id, for "identity", and to N-A2 as LL, for "Leibniz's Law". Given Id and LL, the symmetry of '=' (i.e.: $x = y \rightarrow y = x$) was derived as the first theorem in The Theory of Equality (EQ), and transitivity (i.e.: $x=y \land y=z \rightarrow x=z$) was derived as the second theorem. In this chapter, we continue to use Id and LL, and we begin using Symm and Trans for those two theorems.

In the derivations that follow, Symm and Trans so frequently occur together that it seems like a good idea to combine them into a further principle. First, let us look at two examples:

    1. $a = b$ 1'. $b = a$
    2. $c = b$ 2'. $b = c$
    3. ∴ $a = c$ 3'. ∴ $a = c$

Symm applies to 2 in the first argument and Symm applies to 1' in the second argument. After Symm has been applied, both conclusions follow by Trans. We will write Symm-Trans for arguments like these, and in parentheses we will write first the number of the line to which Symm is applied, followed by the line number of the other sentence. So, for the first argument we would write 'Symm-Trans(2,1)' to the right of line 3 and for the second 'Symm-Trans(1',2')' to the right of line 3'.

We use LL (Leibniz's Law) more liberally here than as a mere abbreviation of *N*-A2. The "idea" of Leibniz's Law is that anything that can be said (in a first-order theory) to be true of an object under one description of it can also be said to be true under any other description of it. For example, suppose we have on a line of a derivation:

$$b = b + c$$

Then, intuitively, any sentence that is true containing 'b' remains true when 'b+c' is substituted for it. Here are some examples (using only '+'):

1) $b + c = (b + c) + c$
2) $a + b = a + (b + c)$
3) $(a + b) + b = (a + b) + (b + c)$
4) $(a + b) + b = (a + (b + c)) + c$

To obtain 1) in a derivation, first take this instance of *N*-A2:

$$\forall x \forall y (x = y \rightarrow (\forall z [x + z = x + z] \rightarrow \forall z [x + z = y + z]))$$

Then instantiate 'x' to 'b' and 'y' to 'b + c'. From the instantiated line, together with the original sentence ('b = b + c'), use MP to get:

$$\forall z [b + z = b + z] \rightarrow \forall z [b + z = (b + c) + z]$$

The antecedent of that line can be obtained by Id, which gives us:

$$\forall z [b + z = (b+c) + z]$$

Finally, instantiate 'z' to 'c', and we get 1), which is what we wanted.

In the previous chapter, derivations involving LL were done more formally. In this chapter, we use LL in a more informal manner, concentrating on the facts about numbers rather than the underlying equality principles.

Exercises:

These exercises are provided in case you wish to assure yourself that the informal uses of LL above can all be turned into derivations. (You may also wish to return to Chapter 8 if you are not certain how to do these exercises.)
1. Turn the above reasoning that shows how to arrive at 1) from 'b = b + c' into an actual derivation.
2-3. Derive any two of 2) - 4) from 'b = b + c'.

9.2 THE MATH-IND RULE

Axiom N-A9, as explained in Chapter 8, is really infinitely many axioms. It "says" that for any formula φx of L_N having only 'x' free, $\forall x \varphi x$ is true provided two conditions are met. One condition is that $\varphi(\underline{0})$ is true, meaning that the sentence resulting from φx by replacing all free occurrences of 'x' by '$\underline{0}$' is true. The second condition is that $\forall x(\varphi x \rightarrow \varphi S(x))$ holds. Thus N-A9 is a principle of mathematical induction <u>within</u> N <u>itself.</u>

From N-A9, we can very easily derive the rule **MATH-IND** for any formula φx of L_N with only 'x' free.

If on one line of a derivation from N, we have:

$$\varphi(\underline{0})$$

and on another line of that derivation, we have:

$$\forall x(\varphi x \rightarrow \varphi S(x))$$

then, on a still later line, we may write:

$$\forall x \varphi x$$

whose assumptions are those of the earlier two lines.

Exercises:

1. Derive MATH-IND from N-A9. (If you are unclear about "deriving a rule", return to Chapter 2, where that concept was introduced.)
2. Show that the following rule (which we will use frequently) is just a special case of MATH-IND (for φx in L_N):

If: $N \vdash \varphi(\underline{0})$, and $N \vdash \forall x(\varphi x \rightarrow \varphi S(x))$,
then: $N \vdash \forall x \varphi x$.

9.3 MATH-IND vs. Mathematical Induction

MATH-IND is an induction-type rule whose use <u>within</u> derivations in N is justified by N-A9. We will also use, however, a principle of Mathematical Induction <u>outside</u> <u>of</u> N to argue <u>that</u> <u>a</u> <u>certain</u> <u>derivation</u> <u>in</u> N <u>exists</u>. For example, suppose we assert that for any <u>number</u> n there's at least one theorem of N having 5+2n official symbols of L_N. We can prove this by Mathematical Induction on n, where n is the number of symbols in a sentence of L_N. So, n can be 0, 1, 2, and so on. When finished, we will have a proof that for 0, 1, 2,...,n there is a theorem having 5 + 2(0), 5 + 2(1), 5 + 2(2)..., 5 + 2(n) symbols. Note that this is a <u>proof</u>, *not* a derivation. The proof establishes that <u>there</u> <u>is</u> a derivation for each natural number, but is not itself a derivation. Here's the theorem:

A THEOREM: For every natural number n, there's at least one theorem χ in N whose official symbols number 5+2n.

> PF: Case 0: n = 0. Let χ be '$\forall x x = x$', which is N-A1 (the "official" sentence for N-A1 contains the universal quantifier, but no parentheses or brackets). '$\forall x x = x$' has 5 symbols. 5+2(0) = 5, so this case works.
>
> Induction Hypothesis: A THEOREM is true for case k. Thus, there is an official sentence ψ of L_N such that $N \vdash \psi$, where ψ has 5+2k symbols.
>
> TO PROVE: There is some official sentence χ such that $N \vdash \chi$ and χ has 5+2(k+1) symbols. By the Induction Hypothesis, we know that $N \vdash \psi$, where ψ has 5 + 2k symbols. The sentence we take for χ is $\neg\neg\psi$, which has 5+2k+2 official symbols, and $N \vdash \neg\neg\psi$ (since $N \vdash \psi$). Furthermore, 5+2k+2 = 5+2(k+1). So we're done. ∎

The purpose of A THEOREM was only to demonstrate the use of Mathematical Induction <u>outside</u> of N so it could be contrasted with the use of MATH-IND <u>within</u> N. Again, note that the above use of Mathematical Induction constitututed a <u>proof</u> <u>that</u> <u>certain</u> <u>theorems</u> <u>exist</u>, and is <u>not</u> a derivational rule. By contrast, MATH-IND is a derivational rule that is used to obtain theorems. Note the difference between our use of MATH-IND in the following theorems and our use of Mathematical Induction in the above proof of A THEOREM. (Later, we will again use Mathematical Induction outside of N.)

We now use MATH-IND to derive the theorem '$\forall x(x = \underline{0} + x)$', which is $\forall x \varphi x$ in our schema above. That is, we now wish to use MATH-IND to derive:

$\forall x \varphi x$: $\qquad\qquad \forall x(x = \underline{0} + x).$

As indicated above, we must first let x be $\underline{0}$ and then derive the sentence that results from Universally Instantiating 'x' to '$\underline{0}$'. That sentence is our $\varphi(\underline{0})$. So, we first derive the sentence:

$\varphi(\underline{0})$: $\qquad\qquad \underline{0} = \underline{0} + \underline{0}.$

Then, we derive the sentence that "says" whenever our original sentence holds for something, it holds for "S of" that thing. That sentence is our $\forall x(\varphi x \rightarrow \varphi S(x))$. So next we derive:

$\forall x(\varphi x \rightarrow \varphi S(x))$: $\qquad \forall x(x = \underline{0} + x \rightarrow S(x) = \underline{0} + S(x)).$

Once we derive $\varphi(\underline{0})$ and $\forall x(\varphi x \rightarrow \varphi S(x))$, we use MATH-IND to complete the derivation of $\forall x \varphi$, which is the desired sentence: $\forall x(x = \underline{0} + x)$.

9.4 THEOREMS OF N

N-T1 $x = \underline{0}+x$

$\qquad\qquad$ √SHOW $\forall x(x = \underline{0} + x)$
$\qquad\qquad$ √SHOW $\underline{0} = \underline{0}+\underline{0}$
1. $\underline{0}+\underline{0} = \underline{0}$ $\qquad\qquad\qquad\qquad\qquad\qquad$ N-A5
2. $\underline{0} = \underline{0}+\underline{0}$ $\qquad\qquad\qquad\qquad\qquad\qquad$ Symm(1)
$\qquad\qquad$ √SHOW $\forall x(x = \underline{0} + x \rightarrow S(x) = \underline{0} + S(x))$
$\qquad\qquad$ √SHOW $a = \underline{0}+a \rightarrow S(a) = \underline{0} + S(a)$
3. $a = \underline{0}+a$ $\qquad\qquad\qquad\qquad\qquad\qquad$ 3,A
$\qquad\qquad$ √SHOW $S(a) = \underline{0}+S(a)$
4. $S(a) = S(\underline{0}+a)$ $\qquad\qquad\qquad\qquad\qquad$ 3,LL
5. $\underline{0}+S(a) = S(\underline{0}+a)$ $\qquad\qquad\qquad\qquad\;$ N-A6
6. $S(a) = \underline{0}+S(a)$ $\qquad\qquad\qquad\qquad\quad$ 3,Symm-Trans(5,4)

```
7.  a = 0 + a → S(a) = 0 + S(a)        SL(3,6)
8.  ∀x(x = 0 + x → S(x) = 0 + S(x))    UG
9.  ∀x(x = 0 + x)                      MATH-IND(2,8)
```

The above derivation of *N*-T1 includes all SHOW lines to make it clear that the MATH-IND Rule has been used correctly. In the derivation of *N*-T2 (below) we include all the steps in the derivation, including SHOW lines. In the terminology of Chapter 2, it is a "full" derivation, showing explicitly the use of MATH-IND. After the following derivation, however, we return to short derivations, whose missing steps can be easily filled in.

Exercise:

Explain in your own words the difference between Mathematical Induction and MATH-IND.

In our use of MATH-IND in the derivation of *N*-T1, we inducted on 'x', since 'x' was the only variable occurring in the sentence. *N*-T2 has two variables, 'x' and 'y', so we have a choice of which variable to induct on. We choose to induct on 'y' because 'y' does not have an 'S' directly in front of it in either occurrence, whereas there is an occurrence of 'S(x)' in *N*-T2. So, inducting on 'y' may result in an easier derivation, though this consideration is not conclusive.

To induct on 'y' requires us to first reverse the prefixed universal quantifiers '∀x∀y' to get get the '∀y' in front. So, instead of deriving '∀x∀y(S(x) + y = S(x+y))', we actually derive '∀y∀x(S(x) + y = S(x+y))', which, by T5.25 ('∀x∀yFxy → ∀y∀xFxy'), yields the desired result. (In the future we take this quantifier reversal for granted.) Now, just as we did in the derivation of *N*-T1, we break the derivation of *N*-T2 into two sub-derivations. First, we derive:

φ(0):   ∀x(S(x) + 0 = S(x + 0)).

We now wish to SHOW ∀y(φy → φS(y)). That is:

φ(y):   ∀x(S(x) + y = S(x + y)).

Hence, to derive $\forall y(\varphi y \rightarrow \varphi S(y))$, we must derive:

$\forall y[\forall x(S(x)+y = S(x+y)) \longrightarrow \forall x(S(x) + S(y) = S[x + S(y)])]$

Then, MATH-IND will give us:

$\forall y \forall x(S(x) + y = S(x+y))$.

We then reverse the universal quantifiers in front, to get:

$\forall x \forall y(S(x) + y = S(x + y))$.

N-T2 $S(x)+y = S(x+y)$

          √SHOW $\forall x \forall y(S(x) + y = S(x+y))$
          √SHOW $\forall y \forall x(S(x) + y = S(x+y))$
          √SHOW $\forall x(S(x) + \underline{0} + S(x+\underline{0}))$
          √SHOW $S(a)+\underline{0} = S(a+\underline{0})$

1. $S(a)+\underline{0} = S(a)$ N-A5
2. $a+\underline{0} = a$ N-A5
3. $S(a+\underline{0}) = S(a)$ LL(2)
4. $S(a)+\underline{0} = S(a+\underline{0})$ Symm-Trans(3,1)
5. $\forall x(S(x) + \underline{0} = S(x+\underline{0}))$ UG

          √SHOW $\forall y[\forall x(S(x)+y = S(x+y)) \longrightarrow$
                      $\forall x(S(x) + S(y) = S[x + S(y)])]$
      √SHOW $\forall x(S(x)+b = S(x+b)) \longrightarrow \forall x(S(x)+S(b) = S(x+S(b)))$

6. $\forall x(S(x)+b = S(x+b))$ 6,A

         √SHOW $\forall x(S(x)+S(b) = S[x+S(b)])$
         √SHOW $S(c)+S(b) = S(c+S(b))$

7. $S(c)+b = S(c+b)$ 6,UI
8. $S(c)+S(b) = S(S(c)+b)$ N-A6
9. $S(S(c)+b) = S(S(c+b))$ 6,LL(7)
10. $S(c)+S(b) = S(S(c+b))$ 6,Trans(8,9)
11. $c+S(b) = S(c+b)$ N-A6
12. $S(c+S(b)) = S(S(c+b))$ LL(10)
13. $S(c)+S(b) = S(c+S(b))$ 6,Symm-Trans(12,10)
14. $\forall x(S(x)+S(b) = S(x+S(b)))$ 6,UG(13)
15. $\forall x(S(x)+b = S(x+b)) \longrightarrow \forall x(S(x)+S(b) = S[x+S(b)])$
                                        SL(6,14)
16. $\forall y[\forall x(S(x)+y = S(x+y)) \longrightarrow \forall x(S(x)+S(y) = S[x+S(y)])]$
                                        UG(15)
17. $\forall y \forall x(S(x) + y = S(x+y))$ MATH-IND(5,16)

CHAPTER 9

Exercise:

Try to derive *N*-T2 by inducting on 'x' rather than 'y'. Does this derivation look like it will be easier or harder than the one given? What considerations seem to affect the relative difficulty of inducting on 'x' or 'y' (in addition to the simple fact that 'y' does not occur immediately after 'S', whereas 'x' does)? Explain your answer.

N-T3   x+y = y+x

We now return to using short derivations where the use of MATH-IND is implicit. For example, the next theorem is 'x+(y+z) = (x+y)+z', which is an abbreviation of the sentence '∀x∀y∀z(x+(y+z) = (x+y)+z)'. We actually derive the sentence '∀z∀x∀y(x+(y+z) = (x+y)+z)', since we will induct on 'z'. In the derivation below, instead of first writing:

$$\text{SHOW } \forall x \forall y (x + (y + \underline{0}) = (x + y) + \underline{0}),$$

we pass to:

$$\text{SHOW } a + (b + \underline{0}) = (a+b) + \underline{0}.$$

Similarly, for the induction step, we skip:

SHOW   ∀z∀x∀y[x + (y+z) = (x+y) + z
              → x + (y + S(z)) = (x+y) + S(z)],

and go directly to:

SHOW a + (b+c) = (a+b) + c → a + (b+S(c)) = (a+b) + S(c).

Then, once we obtain the consequent, 'a+(b+S(c)) = (a+b)+S(c)', we conclude the derivation, leaving the use of MATH-IND (which would be at the end of a full derivation) implicit.

N-T4   x+(y+z) = (x+y)+z

		√SHOW a+(b+$\underline{0}$) = (a+b)+$\underline{0}$
1.	b = b+$\underline{0}$	Symm, *N*-A5
2.	a+b = a+b	Id
3.	a+b = a+(b+$\underline{0}$)	LL(1,2)

```
 4.  a+b = (a+b)+0                              Symm, N-A5
 5.  a+(b+0) = (a+b)+0                          Symm-Trans(3,4)
        SHOW a+(b+c) = (a+b)+c  →  a+(b+S(c)) = (a+b)+S(c)
 6.  a+(b+c) = (a+b)+c                          6, A
        √SHOW a+(b+S(c)) = (a+b)+S(c)
 7.  S(a+(b+c)) = a+S(b+c)                      Symm, N-A6
 8.  S(b+c) = b+S(c)                            Symm, N-A6
 9.  a+S(b+c) = a+(b+S(c))                      LL(8)
10.  S((a+b)+c) = (a+b)+S(c)                    Symm, N-A6
11.  a+(b+S(c)) = S(a+(b+c))                    Symm-Trans(7,9)
12.  ((a+b)+S(c)) = S((a+b)+c)                  Trans(11,10)
13.  S(a+(b+c)) = S((a+b)+c)                    LL(6)
14.  a+(b+S(c)) = S((a+b)+c)                    Trans(11,13)
15.  a+(b+S(c)) = (a+b)+S(c)                    Symm-Trans(12,14)
```

Exercise:

1. Restore all the necessary SHOW lines and add all the extra lines in the above derivation of *N*-T4 so that it is a full derivation and the use of MATH-IND is explicit.

N-T5 $x = \underline{0} \circ x$

N-T6 $S(x) + y = x + S(y)$

Exercise:

1. Derive *N*-T6 again, this time inducting on the other variable. Is one derivation more difficult than the other? Why?

N-T7 $S(x) \circ y = (x \circ y) + x$

N-T8 $x \circ y = y \circ x$

N-T9 $x \circ (y \circ z) = (x \circ y) \circ z$

Each of *N*-T7 — *N*-T9 is analogous to *N*-T2 — *N*-T4, and can be derived similarly.

Exercises:

1. — 3. Derive *N*-T7 — *N*-T9.

Here are some more addition and multiplication principles:

N-T10   x∘(y+z) = x∘y+(x∘z)

N-T11   (x+y)∘z = (x∘z)+(y∘z)

N-T12   ((x ≠ 0) ∧ (x = y ∘ z)) → ((y ≠ 0) ∧ (z ≠ 0))

N-T13   x+z = y+z → x=y

```
                              SHOW a+0 = b+0  →  a=b
   1.   a+0 = b+0                 1,A
                             √SHOW a=b
   2.   a+0 = a                   N-A5
   3.   a = b+0                   1,Symm-Trans(2,1)
   4.   b+0 = b                   N-A5
   5.   a=b                       1,Trans(3,4)
             SHOW ((a+c = b+c) → a=b) → a+S(c) = b+S(c) → a=b
   6.   a+c = b+c → a=b           6,A
                             SHOW a+S(c) = b+S(c → a=b
   7.   a+S(c) = b+S(c)
                             √SHOW a=b
   8.   a+S(c) = S(a+c)           N-A6
   9.   b+S(c) = S(b+c)           N-A6
   10.  S(a+c) = b+S(c)           Symm-Trans(8,7)
   11.  S(a+c) = S(b+c)           Trans(10,9)
   12.  a+c = b+c                 N-A4(11)
   13.  a=b                       MP(12,6)
```

We call '$\underline{0}$', '$S(\underline{0})$', '$S(S(\underline{0}))$', '$S(S(S(\underline{0})))$', ... **numeral terms** and we write them this way: $\underline{0}, \underline{1}, \underline{2}, \ldots, \underline{n}$, where '$\underline{n}$' abbreviates '$\underline{0}$' preceded by n occurrences of 'S' (together with parentheses). Note that '$\underline{n}$' is not itself a term; '$\underline{n}$' signifies the term '$S(S(\ldots S(\underline{0})\ldots))$' that has n occurrences of 'S'. We also write '$\underline{n+1}$' to mean '$\underline{0}$' preceded by n+1 occurrences of 'S'. '$\underline{n+1}$' can also be written as '$S(\underline{n})$' (as in the definition above), and '$\underline{n+1+1}$' and '$\underline{n+2}$' can both be written as '$S(S(\underline{n}))$. In fact, if $\underline{k}$ = $S(\underline{n})$, meaning that $\underline{k}$ has one more 'S' in it than $\underline{n}$ has, then obviously $\underline{n}$ has one less 'S' than $\underline{k}$ has. We can write that $\underline{n}$ has one less 'S' in it than $\underline{k}$ has this way: $\underline{n}$ = $\underline{k-1}$. Pictorially:

n is: $\qquad$ S(S(...S(0)...))
n occurrences of 'S'

k-1 is: $\qquad$ S(S(...S(0)...))
k-1 occurrences of 'S'

Since n = k - 1, the expressions on the right of the colons above are identical. Hence, when n = k - 1, *N* ⊢ **n** = **k-1** because '**n**' and '**k-1**' are descriptions of the same expression. We can derive '**n** = **k-1**' in *N* by instantiating 'x' in *N*-T1 to the term with n occurrences of 'S' preceding '**0**', which is the same term as the one with k-1 occurrences of 'S' preceding '**0**'.

We now derive some number-theoretic facts, using numeral terms (sometimes rewriting '**n**' as 'S(S(...S(0)...))', with n occurrences of 'S', and sometimes keeping '**n**'):

N-T14   x+**1** = S(x)

          √SHOW a+S(**0**) = S(a)
1. a+S(**0**) = S(a+**0**)   *N*-A6
2. a+**0** = a   *N*-A5
3. S(a+**0**) = S(a)   LL(2)
4. a+S(**0**) = S(a)   Trans(1,3)

N-T15   x∘**1** = x
N-T16   x∘**2** = x+x
 [Hint: Rewrite 'a∘**2**' as 'a∘S(**1**)' and use *N*-A8 and *N*-T14.]

N-T17   x+y = **0** → x=**0** ∧ y=**0**

         SHOW (a+**0** = **0** → a = **0** ∧ **0** = **0**)
1. a+**0** = **0**   1,A
        √SHOW a=**0** ∧ **0** = **0**
2. a+**0** = a   *N*-A5
3. a=**0**   1,Symm-Trans(2,1)
4. **0** = **0**   *N*-A1
5. a=**0** ∧ **0** = **0**   1,SL(3,4)
       SHOW (a+b = **0** → a=**0** ∧ b=**0**) →
        (a+S(b) = **0** → a=**0** ∧ S(b) = **0**)
6. a+b = **0** → a=**0** ∧ b=**0**   6,A
       √SHOW a+S(b) = **0** → a=**0** ∧ (b)=**0**
7. a+S(b) = **0**   7,A

8.	a+S(b) = S(a+b)	N-A6
9.	S(a+b) = 0	7,Symm-Trans(8,7)
10.	0 ≠ S(a+b))	N-A3
11.	0 = S(a+b)	7,Symm(9)

(The above derivation is slightly unusual, as the assumption on line 7 above leads to a contradiction, but it is correct. For a more satisfying and elegant derivation of N-T17, first derive N-T22 (without using N-T17 — N-T21) and then use it to derive N-T17 <u>without</u> MATH-IND.)

Exercise:

1. Derive N-T17 using alternative specified above.

N-T18 $x \neq \underline{0} \rightarrow (y \circ x = \underline{0} \rightarrow y = \underline{0})$

N-T19 $x+y = \underline{1} \rightarrow ((x=\underline{0} \land y=\underline{1}) \lor (x=\underline{1} \land y=\underline{0}))$
[Hint: First SHOW '(a+$\underline{0}$ = $\underline{1}$) → ((a = $\underline{0}$ ∧ $\underline{0}$ = $\underline{1}$) ∨ (a = $\underline{1}$ ∧ $\underline{0}$ = $\underline{0}$))', then SHOW the induction step, and use MATH-IND.]

N-T20 $x \circ y = \underline{0} \rightarrow (x = \underline{0}) \lor (y = \underline{0})$

N-T21 $x \neq \underline{0} \rightarrow (y \circ x = z \circ x \rightarrow y = z)$

To derive N-T21 by MATH-IND, we first derive:

(1) $\forall x \forall y (x \neq \underline{0} \rightarrow (y \circ x = \underline{0} \circ x \rightarrow y=\underline{0}))$.

And then we derive:

(2) $\forall z (\forall x \forall y [x \neq \underline{0} \rightarrow (y \circ x = z \circ x \rightarrow y=z)] \rightarrow \forall x \forall y [x \neq \underline{0} \rightarrow (y \circ x = S(z) \circ x \rightarrow y=S(z))])$.

From which the conclusion that follows <u>strictly</u> is:

(3) ∴ $\forall z \forall x \forall y [x \neq \underline{0} \rightarrow (y \circ x = z \circ x \rightarrow y=z)]$.

But, as we have seen before, the universal quantifiers preceding a sentence may be in any order.

SHOW $\forall x \forall y \forall z (x \neq \underline{0} \rightarrow (y \circ x = z \circ x \rightarrow y=z))$
SHOW $\forall x \forall y [x \neq \underline{0} \rightarrow (y \circ x = \underline{0} \circ x \rightarrow y=\underline{0})]$
SHOW $a \neq \underline{0} \rightarrow (b \circ a = \underline{0} \circ a \rightarrow b=\underline{0})$

FIRST-ORDER NUMBER THEORY

1.	$a \neq \underline{0}$	1, A
	SHOW $b \circ a = \underline{0} \circ a \rightarrow b = \underline{0}$	
2.	$b \circ a = \underline{0} \circ a$	
	√SHOW $b = \underline{0}$	
3.	$a \neq \underline{0} \rightarrow (b \circ a = \underline{0} \rightarrow b = \underline{0})$	N-T17
4.	$b \circ a = \underline{0} \rightarrow b = \underline{0}$	SL(1,3)
5.	$b = \underline{0}$	SL(2,4)
	SHOW $\forall z(\forall x \forall y[x \neq \underline{0} \rightarrow (y \circ x = z \circ x \rightarrow y = z)]$	
	$\rightarrow \forall x \forall y[x \neq \underline{0} \rightarrow (y \circ x = S(z) \circ x \rightarrow y = S(z))])$	
	SHOW $\forall x \forall y[x \neq \underline{0} \rightarrow (y \circ x = c \circ x \rightarrow y = c)]$	
	$\rightarrow \forall x \forall y[x \neq \underline{0} \rightarrow (y \circ x = S(c) \circ x \rightarrow y = S(c))]$	
6.	$\forall x \forall y(x \neq \underline{0} \rightarrow (y \circ x = c \circ x \rightarrow y = c))$	
	SHOW $\forall x \forall y(x \neq \underline{0} \rightarrow (y \circ x = S(c) \circ x \rightarrow y = S(c)))$	
	SHOW $a \neq \underline{0} \rightarrow (b \circ a = S(c) \circ a \rightarrow b = S(c))$	
7.	$a \neq \underline{0}$	
	SHOW $b \circ a = S(c) \circ a \rightarrow b = S(c)$	
8.	$b \circ a = S(c) \circ a$	
	√SHOW $b = S(c)$	
9.	$S(c) \neq \underline{0}$	N-A3
10.	$a \neq \underline{0} \rightarrow (S(c) \circ a = \underline{0} \rightarrow S(c) = \underline{0})$	N-T18
11.	$S(c) \circ a = \underline{0} \rightarrow S(c) = \underline{0}$	SL(7,10)
12.	$S(c) \circ a \neq \underline{0}$	SL(9,11)
13.	$b \circ a \neq \underline{0}$	LL(8,12)
	√SHOW $b \neq \underline{0}$	
14.	$b = \underline{0}$	14, A
	√SHOW A CONTRADICTION	
15.	$a \circ b = \underline{0}$	14, LL, N-T7(14)
16.	$b \circ a = \underline{0}$	14, Symm
17.	$b \neq \underline{0}$	CONTRA(13,16)
18.	$b \neq \underline{0} \rightarrow \exists y(b = S(y))$	T20
19.	$\exists y(b = S(y))$	SL(17,18)
20.	$b = S(d)$	20, ES
21.	$S(d) \circ a = S(c) \circ a$	LL(20,8)
22.	$S(d) \circ a = a \circ d + a$	N-A8, N-T7
23.	$S(c) \circ a = a \circ c + a$	N-A8, N-T7
24.	$a \circ d + a = a \circ c + a$	LL, twice(22,23,21)
25.	$a \circ d = a \circ c$	N-T12
26.	$a \neq \underline{0} \rightarrow (d \circ a = c \circ a \rightarrow d = c)$	UI(6)
27.	$d \circ a = c \circ a \rightarrow d = c$	SL(7,26)
28.	$d \circ a = c \circ a$	N-T12 twice(25)
29.	$d = c$	SL(28,27)
30.	$S(d) = S(c)$	LL(29)
31.	$b = S(c)$	Trans(20,30)

N-T22 $x \neq \underline{0} \rightarrow \exists y(x = S(y))$
[This is (7′) and also Q-A7 in *First-Order Theories*. Deriving N-T22 shows that theory $\mathcal{P}$, introduced in Chapter 8, is at least as strong as theory Q.]

N-T23 $x \neq \underline{0} \rightarrow (x \neq \underline{1} \rightarrow \exists y(x = S(S(y))))$

We now provide some definitions to abbreviate concepts in L_{N^+} (i.e., the Language of Number Theory together with logical names).

In what follows, we use 'r', 's', 't' with or without primes or subscripts to stand for terms of L_{N^+}. In definition (1) below 'x' is the <u>first</u> variable not occurring in the formula on the left.

9.5   DEFINITIONS OF FURTHER CONCEPTS

Here, we define '<' ("less-than"), '≤' ("less-than or equal"), '>' ("greater-than"), and '≥' ("greater-than or equal") in standard ways within N:

(1) $r < s$ $\exists x(S(x) + r = s)$
(2) $r \leq s$ $(r < s) \vee (r = s)$
(3) $r > s$ $s < r$
(4) $r \geq s$ $s \leq r$

On the basis of definitions (1)-(4) many facts about numbers can be derived in N (the "facts" are "about numbers" when we interpret our non-logical symbols to have their customary meanings).

N-T24 $x \not< x$
[Hint: assume N-T24 is false, use definition (1) above and obtain a contradiction.]

N-T25 $\underline{0} < S(x)$

        √SHOW $\forall x(\underline{0} < S(x))$
        √SHOW $\underline{0} < S(a)$
        √SHOW $\exists x(S(x) + \underline{0} = S(a))$
1. $S(a) + \underline{0} = S(a)$ N-A5
2. $\exists x(S(x) + \underline{0} = S(a))$ EG

N-T26 $\neg \exists x(x < \underline{0})$

N-T27 $x = S(y) \rightarrow y < x$

N-T28   x < y → S(x) < S(y)

N-T29   S(x) < S(y) → x < y

We now sketch a derivation of the Law of Trichotomy, using MATH-IND. For this derivational sketch,

∀xφx is: ∀x∀y[(x < y) ∨ (x = y) ∨ (y < x)]

φ is :   ∀y[(x < y) ∨ (x = y) ∨ (y < x)]

φ(0) is: ∀y[(0 < y) ∨ (0 = y) ∨ (y < 0)]

∀x(φx → φS(x)) is:
 ∀z(∀y[(z < y) ∨ (z = y) ∨ (y < z)] →
   ∀y[(S(z) < y) ∨ (S(z) = y) ∨ (y < S(z))])

N-T30   (x < y) ∨ (x = y) ∨ (y < x)
    √SHOW ∀y[(0 < y) ∨ (0 = y) ∨ (y < 0)]
    √SHOW (0 < a) ∨ (0 = a) ∨ (a < 0)
We know by N-T29 that nothing is less than 0, so we concentrate on the other two disjuncts:
    √SHOW (0 < a) ∨ (0 = a)
1. 0 ≠ a   1,A
    √SHOW 0 < a
    √SHOW ∃x(S(x) + 0 = a)
2. ∃x(S(x) + 0 = a)   by N-T22
    √SHOW ∀z(∀y[(z < y) ∨ (z = y) ∨ (y < z)] →
      ∀y[(S(z) < y) ∨ (S(z) = y) ∨ (y < S(z))])
    √SHOW ∀y[(b < y) ∨ (b = y) ∨ (y < b)] →
      ∀y[(S(b) < y) ∨ (S(b) = y) ∨ (y < S(b))]
3. ∀y[(b < y) ∨ (b = y) ∨ (y < b)]   3,A
    √SHOW ∀y[(S(b) < y) ∨ (S(b) = y) ∨ (y < S(b))]
    √SHOW (S(b) < c) ∨ (S(b) = c) ∨ (c < S(b))
Assume c = 0. Then, S(b) > c, by N-T25. So, assume c ≠ 0. Then, by N-T22:
4. S(d) = c   essentially by N-T22
We now can use our inductive hypothesis (on line 3) to give us three alternatives:
5. (b < d) ∨ (b = d) ∨ (d < b)   from 3.
If b < d, then S(b) < S(d), and
if b = d, then S(b) = S(d), and
if d < b, then S(d) < S(b), by previous theorems. Since c = S(d), we get by Sentential Logic:
6. (S(b) < c) ∨ (S(b) = c) ∨ (c < S(b))

Exercise:

1. Provide a full derivation for the sketch of *N*-T30 above.

N-T31 $x < y \rightarrow y \not< x$

N-T32 $x = y \rightarrow x \not< y$

N-T33 $y \neq \underline{0} \rightarrow \exists z(x < y \circ z)$

N-T34 $x < y \circ z \rightarrow z \neq \underline{0}$

N-T35 $x \neq \underline{0} \rightarrow (y < z \leftrightarrow x \circ y < x \circ z)$

We now define **t divides r** for terms t and r:

$t|r$ is defined as: $\exists x(r = t \circ x)$,
where 'x' is the first variable not in t or r.

Here are a number of theorems about divisibility that are left as *exercises*:

N-T36 $x | x \circ y$

N-T37 $x | x$

N-T38 $x | \underline{0}$

N-T39 $\underline{0} | x \leftrightarrow x = \underline{0}$

N-T40 $\underline{1} | x$

N-T41 $x | \underline{1} \rightarrow x = \underline{1}$

N-T42 $(x|y \land y|z) \rightarrow x|z$

N-T43 $(x|y \land y|x) \rightarrow x = y$

N-T44 $x|y \rightarrow x|(z \circ y)$

N-T45 $x|y \land x|z \rightarrow x|(y + z)$

FIRST-ORDER NUMBER THEORY

It should be clear that we can extend divisibility further. For example, we can define **r is a prime** this way:

Prime(r) is: $(\underline{1} < r) \land \forall x[x | r \rightarrow ((x = \underline{1}) \lor (x = r))]$
(where 'x' is the first variable not in r)

And, then we can derive many theorems about primes. But, instead of developing primes further, we wish to take up an issue that was begun early in this chapter.

In the beginning of this chapter, we proved a theorem called A THEOREM just to provide an example of the use of Mathematical Induction <u>outside</u> N. Then, we began using the derivational rule MATH-IND extensively <u>inside</u> N, in order to derive many of the previous theorems. Now, we need to return to using Mathematical Induction <u>outside</u> of N (i.e., in the language used to talk <u>about</u> theorems — the "metalanguage of N") in order to prove some more theorems. The first theorem we prove "says" whenever m is a different natural number from n, there is a derivation of the sentence '$\neg(\underline{m} = \underline{n})$' in N. Since we are stating that for each distinct m and n, such a derivation exists, we are expressing a fact <u>about</u> infinitely many derivations in N. We establish the truth of this statement by a proof (<u>not</u> a derivation) that these infinitely many derivations exist.

N-T46 REPRESENTATION THEOREM FOR EQUALITY:
(A) If m and n are natural numbers such that m ≠ n, then $N \vdash \neg(\underline{m} = \underline{n})$.
(B) If m and n are natural numbers such that m = n, then $N \vdash (\underline{m} = \underline{n})$.

PF OF (A): This proof (<u>not</u> a derivation) is conducted <u>outside</u> of N ("in the metalanguage of N") and <u>describes</u> derivations in N. The proof is by Mathematical Induction on the natural number (<u>not</u> the numeral term) n.

Case 0: n = 0.
PROVE: $N \vdash \neg(\underline{m} = \underline{0})$

Since, by hypothesis m ≠ n, m > 0. Thus, there is some natural number k, such that k + 1 = m.

PROVE: $N \vdash \neg(\underline{k+1} = \underline{0})$

k+1 has one more 'S' in it than k. By definition of 'numeral term': k+1 is S(k). So,

$$\sqrt{\text{PROVE}}: \mathcal{N} \vdash \neg(S(\underline{k}) = \underline{0})$$

By instantiating 'x' in $\mathcal{N}$-T3 to 'k', we get immediately:

1. $\neg(S(\underline{k}) = \underline{0})$

Our Induction Hypothesis (IH) is the following: for n and any m such that m ≠ n, it is true that $\mathcal{N} \vdash \neg(\underline{m} = \underline{n})$. So, we wish to prove the case for n + 1:

$$\text{PROVE: If } p \ne n + 1, \text{ then } \mathcal{N} \vdash \neg(\underline{p} = \underline{n+1})$$

Assume that p ≠ n + 1. Now,

$$\text{PROVE: } \mathcal{N} \vdash \neg(\underline{p} = \underline{n+1})$$

As before, 'n+1' stands for the same thing that 'S(n), does (namely '0' preceded by n + 1 occurrences of 'S'). So,

$$\text{PROVE: } \mathcal{N} \vdash \neg(\underline{p} = S(\underline{n}))$$

If p = 0, we get the desired result by assuming '$\underline{p} = S(\underline{n})$' and getting a contradiction from $\mathcal{N}$-T3 (and Symm). So, we may as well assume p ≠ 0. Thus, there is some natural number k, such that k + 1 = p. k+1 is S(k). So,

$$\sqrt{\text{PROVE}}: \mathcal{N} \vdash \neg(S(\underline{k}) = S(\underline{n}))$$

Since, by assumption, p ≠ n + 1, and k + 1 = p, it follows that k + 1 ≠ n + 1. Hence, k ≠ n. But, then, k is one of the numbers satisfying our Induction Hypothesis. So, we have:

2. $\neg(\underline{k} = \underline{n})$ by IH.
3. $S(\underline{k}) = S(\underline{n}) \rightarrow \underline{k} = \underline{n}$ by $\mathcal{N}$-A4.
4. $\neg(S(\underline{k}) = S(\underline{n}))$ by SL(2,3).

This concludes the proof of (A).

PF OF (B): We explained why (B) holds when numeral terms were introduced, but it bears repeating. When m = n,

'$\underline{m} = \underline{n}$' is an instance of N-A1, since $\underline{m}$ is the same term as $\underline{n}$ is (namely the term with '$\underline{0}$' preceded by m occurrences of 'S', which is the same term as the one with '$\underline{0}$' preceded by n occurrences of 'S').
This concludes the proof of the theorem. ∎

Exercises:

Use the "pattern of reasoning" in the proof of the
REPRESENTATION THEOREM FOR EQUALITY to construct
derivations in N of the following results (but do not use the theorem itself):

1. $\underline{0} = \underline{3}$
2. $\underline{5} = \underline{5}$
3. $\underline{7} \neq \underline{8}$

The reason the above theorem is called the
REPRESENTABILITY THEOREM FOR EQUALITY is that it establishes that equality can be "represented" in N. The "representability" of numeric relations in N is an important and interesting topic in its own right.

9.6 THE REPRESENTABILITY OF NUMERIC RELATIONS IN N

Let's look at an example first. Take some property of the natural numbers. Being a prime number, for instance. We know that 2 is prime, 3 is prime, 5 is prime, and so on. Suppose that corresponding to the numeric property "being a prime", there is a formula φx (with only 'x' free) in L_N such that:

(1) whenever n *is* prime,
  the sentence $\varphi(\underline{n})$ is derivable in N, and,
(2) whenever n is *not* prime,
  the sentence $\neg\varphi(\underline{n})$ is derivable in N.

Intuitively speaking, we may wish to say that when (1) and (2) are satified, the formula φx <u>uniquely</u> <u>characterizes</u> the prime numbers. This is so because when we substitute the numeral term of a prime for 'x', φ of that numeral term is derivable in N, and when we substitute the numeral term of a non-prime for 'x', $\neg\varphi$ of that numeral term is derivable in N. (And, of course, we <u>do</u> have a candidate for φx in this case — it is the formula Prime(x) that was defined above.)

When such a formula φ exists for some numeric property, that property is said to be **representable in** N.

The notion of "representability in N" (or in any theory with numeral terms) can be extended to cover relations. For example, consider the binary relation "less-than". If "less-than" is representable in N, then there is a formula φxy of L_N (whose only free variables are 'x' and 'y') such that both of these clauses are true:

(1) if n < m, then $N \vdash \varphi(\underline{n},\underline{m})$, and
(2) if n ≮ m, then $N \vdash \neg\varphi(\underline{n},\underline{m})$.

We will soon prove that "less-than" <u>is</u> in fact representable in N. So far, we have established that only equality is representable in N. The fact that certain numeric relations are representable in N will have important theoretical consequences in Chapter 11, *Gödel's Theorem*.

Exercises:

1. Provide your own definition of the representability in N of any k-ary relation R^k. (Representability is explained more thoroughly in Chapter 11, and you may wish to look ahead for this definition.)
2. Prove that if R^k is representable in N, there is a decision procedure for determining for any k natural numbers $n_1, n_2, \ldots, n_k$, whether $(n_1, n_2, \ldots, n_k) \in R^k$ or not.

We now wish to prove that for any <u>numbers</u> m and n, $N \vdash \underline{m+n} = \underline{m}+\underline{n}$. In words, this means that the result of adding two numbers <u>first</u>, then getting the numeral term of that sum, is the same as <u>first</u> getting the numeral terms for each of the numbers separately, then adding-in-N the N-sum of those two numeral terms. Thus, we have two kinds of addition, number addition and numeral-term addition. As before, we use an underline to indicate the result of taking the numeral term of whatever is underlined. Hence, 'm+n' is the sum of the numbers m and n, and '<u>m+n</u>' is the <u>numeral</u> <u>term</u> of that sum. On the other hand, '+' signifies addition-of-numeral-terms-in-N. Hence, '<u>m</u>+<u>n</u>' is the sum-in-N of the numeral terms <u>m</u> and <u>n</u>. Also, (outside of N) we write that two <u>numbers</u> are equal by using the normal equals sign '=', and we write that two <u>numeral</u> <u>terms</u> are equal-in-N

FIRST-ORDER NUMBER THEORY

by using the sign '=' that looks almost the same as the real equals sign.

Again, this proof uses Mathematical Induction. <u>First</u>, we prove:

(1) $N \vdash \underline{m+0} = \underline{m+0}$. This is our base case, when n = 0.

<u>Next</u>, we prove:

(2) if $N \vdash \underline{m+n} = \underline{m+n}$, then $N \vdash \underline{m+(n+1)} = \underline{m+n+1}$.

N-T47 $\underline{m+n} = \underline{m+n}$

$\sqrt{}$PROVE $N \vdash \underline{m+0} = \underline{m+0}$

1. $\underline{m+0} = \underline{m}$ the m+0th numeral term is the mth numeral term, (i.e., $\underline{m}$)
2. $\underline{m} = \underline{m+0}$ N-A5, Symm
3. $\underline{m+0} = \underline{m+0}$ Trans(1,2)

PROVE if $N \vdash \underline{m+n} = \underline{m+n}$, then $N \vdash \underline{m+(n+1)} = \underline{m+n+1}$

4. $\underline{m+n} = \underline{m+n}$ by IH

PROVE $N \vdash \underline{m+(n+1)} = \underline{m+n+1}$

But, since the numeral term for m+(n+1) is the same as the numeral term for (m+n)+1:

PROVE $N \vdash \underline{(m+n)+1} = \underline{m+n+1}$

5. $S(\underline{m+n}) = S(\underline{m+n})$ LL(4)
6. $S(\underline{m+n}) = \underline{(m+n)+1}$ by definition of numeral term
7. $S(\underline{m+n}) = \underline{m}+S(\underline{n})$ N-A6, Symm
8. $\underline{(m+n)+1} = S(\underline{m+n})$ Symm-Trans(6,5)
9. $\underline{(m+n)+1} = \underline{m}+S(\underline{n})$ Trans(7,8)
10. $\underline{(m+n)+1} = \underline{m+n+1}$ by definition of numeral term

N-T48 $\underline{m \circ n} = \underline{m} \circ \underline{n}$

<u>Exercises</u>:

1. Using the reasoning of the proof of the above theorem, but not the theorem itself, derive the following results:

A) $\underline{2+0} = \underline{2} + \underline{0}$
B) $\underline{3+5} = \underline{3} + \underline{5}$

2. Prove N-T48 by Mathematical Induction.

Here are two more theorems whose proofs require the use of Mathematical Induction.

N-T49 For every n,
$\forall x[((x=\underline{0}) \vee (x=\underline{1}) \vee \ldots \vee (x=\underline{n})) \longleftrightarrow x \leq \underline{n}]$

N-T50 For every n and every φ,
$\forall x[((\varphi(\underline{0}) \wedge \varphi(\underline{1}) \wedge \ldots \wedge \varphi(\underline{n})) \longleftrightarrow \forall x(x \leq \underline{n} \rightarrow \varphi(\underline{n}))]$

We now prove that "less-than" is representable in N.

N-T51: (A) If n < m, then $N \vdash \underline{n} < \underline{m}$, and
(B) If n $\not<$ m, then $N \vdash \neg(\underline{n} < \underline{m})$.

PF OF (A): By Mathematical Induction on n.

PROVE: If 0 < m, then $N \vdash \underline{0} < \underline{m}$

Assume 0 < m. Then there is some number k, such that k + 1 = m. So, m = k+1, and, by definition of numeral term, $\underline{k+1} = S(\underline{k})$. Hence:

PROVE: $N \vdash \underline{0} < S(\underline{k})$

By definition of '<' (in clause (1) above):

√PROVE: $N \vdash \exists x(S(x) + \underline{0} = S(\underline{k}))$

1. $S(\underline{k}) + \underline{0} = S(\underline{k})$ N-A5
2. $\exists x(S(x) + \underline{0} = S(\underline{k}))$ EG

Exercises:

1. Finish the proof of (A) above by proving the induction step.
2. Prove (B) [Hint: If n $\not<$ m, then n = m or m < n. If n = m, use the earlier result of N-T46 to get $N \vdash \underline{n} = \underline{m}$. If n < m, use (A).]

N-T51 is now complete (assuming the completion of the above *exercises*). ∎

It is clear at this point that N can be developed much further. But, it becomes arduous to establish more complicated numerical results by deriving them in a first-order axiomatic theory such as N. To establish more facts about numbers, it is probably better at this point to turn to a standard book on number theory, where the proofs are carried out informally, with little or no mention of the underlying logic. A few of the results in those books may depend on concepts that are not translatable into first-order sentences. First-order sentences, you will recall, contain quantifiers that range over only the objects in a given domain of discourse. In first-order Number Theory those objects are numbers. If we wish to make a true mathematical statement not only about numbers but about sets of them as well, we may need another <u>kind</u> of quantifier — a second-order quantifier — to create a formal sentence stating that fact. Some true second-order sentences about numbers (those containing quantifiers over sets of numbers, in addition to having quantifiers over the numbers themselves) can not be fully translated into true first-order sentences. Thus, many facts about numbers, sets of numbers, sets of sets of numbers, and so on may not be expressible using quantifiers ranging over only the numbers themselves. Clearly, mathematical facts that are not even <u>expressible</u> in N cannot be derived in N.

Moreover, some mathematical facts of arithmetic that <u>can</u> be translated into sentences of N have proofs using expressions that may not be translatable into sentences of N. Two questions then arise: (1) whether concepts in these proofs <u>can</u> be expressed in N (however indirectly), and (2) whether, even if (1) fails, there may be <u>other</u> derivations in N of these same results.

In Chapter 11, *Gödel's Theorems*, we investigate whether any first-order theory like N can be used to obtain all the truths of arithmetic that can be expressed in L_N. That is, we investigate whether N is mathematically complete. In Chapter 10, *Models For First-Order Theories*, we present some additional concepts that will assist that investigation and also enable us to understand first-order theories from the perspective of the abstract structures that the first-order theories are "about".

CHAPTER 10

MODELS FOR FIRST-ORDER THEORIES

[* ADVANCED *][1]

TOPICS: *Structure, vs. interpretation, homomorphism, monomorphism, epimorphism, isomorphism, elementary equivalence, substructure, extension, elementary substructure, elementary extension, cardinality of first-order languages, Löwenheim-Skolem (-Tarski & Vaught) theorems (Downward and Upward), Capture & Escape theorems, being a natural number is not capturable, categoricity & (mathematical) completeness, k-categoricity, sketch of ultraproduct construction, Łoś' Theorem, and model-theoretic proof of compactness.*

Chapter 8, *First-Order Theories*, dealt almost exclusively with syntactic issues. In that chapter a first-order theory $\mathcal{T}$ (with equality) was defined as consisting of a language L_T, an intial set of sentences Δ_T, and its set of theorems $Thms_T$. The sentences of a theory $\mathcal{T}$ are syntactic objects: strings arranged in certain specified patterns. Similarly, derivations are syntactic objects: sequences of sentences (and numbers) whose correctness can be checked by visual examination of the shapes of the sentences in the sequences. <u>After</u> a theory has been constructed, we can use interpretations to determine whether or not the sentences of that theory are true. When interpretations arise in this way, they seem dependent on the existence of sentences that they are to interpret. Without the sentences there would be nothing for the interpretations to interpret.

There is a completely different perspective in modern logic according to which interpretations can be looked at as providing algebraic structures that can be investigated in their own right. The properties of algebraic structures can be studied <u>independently</u> of any particular language. From <u>this</u> perspective, then, structures are presented first.

---

[1] Most of the principles explained in this chapter can be understood on an intuitive level, but many of the proofs require some familiarity with set theory, such as could be gotten, for example, from Halmos [1960].

Then, <u>afterwards</u>, we can investigate sentences of a first-order language concocted to express facts about these structures (though the particular language used to express these facts is independent of the properties of the structures themselves). Thus, the central characters in this chapter will be first-order structures themselves.

10.1 STRUCTURES

Let's begin with an example. The following is a first-order structure:

$$S = \langle \{0,1,2,3,4\}, \cdot, *, 0 \rangle$$

S consists of the set of the first five natural numbers, together with two operations and a distinguished element. The set $\{0,1,2,3,4\}$ is the universe of S, the binary operation $\cdot$ is "plus MOD 5", $*$ is the "inverse operation" for $\cdot$, and 0 is the distinguished element for the set. "Plus MOD 5" is the operation such that $(e \cdot e') =$ the remainder left over when e is added to e' and the sum is divided by 5. So, for example, $(3 \cdot 4) = 2$, since $3 + 4 = 7$, which, divided by 5 leaves the remainder 2. 0 is the identity element for $\cdot$, which means that for any element e, $(e \cdot 0) = e$, and $(0 \cdot e) = e$ too. We can see that this is true: any element added-MOD-5 to 0 returns that element. For each element e, its inverse e^* is the element such that $(e \cdot e^*) = 0$. For example, 1^* is 4, since $(1 \cdot 4) = 0$ (and $(4 \cdot 1) = 0$). It happens in this case that S is an example of a group. That means that S satisfies all the properties that make it a group. At this moment we do not care in what language those properties are expressed, just that S must have them in order for it to be a group (the only property of groups not yet mentioned is that $\cdot$ must be associative).

Of course, there are many other examples of groups besides the example just presented. And, there are many other <u>kinds</u> of structures besides group structures. Now that we have seen one example of a first-order structure, let us define them in general.

A **first-order structure** S has three things: (1) a non-empty set S called the universe of S, (2) some number of relations on elements of S, and (3) some number of operations or functions. In general, a first-order structure looks like this:

$$S = \langle S, R^k, O^k \rangle$$

(Note that 'R^k' represents <u>all</u> the relations in S and that 'O^k' represents <u>all</u> the operations. We sometimes write 'R^k' and 'O^k' for a particular relation or operation. These two distinct uses should be clear from context.)

In the group presented above, S happens to have no relations, $S = \{0,1,2,3,4\}$, and $\cdot$, $*$, and 0 are all operations. (0 is technically a 0-ary operation, as are all distinguished elements of a structure.)

Now that we have defined first-order structures, we can explain how a first-order language can be used to express facts about structures. Suppose we begin with the following first-order structure S, where S is the universe of S, R_1^2 is a binary relation on S, O_1^2 and O_2^2 are two binary operations on S, and O_1^0 is a 0-ary operation on S (i.e., O_1^0 is a distinguished element of S):

$$S = \langle S,\ R_1^2,\ O_1^2,\ O_2^2,\ O_1^0 \rangle$$

To express facts about structure S in a first-order language requires an "appropriate" language for S. We need a first-order language L_T containing the following non-logical symbols: one binary relation symbol 'R_1^2' to be interpreted as the relation R_1^2 in structure S, two binary operation symbols 'O_1^2' and 'O_2^2' to be interpreted as O_1^2 and O_2^2 in S, and one 0-ary operation symbol 'O_1^0' to be interpreted as O_1^0 in S. It is not necessary to create a non-logical symbol to be interpreted as S in S, since '$\forall$' will now be interpreted as ranging over the elements of S. Thus, an **appropriate language for** S has a unique k-ary relation symbol and operation symbol for each k-ary relation and operation in structure S. When two structures S_1 and S_2 have the same number of k-ary relations and k-ary operations for any k, we say they are of the same **type**. This means a language appropriate for the one is appropriate for the other.

Now, suppose that the universe S of structure S is $\mathbb{N}$, the set of natural numbers, and R_1^2, O_1^2, O_2^2, O_1^0 are, respectively, the less-than relation, $<$, the binary operation of addition, $+$, the binary operation of multiplication, $\cdot$, and the distinguished element 0. Then the structure is more clearly portrayed as:

$$\mathcal{S} = \langle \mathbb{N}, <, +, \cdot, 0 \rangle.$$

We often use perspicuous symbols in a language appropriate for a structure $\mathcal{S}$, such as the symbol '<' for the less-than relation <, the symbol '+' for the addition operation +, the symbol '∘' for the multiplication operation ·, and the symbol '$\underline{0}$' for the natural number 0. Strictly speaking, relation and operation letters are selected from the infinite stock available in L_T, as specified at the beginning of Chapter 8. Thus, the easier-to-read symbols we actually use can be thought of as stand-ins for the official symbols from L_T.

Structure vs. Interpretation

An interpretation can now be seen as a mapping whose range is a structure. Formally, we can identify the domain of the mapping as the following set {'domain'} ∪ {the non-logical symbols of a language appropriate for that structure}. The interpretation maps the word 'domain' to the universe **S** of the structure, and the interpretation maps each relation letter 'R_j^k' from the language to the relation $\mathbf{R}_j^k$ of the structure and each operation letter 'O_j^k' to the operation $\mathbf{O}_j^k$ of the structure. We can state this more succinctly by saying that an interpretation $\mathcal{I}$ is a function such that:

$$\mathcal{I}: (\{\text{'domain'}\} \cup \{\text{non-logical symbols of } L_T\}) \longrightarrow \{\mathbf{S}, \mathbf{R}_j^k, \mathbf{O}_j^k\},$$

where $\mathcal{I}(\text{'domain'}) = \mathbf{S}$, and for any relation letter R_j^k, $\mathcal{I}(R_j^k) = \mathbf{R}_j^k$, and for any O_j^k, $\mathcal{I}(O_j^k) = \mathbf{O}_j^k$. So, the elements in the range of the mapping are the elements of a structure (the map being onto, of course).

Given any structure, we say that the interpretation whose range is that structure is **the associated interpretation**. From the other direction, we say that an interpretation **specifies** a structure. To evaluate whether a sentence is true for a given structure requires us to use the associated interpretation, rather than the structure itself, since it is the interpretation that establishes the formal link between the language and the structure. When we say that a sentence is true (or false) in a structure, what is meant strictly is that the sentence is true (or false) for the associated interpretation. The same goes for an entire set of sentences. When it is said, for example, that a structure $\mathcal{S}$ models a set of sentences Γ (i.e., $\mathcal{S}(\Gamma) = T$), what is strictly the case is that the associated interpretation $\mathcal{I}$

models Γ (i.e., $\mathcal{I}(\Gamma) = \mathbf{T}$). When an interpretation $\mathcal{I}$ models a set of sentences Γ we often write the letter '$\mathfrak{M}$' instead of '$\mathcal{I}$' and say that $\mathfrak{M}$ models Γ or $\mathfrak{M}$ is a model of Γ. (If $\mathfrak{M}$ models a singleton set, $\{\varphi\}$, we normally drop the braces and say that $\mathfrak{M}$ models φ.)

There is another terminological issue that should be clarified. The subject of this chapter is usually identified as "Model Theory", rather than "Structure Theory", even when there are no sentences either explicitly or implicitly "modeled" (i.e., true under the associated interpretation). Recall that we have used 'model' up to this point as an interpretation for which a given sentence or set of sentences is true. But, it has another meaning as well. It is often used in a "neutral" sense to mean structure. We defer to this almost-universally accepted usage in this chapter, while retaining the earlier meaning as well. Thus, in this chapter, "models" are, almost always, structures. Our definition of the associated interpretation for a structure renders the ambiguity between a structure in and of itself and one that models something (via its associated interpretation) harmless. But, the ambiguity is worth noting.

Exercise:

Give a precise explanation in your own words of the difference between a structure and its associated interpretation.

10.2 FIRST-ORDER THEORY (model-theoretic version)

To obtain a first-order theory model-theoretically often begins with a class of models (of the same type) that have certain properties in common. Then, using an appropriate language for those models, we look for a set of axioms that "characterize" those models. That is, we look for axioms that are true for exactly that class of models (via the associated interpretations). The resultant theory $\mathcal{T}$ consists of (1) the language L_T; (2) the axioms Δ_T; and (3) all sentences $\mathbf{Cns}_T$ that are consequences of Δ_T. This model-theoretic approach provides a different perspective on a first-order theory than that provided in Chapter 8. Recall from Chapter 8 that a first-order theory $\mathcal{T}$ was defined as an ordered triple consisting of (1) a language, L_T; (2) a set of initial sentences, Δ_T; and (3) the set of all theorems,

Thms$_T$, that can be derived from Δ_T. We said in that chapter that a sentence φ <u>belongs to</u> $\mathcal{T}$ when $\varphi \in$ Thms$_T$, and we also mentioned that a theory is sometimes identified in a loose sense with its theorems.

In this chapter we can identify a theory loosely with the <u>consequences</u> of Δ_T rather than its <u>theorems</u>. We write **Cns(Γ)** as the set of all consequences of the set of sentences Γ. When Γ is the set of initial sentences of a theory — i.e., when $\Gamma = \Delta_T$ — we usually write Cns$_T$ (as indicated above) rather than Cns(Δ_T). Since Thms$_T$ = Cns$_T$ (via completeness and soundness), exactly the same sentences are arrived at model-theoretically as are obtained proof-theoretically (from derivations). So, we now say that a sentence φ <u>belongs to</u> $\mathcal{T}$ if and only if $\varphi \in$ Cns$_T$.

From the perspective of model theory, then, it is more appropriate to consider a theory $\mathcal{T}$ to be the following ordered triple:

$$< L_T, \Delta_T, Cns_T >,$$

which is justified by the earlier proof that Thms$_T$ = Cns$_T$. When thinking of a theory $\mathcal{T}$ deductively — in terms of the <u>theorems</u> that can be derived — we referred to the <u>theses</u> of $\mathcal{T}$. In thinking of a theory model-theoretically — in terms of its logical <u>consequences</u> — we will speak of the <u>assertions</u> of $\mathcal{T}$. A sentence φ is **an assertion of** $\mathcal{T}$ just in case $\varphi \in$ Cns$_T$. So, obviously, a sentence is an assertion of a theory just in case it is a thesis of that theory. Later in this chapter we connect the assertions of a theory with properties (of a class of models) that the theory characterizes.

One way to obtain a theory $\mathcal{T}$ is to begin with a structure and take the set of all sentences (of an appropriate language) that are **True** in that structure. We write $\mathcal{T}h(\mathcal{S}_1)$ for: $\{\varphi|\ \varphi$ is **True** in $\mathcal{S}_1\}$.

CLAIM: $\mathcal{T}h(\mathcal{S}_1)$ is a first-order theory.

To prove the above claim that $\mathcal{T}h(\mathcal{S}_1)$ is a first-order theory requires us

TO PROVE: $\mathcal{T}h(\mathcal{S}_1) = Cns(\mathcal{T}h(\mathcal{S}_1))$.

PF of CLAIM: ($\Longrightarrow$) This is the easy direction. For <u>any</u> set of sentences Γ, $\Gamma \subseteq Cns(\Gamma)$, since for any sentence $\varphi \in \Gamma$, $\Gamma \models \varphi$.

($\Longleftarrow$) In this direction, we must prove that $Cns(Th(S_1))$ has no <u>more</u> sentences than $Th(S_1)$ itself. Take any sentence ψ such that $\psi \in Cns(Th(S_1))$. That means that $Th(S_1) \models \psi$. We will prove that ψ must be true in S_1. Since $Th(S_1) \models \psi$, ψ is true in <u>any</u> <u>model</u> of $Th(S_1)$. But, S_1 is itself one of those models. Hence, ψ is true in S_1 also. Thus, $\psi \in Th(S_1)$. ∎

Actually, $Th(S_1)$ is a special case of a more general fact. Let $\mathcal{C}$ be a <u>class</u> of models of the same type. Then: $Th(\mathcal{C}) = \{\varphi |\ \varphi$ is **True** for <u>every</u> model in $\mathcal{C}\}$. (Note in this case that the first-order language containing φ is not explicitly mentioned. Here and elsewhere, we assume φ belongs to some <u>appropriate</u> language for the models in $\mathcal{C}$.)

CLAIM: $Th(\mathcal{C})$ is a theory.

Exercise:

Prove the above claim.

In words, $Th(\mathcal{C})$ is **the theory of the class of structures** $\mathcal{C}$ or just **the theory of** $\mathcal{C}$. Let us explore this notion and relate it to some earlier ones.

Let $MOD(\Gamma)$ be **the class of all models of the set of sentences** Γ. Suppose that a sentence φ is <u>not</u> a consequence of Γ and $\neg\varphi$ is <u>not</u> a consequence of Γ either. Let's further suppose that Γ has at least one model. Now, let's look at the set of sentences $\Gamma \cup \{\varphi\}$ and ask what the relationship is between $MOD(\Gamma)$ and $MOD(\Gamma \cup \{\varphi\})$. One way to think of this is to notice that the models in $MOD(\Gamma)$ (we assumed there was at least one) can be split into two categories: those that <u>do</u> model the sentence φ and those that do <u>not</u>. The models in $MOD(\Gamma \cup \{\varphi\})$, however, are just the ones in $MOD(\Gamma)$ that <u>do</u> model φ. Hence, $MOD(\Gamma \cup \{\varphi\})$ contains exactly "half" of the models of $MOD(\Gamma)$. More generally, if $\Gamma \subseteq \Gamma'$, then $MOD(\Gamma') \subseteq MOD(\Gamma)$.

Now, let us look at two classes of models (or structures) $\mathcal{C}$ and $\mathcal{C}'$, where $\mathcal{C} = \{\mathfrak{M}\}$ and $\mathcal{C}' = \{\mathfrak{M}, \mathfrak{M}'\}$. That is, $\mathcal{C}$ is a singleton containing only model $\mathfrak{M}$, and $\mathcal{C}'$ contains $\mathfrak{M}$ and $\mathfrak{M}'$, where $\mathfrak{M}'$ is different from $\mathfrak{M}$. Clearly, $\mathcal{C} \subseteq \mathcal{C}'$. Let's suppose that for some sentence ψ, $\mathfrak{M}'(\psi) = \mathbf{T}$, whereas

$\mathfrak{M}(\neg\psi)$ = **T**. Then the <u>theory</u> of $\mathcal{C}'$ does <u>not</u> contain <u>either</u> ψ <u>or</u> $\neg\psi$ (since <u>neither</u> ψ <u>nor</u> $\neg\psi$ is true for <u>all</u> models in $\mathcal{C}'$). <u>But</u>, it <u>is</u> true that $\neg\psi \in \mathcal{T}h(\mathcal{C})$, since $\mathfrak{M}(\neg\psi)$ = **T**. Hence, in this special case, $\mathcal{T}h(\mathcal{C}') \subseteq \mathcal{T}h(\mathcal{C})$. In general, if $\mathcal{C} \subseteq \mathcal{C}'$, then $\mathcal{T}h(\mathcal{C}') \subseteq \mathcal{T}h(\mathcal{C})$.

Exercises:

Prove the following general facts about models (extending the intuitive arguments for the special cases explained above):
1. If $\Gamma \subseteq \Gamma'$, then $\mathcal{MOD}(\Gamma') \subseteq \mathcal{MOD}(\Gamma)$.
2. If $\mathcal{C} \subseteq \mathcal{C}'$, then $\mathcal{T}h(\mathcal{C}') \subseteq \mathcal{T}h(\mathcal{C})$.

Now, let Σ = {all sentences of L_T}, where $\mathcal{T}$ is any theory, and then $\mathcal{MOD}(\Sigma)$ is the class of all models of <u>all</u> sentences of L_T. Of course, $\mathcal{MOD}(\Sigma) = \emptyset$, since there are <u>no</u> models of <u>all</u> sentences of L_T. Let $\mathcal{MOD}_L$ be the class of <u>all</u> models for the language L_T. Then, the <u>theory</u> of $\mathcal{MOD}_L$ (i.e., $\mathcal{T}h(\mathcal{MOD}_L)$) is empty, since there cannot be a single sentence that is true for <u>all</u> models of the language.

Exercises:

Prove the following facts about models discussed above:
1. $\mathcal{MOD}(\Sigma) = \emptyset$.
2. $\mathcal{T}h(\mathcal{MOD}_L) = \emptyset$.
   Now, prove the following:
3. $\mathcal{MOD}(\emptyset) = \mathcal{MOD}_L$.
4. $\mathcal{T}h(\emptyset) = \Sigma$.

A number of facts about models, classes of models, and theories of classes of models should become clear. For example:

$$\text{Cns}(\Gamma) = \mathcal{T}h(\mathcal{MOD}(\text{Cns}(\Gamma))),$$

since the sentences true in all models of the consequences of Γ are just those consequences of Γ. Since $\mathcal{MOD}(\text{Cns}(\Gamma)) = \mathcal{MOD}(\Gamma)$,

$$\text{Cns}(\Gamma) = \mathcal{T}h(\mathcal{MOD}(\Gamma)).$$

We also get these "cancellation properties":

$$MOD(Th(MOD(\Gamma))) = MOD(\Gamma), \text{ and}$$
$$Th(MOD(Th(\mathcal{C}))) = Th(\mathcal{C}).$$

Exercises:

Prove the above-mentioned facts, namely:
1. $Cns(\Gamma) = Th(MOD(Cns(\Gamma)))$.
2. $MOD(Cns(\Gamma)) = MOD(\Gamma)$.
3. $Cns(\Gamma) = Th(MOD(\Gamma))$.
4. $MOD(Th(MOD(\Gamma))) = MOD(\Gamma)$.
5. $Th(MOD(Th(\mathcal{C}))) = Th(\mathcal{C})$.

Given the above facts, these principles should not be difficult to prove, for any theory T:

6. $Cns(\emptyset) \subseteq T \subseteq Cns(\Sigma)$.
7. $Th(MOD_L) \subseteq T \subseteq Th(\emptyset)$.
8. $Cns(\Sigma) = \Sigma$.
9. $Cns(\emptyset) \subseteq T \subseteq Th(\emptyset)$.
10. $Th(MOD_L) \subseteq T \subseteq Cns(\Sigma)$.

The following two exercises concern only the union and intersection of classes of models:

11. Prove $MOD(\Gamma) \cap MOD(\Delta) = MOD(\Gamma \cup \Delta)$.
12. Prove or disprove: $MOD(\Gamma) \cup MOD(\Delta) = MOD(\Gamma \cap \Delta)$. [Hint: one direction works. Find a counter-example to the other.]

10.3 ISOMORPHISM & ELEMENTARY EQUIVALENCE

We define a **homomorphism** from $\mathcal{S}_1$ to $\mathcal{S}_2$, where $\mathcal{S}_1$, $\mathcal{S}_2$ are structures of the same type. The intuitive idea is that a homomorphism from one structure to another preserves the relations and operations from the first structure to the second.

Formally, a **homomorphism** from $\mathcal{S}_1$ to $\mathcal{S}_2$ is a function h such that $h: \mathbf{S}_1 \to \mathbf{S}_2$ and for every $e_1, e_2, \ldots, e_k \in \mathbf{S}_1$,

(i) $h(O_1^k \langle e_1, e_2, \ldots, e_k \rangle) = O_2^k \langle h(e_1), h(e_2), \ldots, h(e_k) \rangle$, and

(ii) $R_1^k \langle e_1, e_2, \ldots, e_k \rangle$ iff $R_2^k \langle h(e_1), h(e_2), \ldots, h(e_k) \rangle$.

For example, let $\mathbf{S}_1 = \mathbb{N}$ (i.e., $\{0, 1, 2 \ldots\}$) and $\mathbf{S}_2 = \{\ldots -4, -2, 0, 2, 4 \ldots\}$. Then, define $+_1$, $<_1$, as ordinary "plus" and "less than" on $\mathbf{S}_1$. Similarly $+_2$, $<_2$ are "plus" and "less than" on $\mathbf{S}_2$. We now have two structures:

$$\mathcal{S}_1 = \langle S_1, +_1, <_1, \rangle, \text{ and}$$

$$\mathcal{S}_2 = \langle S_2, +_2, <_2, \rangle.$$

Now, define $h_1(e) = 2e$ for each $e \in S_1$. So, $h_1(0) = 0$, $h_1(1) = 2$, $h_1(2) = 4$, etc. Let us look at two elements in S_1, say 0 and 3. For these two elements, clauses (i) and (ii) entail that:

(a) $h_1(+_1(0,3)) = +_2(0,6)$, and

(b) $<_1(0,3)$ iff $<_2(0,6)$

We can rewrite (a) and (b) to make them more readable:

(a)' $h_1(0 +_1 3) = (0 +_2 6)$, and

(b)' $0 <_1 3$ iff $0 <_2 6$

(a)' says that the image under h_1 of $(0 +_1 3)$ is the same as $0 +_2 6$. $0 +_1 3 = 3$ and $h_1(3) = 6$. Hence, $h_1(0 +_1 3) = 6$. And, on the right side, $0 +_2 6 = 6$, as well. So, (a)' holds. Clearly (b)' holds too, since $0 < 3$ and $0 < 6$.

More generally, we can see that clauses (i) and (ii) hold for <u>any</u> two elements δ_1 and δ_2 in S_1. To Prove (i), we need to prove $h_1(\delta_1 +_1 \delta_2) = h_1(\delta_1) +_2 h_1(\delta_2)$. By the definition of h_1, the left side of that equation is $2(\delta_1 +_1 \delta_2)$, which equals $2\delta_1 +_1 2\delta_2$. The right side of the equation is (also by the definition of h_1) $2\delta_1 +_2 2\delta_2$. Since $+_1$ and $+_2$ are defined as ordinary "plus" on the elements for which they are defined, the left side equals the right side and (i) has been proved. It is straightforward to prove (ii) for this example. So, h_1 is a homomorphism in this case.

Exercises:

1. Prove that clause (ii) holds for h_1 in the example above.
2. For h_1 defined as above from $\mathcal{S}_1$ to $\mathcal{S}_3$, where the only difference here is that S_3, the universe of $\mathcal{S}_3$, is the set $\{0,2,4...\}$, which is a proper subset of S_2. Everything else is defined the same as above. Prove (i) and (ii) as before, <u>and</u> prove that h_1 is both 1-1 and onto in this case.

Again, the idea behind a homomorphism is to preserve the structural relations and operations in going from one structure to another.

Here is another example:

$$S_3 = \langle \mathbb{N}, \cdot_3 \rangle \text{ and}$$
$$S_4 = \langle \{0,1,2,3,4\}, \cdot_4 \rangle$$

where "$\cdot_3$" is ordinary multiplication on the natural numbers and "$\cdot_4$" is the remainder obtained when the ordinary product of any two numbers in $\{0,1,2,3,4\}$ is divided by 5 (i.e., $\cdot_4(e,e') = e \times e'$ MOD 5). So, $\cdot_4 (0,1) = 0$, $\cdot_4 (1,3) = 3$, $\cdot_4 (2,3) = 1$, $\cdot_4 (3,4) = 2$, etc.

Now, we define $h_2: \mathbb{N} \rightarrow \{0, 1, 2, 3, 4\}$ such that $h_2(e)$ is the remainder obtained when e is divided by 5 (i.e., $h_2(e) = $ e MOD 5). We claim h_2 is a homomorphism. (To take just one example, let us check whether $h_2(3 \cdot_3 17) = h_2(3) \cdot_4 h_2(17)$. To get the left side, we divide 51 by 5, and get 1 as the remainder. Now let us test the other side. Since $h_2(3) = 3$, and $h_2(17) = 2$, we must consider $(3 \cdot_4 2)$, which is 1. So both sides match.)

Exercise:

Prove that h_2 in the above example is a homomorphism.

Notice that h_1 is not onto and h_2 is not 1-1 in the two examples provided. So, obviously a homomorphism does not need to be 1-1 or onto. A homomorphism that is 1-1 is a **monomorphism** and a homomorphism that is onto is an **epimorphism**.

For us, the most important special case of a homomorphism is an **isomorphism**. When the mapping h is both 1-1 and onto, then the homomorphism is an **isomorphism**. The reason isomorphisms are so important to us is that they represent the notion of "identity of structure". For example, consider the following structures:

$$S_1 = \langle \{1,2,3,\ldots\}, +_1 \rangle$$
$$S_2 = \langle \{2,4,6,\ldots\}, +_2 \rangle$$

Even though $S_1 - S_2 = \{1,3,5,\ldots\}$, we can consider S_1 and S_2 to be "essentially the same structure". Why? Formally, the answer is that there is a 1-1 correspondence $h: S_1 \rightarrow S_2$, such that $h(e) = 2e$ *and* h is a homomorphism. In short, the two structures are isomorphic, written: $S_1 \simeq S_2$.

Intuitively speaking, S_1 is essentially the same structure as S_2 because we can think of $2, 4, 6, \ldots$, as just being disguised to look different from $1, 2, 3, \ldots$. That is, we can say that as far as the operation $+_2$ is concerned — and this is the only operation in S_2 — there is no way to distinguish elements of S_2 from their counterparts in S_1.

Exercises:

1. Prove that h in the above example is an isomorphism from S_1 to S_2.

Let $S_1 = \langle \{a,b,c,d\}, +, \cdot \rangle$, and let $S_2 = \langle \{i,j,k,l\}, *, \circ \rangle$. Define h from S_1 to S_2 such that $h(a) = k$, $h(b) = i$, $h(c) = j$, and $h(d) = l$. And the following tables define the two operations in the two structures:

+	a	b	c	d
a	a	b	c	d
b	b	a	d	c
c	c	d	a	b
d	d	c	b	a

·	a	b	c	d
a	a	a	a	a
b	a	b	c	d
c	a	a	a	a
d	a	b	c	d

*	i	j	k	l
i	k	l	i	j
j	l	k	j	i
k	i	j	k	l
l	j	i	l	k

∘	i	j	k	l
i	i	j	k	l
j	k	k	k	k
k	k	k	k	k
l	i	j	k	l

2. Prove h is an isomorphism from S_1 to S_2.

When the truths (expressed in a first-order language) of one structure, S_1, are indistinguishable from the truths of another structure, S_2, then S_1 is **elementarily equivalent** to S_2, written: $S_1 \equiv S_2$. More formally, S_1 is **elementarily equivalent** to S_2 if for each sentence φ of a language L_T of the appropriate type,

$$\mathcal{I}_1(\varphi) = \mathcal{I}_2(\varphi).$$

In words, S_1 is elementarily equivalent to S_2, if the interpretation for S_1 of any sentence of the language results in the same truth value as the interpretation for S_2 of that sentence.

It is a fact that if two structures are "essentially the same" — meaning they are isomorphic — then they are elementarily equivalent. We will prove this fact as a theorem. But, before embarking on the exact statement of the theorem and its very technical proof, let us look at the idea behind the theorem.

<u>Idea</u> <u>of</u> <u>The</u> <u>Isomorphism</u> <u>Theorem</u>: According to the Isomorphism Theorem, when two structures are essentially the same structure, then any sentence that expresses something that is true for one structure also expresses something that is true for the other one as well. Put this way the theorem is not at all surprising. The theorem establishes a bridge between the purely algebraic properties of a structure and what a first-order language can "say" about the structure. In particular: no first-order language can say something true about one structure while using the same sentence to say something false about an isomorphic one.

The following proof of the Isomorphism Theorem is quite technical and lengthy, though as we can see from the paragraph above, the idea behind it is easy to grasp. The idea is, again: a sentence of a first-order language that says something true about a given structure must say something true about all isomorphic copies of that structure. You may wish to skip the technical details of the proof for now and return to them later.

Coinciding with our informal statement of the Isomorphism Theorem, we need:

TO PROVE: if $S_1 \simeq S_2$ then for any φ of the appropriate L_T:

$$\mathcal{I}_1(\varphi) = \mathcal{I}_2(\varphi),$$

where $\mathcal{I}_1$ is the interpretation associated with S_1 and $\mathcal{I}_2$ is the interpretation associated with S_2. We actually prove a somewhat stronger version of this fact for L_{T^+}, the extended language that includes logical names. Suppose:

$$S_1 = \langle S_1, O_1^k, R_1^k \rangle,$$
$$S_2 = \langle S_2, O_2^k, R_2^k \rangle,$$

and $e_1, e_2, \ldots, e_k \in S_1$. And, let n be a naming function for S_1. Here is an exact statement of the result to be proved.

ISOMORPHISM THEOREM: Let f be an isomorphism from $\mathcal{S}_1$ to $\mathcal{S}_2$. Then $\langle \mathcal{I}_1, n \rangle(\varphi) = \langle \mathcal{I}_2, f \circ n \rangle(\varphi)$ for any sentence $\varphi \in L_{T^+}$.

What is given to us in the hypothesis of the ISOMORPHISM THEOREM is that f is an isomorphism between $\langle S_1, O_1^k, R_1^k \rangle$ and $\langle S_2, O_2^k, R_2^k \rangle$. Specifically, that means $f: S_1 \to S_2$ is 1-1 and onto and for $e_1, e_2, \ldots, e_k \in S_1$:

(i) $f(O_1^k \langle e_1, e_2, \ldots, e_k \rangle) = O_2^k \langle f(e_1), f(e_2), \ldots, f(e_k) \rangle$,

(ii) $R_1^k \langle e_1, e_2, \ldots, e_k \rangle$ iff $R_2^k \langle f(e_1), f(e_2), \ldots, f(e_k) \rangle$.

((i) and (ii) say that f is a homomorphism). So now, for any sentence φ of L_{T^+} we need:

TO PROVE: $\langle \mathcal{I}_1, n \rangle(\varphi) = \langle \mathcal{I}_2, f \circ n \rangle(\varphi)$.

PF: First, for any variable-free term δ of L_{T^+}, we need:

TO PROVE:
1) $f(\langle \mathcal{I}_1, n \rangle(\delta)) = \langle \mathcal{I}_2, f \circ n \rangle(\delta)$.

Case 1a: δ is a name. Then (for the left side) $f(\langle \mathcal{I}_1, n \rangle(\delta)) = f(n(\delta))$. And (for the right side) $\langle \mathcal{I}_2, f \circ n \rangle(\delta) = f \circ n(\delta) = f(n(\delta))$ also.

Case 1b: δ is a 0-ary operation letter O^0. Then, for the left side, $f(\langle \mathcal{I}_1, n \rangle(O^0)) = f(O_1^0)$. And, for the right side, $\langle \mathcal{I}_2, f \circ n \rangle(O^0) = O_2^0$. But, by condition (i) we have $f(O_1^0) = O_2^0$. Thus, the two sides are equal.

Case 2: δ is a k-ary operation symbol and $\delta_1, \delta_2, \ldots, \delta_k$ are any variable-free terms. We prove this by Complete Induction on the number of symbols in δ that $f(\langle \mathcal{I}_1, n \rangle(\delta)) = \langle \mathcal{I}_1, f \circ n \rangle(\delta)$, assuming it holds for any variable-free term γ with fewer symbols than δ. We suppose that $\langle \mathcal{I}_1, n \rangle(\delta_i) = e_i$ for $1 \le i \le k$. Now, we need:

TO PROVE:
$f(O_1^k) \langle \langle \mathcal{I}_1, n \rangle(\delta_1), \langle \mathcal{I}_1, n \rangle(\delta_2), \ldots, \langle \mathcal{I}_1, n \rangle(\delta_k) \rangle =$
$(O_2^k) \langle \langle \mathcal{I}_2, f \circ n \rangle(\delta_1), \langle \mathcal{I}_2, f \circ n \rangle(\delta_2), \ldots, \langle \mathcal{I}_2, f \circ n \rangle(\delta_k) \rangle$.

Since each δ_i, $1 \leq i \leq k$, has fewer symbols than δ, our induction hypothesis gives us that:

$$f(\langle \mathcal{I}_1, n \rangle(\delta_i)) = \langle \mathcal{I}_2, f \circ n \rangle(\delta_i).$$

Since $\langle \mathcal{I}_1, n \rangle(\delta_i) = e_i$, we have:

$$f(e_i) = \langle \mathcal{I}_2, f \circ n \rangle(\delta_i) \text{ for each } \delta_i.$$

Thus, we need:

TO PROVE:
$$f(O_1^k)\langle e_1, e_2, \ldots, e_k \rangle = (O_2^k)\langle f(e_1), f(e_2), \ldots, f(e_k) \rangle.$$

But, this is given by condition (i). This concludes the proof of **1)**. ∎

Now, for any sentence φ of L_{T^+}, we need:

TO PROVE:
2) $\qquad\qquad\qquad \langle \mathcal{I}_1, n \rangle(\varphi) - \langle \mathcal{I}_2, f \circ n \rangle(\varphi).$

<u>Case 1a:</u> φ is atomic, and $\varphi = R^k \delta_1 \delta_2 \ldots \delta_k$, where R^k is not '='. We need:

TO PROVE:
$R_1^k \langle \langle \mathcal{I}_1, n \rangle(\delta_1), \langle \mathcal{I}_1, n \rangle(\delta_2), \ldots, \langle \mathcal{I}_1, n \rangle(\delta_k) \rangle$ iff
$R_2^k \langle \langle \mathcal{I}_2, f \circ n \rangle(\delta_1), \langle \mathcal{I}_2, f \circ n \rangle(\delta_2), \ldots, \langle \mathcal{I}_2, f \circ n \rangle(\delta_k) \rangle.$

By **1)**, $f(\langle \mathcal{I}_1, n \rangle(\delta_i)) = \langle \mathcal{I}_2, f \circ n \rangle(\delta_i)$ for any variable free term δ_i. So, if $\langle \mathcal{I}_1, n \rangle(\delta_i) = e_i$, then $\langle \mathcal{I}_2, f \circ n \rangle(\delta_i) = f(e_i)$. Hence, we need:

TO PROVE:
$R_1^k \langle e_1, e_2, \ldots, e_k \rangle$ iff $R_2^k \langle f(e_1), f(e_2), \ldots, f(e_k) \rangle.$

But, we get this result immediately from condition (ii).

<u>Case 1b:</u> φ is atomic and φ is $t_1 = t_2$, where t_1, t_2 are variable-free terms. We need:

TO PROVE:
$\langle \mathcal{I}_1, n \rangle(t_1 = t_2)$ iff $\langle \mathcal{I}_2, f \circ n \rangle(t_1 = t_2).$

That is, we need:

TO PROVE:
$\langle \mathcal{I}_1, n\rangle(t_1) = \langle \mathcal{I}_1, n\rangle(t_2)$ iff $\langle \mathcal{I}_2, f\circ n\rangle(t_1) = \langle \mathcal{I}_2, f\circ n\rangle(t_2)$.

Suppose $\langle \mathcal{I}_1, n\rangle(t_1) = e_1$ and $\langle \mathcal{I}_1, n\rangle(t_2) = e_2$. We know by 1) that:

$$\langle \mathcal{I}_2, f\circ n\rangle(t_1) = f(e_1) \quad \text{and}$$
$$\langle \mathcal{I}_2, f\circ n\rangle(t_2) = f(e_2).$$

Thus, we need:

TO PROVE: $e_1 = e_2$ iff $f(e_1) = f(e_2)$.

In one direction there is nothing to prove: if $e_1 = e_2$, then $f(e_1) = f(e_2)$. But, what establishes the other direction? If $f(e_1) = f(e_2)$, then is $e_1 = e_2$? It is established by the fact that f is 1-1. (Thus, this clause would fail if f were only a homomorphism.)

Case 2a: φ is not atomic. Suppose $\varphi = \forall\alpha\psi$. We need:

TO PROVE: $\langle \mathcal{I}_1, n\rangle(\forall\alpha\psi) = \langle \mathcal{I}_2, f\circ n\rangle(\forall\alpha\psi)$.

Our Induction Hypothesis is that this result holds for all sentences of L_{T^+} with fewer symbols than φ.

By the truth clause for universal sentences:

$$\langle \mathcal{I}_1, n\rangle(\forall\alpha\psi) = T \quad \text{iff} \quad \langle \mathcal{I}_1, n(\beta\rightarrow e)\rangle(\psi[e]) = T$$
$$\text{for all } e \in S_1.$$

And, by our Induction Hypothesis:

$$\langle \mathcal{I}_1, n(\beta\rightarrow e)\rangle(\psi[e]) = T \text{ for each } e \in S_1 \text{ iff}$$
$$\langle \mathcal{I}_2, f\circ n(\beta\rightarrow e)\rangle(\psi[f(e)]) = T \text{ for each } f(e) \in S_2.$$

But, since f is onto, every $e' \in S_2$ is such that for some $e \in S_1$, $e' = f(e)$. Thus, if $\langle \mathcal{I}_2, f\circ n(\beta\rightarrow e)\rangle(\psi[f(e)]) = T$ for all $f(e)$, then $\langle \mathcal{I}_2, f\circ n\rangle(\forall\alpha\psi) = T$. This concludes the proof of Case 2a. ∎

Exercises:

Prove that this theorem holds for the rest of the cases, where φ is:
1. $(\psi \vee \chi)$, 2. $(\psi \wedge \chi)$, 3. $\neg\psi$, 4. $(\psi \rightarrow \chi)$,
5. $(\psi \leftrightarrow \chi)$, and 6. $\exists\alpha\psi$.

COROLLARY (OF ISOMORPHISM THEOREM): If $\mathcal{S}_1 \simeq \mathcal{S}_2$ and $\varphi \in L_T$, then $\mathcal{I}_1(\varphi) = \mathcal{I}_2(\varphi)$.

Exercises:

1. Explain in your own words the ISOMORPHISM THEOREM and the overall strategy of the proof.
2. Prove the above COROLLARY.

10.4 SUBSTRUCTURES & ELEMENTARY SUBSTRUCTURES

Consider the following two structures, where $\mathbb{N}$ is the set of natural numbers, and $+_1$ and $<_1$ are ordinary "plus" and "less than" on the natural numbers; $\mathbb{R}$ is the set of real numbers, and $+_2$ and $<_2$ are ordinary "plus" and "less than" on the reals:

$$\mathcal{S}_1 = \langle \mathbb{N}, +_1, <_1 \rangle$$
$$\mathcal{S}_2 = \langle \mathbb{R}, +_2, <_2 \rangle$$

It is clear that $\mathbb{N} \subseteq \mathbb{R}$ (i.e., the set of natural numbers is included in the set of real numbers), and that $+_2$ restricted to $\mathbb{N}$ is the same as $+_1$, and that $<_2$ restricted to $\mathbb{N}$ is the same as $+_1$. That is (using symbols from Chapter 0), $+_1 = +_2|\mathbb{N}$, and $<_1 = <_2|\mathbb{N}$. We say in this case that $\mathcal{S}_1$ is a **substructure** of $\mathcal{S}_2$ and that $\mathcal{S}_2$ is an **extension** of $\mathcal{S}_1$. More generally, whenever $S_1 \subseteq S_2$ and each operation and relation in $\mathcal{S}_1$ is the restriction of the corresponding operation or relation in $\mathcal{S}_2$, then $\mathcal{S}_1$ is a **substructure** of $\mathcal{S}_2$ and $\mathcal{S}_2$ is an **extension** of $\mathcal{S}_1$, written $\mathcal{S}_1 \subseteq \mathcal{S}_2$.

Observe that in creating a substructure $\mathcal{S}_1$ of $\mathcal{S}_2$ we may start with a subset S of the universe of $\mathcal{S}_2$. But, before defining the operations and relations in $\mathcal{S}_1$ as the restrictions of those operations and relations in $\mathcal{S}_2$, we must first ensure that all operations are closed in the universe of $\mathcal{S}_1$. This will normally require the construction of another set S_1, where $S \subseteq S_1$ and S_1 is the closure of S.

SUBSTRUCTURE THEOREM: If $S_1 \subseteq S_2$, then for every <u>atomic</u> sentence $\varphi \in L_{T^+}$, and every n on S_1:

$$\langle \mathcal{I}_1, n \rangle(\varphi) = \langle \mathcal{I}_2, n \rangle(\varphi).$$

Exercise:

Prove the SUBSTRUCTURE THEOREM.

If $S_1 \subseteq S$ and for any sentence $\varphi \in L_{T+}$ and any naming function n on S_1:

$$\langle \mathcal{I}_1, n \rangle(\varphi) = \langle \mathcal{I}_2, n \rangle(\varphi),$$

Then S_1 is an **elementary substructure** of S_2 and S_2 is an **elementary extension** of S_1, written: $S_1 \preceq S_2$. It may be tempting to think that if $S_1 \simeq S_2$ and $S_1 \subseteq S_2$, then $S_1 \preceq S_2$. We show this is not the case by taking an example where $S_1 \simeq S_2$ and $S_1 \subseteq S_2$, but $S_1 \not\preceq S_2$.

Let us define S_1 and S_2 as follows, where $\mathbb{N}^+$ is the set of positive numbers, and $<_1$ is ordinary "less than" on the positive numbers; and $\mathbb{N}$ is the set of natural numbers and $<_2$ is "less than" on the natural numbers:

$$S_1 = \langle \mathbb{N}^+, <_1 \rangle, \qquad S_2 = \langle \mathbb{N}, <_2 \rangle.$$

Let $f: \mathbb{N}^+ \to \mathbb{N}$ be such that for any $e \in \mathbb{N}^+$, $f(e) = e - 1$. f is clearly 1-1 and onto. Furthermore, for any $e_1, e_2 \in \mathbb{N}^+$ it is clear that $e_1 <_1 e_2$ iff $f(e_1) <_2 f(e_2)$. Thus, f is an isomorphism from $\mathbb{N}^+$ to $\mathbb{N}$. That is, $S_1 \simeq S_2$.

It is also clear that $S_1 \subseteq S_2$, since $\mathbb{N}^+ \subseteq \mathbb{N}$ and for $e_1, e_2 \in \mathbb{N}^+$, $e_1 <_1 e_2$ iff $f(e_1) <_2 f(e_2)$. Hence, $S_1 \simeq S_2$ and $S_1 \subseteq S_2$. To see that $S_1 \not\preceq S_2$, consider the sentence '$\neg \exists x F^2 xa$', which says that there is no number less than a. When 'a' stands for 1, this sentence is true for the positive integers but false for the natural numbers (since 0 is a natural number less than 1). To show this more technically, we take $\mathcal{I}_1$ to be the interpretation associated with S_1, and $\mathcal{I}_2$ the interpretation associated with S_2. And we suppose $n(a) = 1$:

```
    𝒥₁                              𝒥₂
┌─────────────────┐          ┌─────────────────┐
│ S₁ = ℕ⁺         │          │ S₂ = ℕ          │
│ F² = < on ℤ⁺    │          │ F² = < on ℕ     │
└─────────────────┘          └─────────────────┘

    n                               n
┌──────────┐                  ┌──────────┐
│ n(a) = 1 │                  │ n(a) = 1 │
│ n(b) = 2 │                  │ n(b) = 2 │
│ n(c) = 3 │                  │ n(c) = 3 │
│    .     │                  │    .     │
│    .     │                  │    .     │
│    .     │                  │    .     │
└──────────┘                  └──────────┘
```

$\langle \mathcal{J}_1, n \rangle(\neg \exists x F^2 x[1]) = T$, since no positive integer is less than 1. But, $\langle \mathcal{J}_2, n \rangle(\neg \exists x F^2 x[1]) = F$ because the natural number 0 *is* less than 1.

What we can say here is that for a given positive integer e, the number e does not play the same "truth role" in the two structures. In particular, 1 is the least element of $\mathbb{N}^+$, but 1 is *not* the least element of $\mathbb{N}$. However, given $f: \mathbb{N}^+ \to \mathbb{N}$ where f is an isomorphic mapping from one structure to the other, e and $f(e)$ *do* play the same "truth role" in their respective structures. We sum this up more precisely (for the two structures previously mentioned, where n is a naming function on S_1, f is the isomorphism, and $\varphi \in L_{T^+}$) by saying:

$$\langle \mathcal{J}_1, n \rangle(\varphi) = \langle \mathcal{J}_2, f \circ n \rangle(\varphi), \quad \text{but}$$
$$\langle \mathcal{J}_1, n \rangle(\varphi) \neq \langle \mathcal{J}_2, n \rangle(\varphi).$$

A very nice characterization for a substructure to be an *elementary* substructure is given by TARSKI'S THEOREM:

TARSKI'S THEOREM: Suppose $\mathcal{S}_1 \subseteq \mathcal{S}_2$ and n is a naming function for $\mathcal{S}_1$. Then $\mathcal{S}_1 \preceq \mathcal{S}_2$ iff for any sentence $\exists \alpha \varphi$ of L_{T^+} such that $\langle \mathcal{J}_2, n \rangle(\exists \alpha \varphi) = T$, there is some element $e' \in S_1$ such that $\langle \mathcal{J}_2, n \rangle(\varphi[e']) = T$.

PF: ($\Rightarrow$.) Suppose $\mathcal{S}_1 \preceq \mathcal{S}_2$ and $\langle \mathcal{J}_2, n \rangle(\exists \alpha \varphi) = T$. Then $\langle \mathcal{J}_1, n \rangle(\exists \alpha \varphi) = T$. Thus $\langle \mathcal{J}_1, n(\beta \to e) \rangle(\varphi[e]) = T$ for some $e \in S_1$. Since $\mathcal{S}_1 \preceq \mathcal{S}_2$, $\langle \mathcal{J}_2, n(\beta \to e) \rangle(\varphi[e]) = T$ as well.

($\Leftarrow$.) We now suppose for any sentence $\exists \alpha \varphi \in L_{T^+}$, such that $\langle \mathcal{J}_2, n \rangle(\exists \alpha \varphi) = T$, there is an element $e \in S_1$ such that $\langle \mathcal{I}_2, n(\beta \to e) \rangle (\varphi[e]) = T$. And we show $\mathcal{S}_1 \leq \mathcal{S}_2$ by showing for any $\psi \in L_{T^+}$

*
$$\langle \mathcal{I}_1, n_1 \rangle (\psi) = \langle \mathcal{I}_2, n_1 \rangle (\psi)$$

The proof is by Complete Induction on the height (number of quantifiers and connectives) of ψ. When ψ is atomic, * is true since $\mathcal{S}_1 \subseteq \mathcal{S}_2$. Assuming * holds for χ_1 and χ_2, it is not difficult to show * holds for $\neg \chi_1$, $(\chi_1 \to \chi_2)$, $(\chi_1 \vee \chi_2)$, $(\chi_1 \wedge \chi_2)$, and $(\chi_1 \leftrightarrow \chi_2)$.

We now show * for $\exists \alpha \chi$. In particular, we show

**
$$\langle \mathcal{I}_1, n_1 \rangle (\exists \alpha \chi) = \langle \mathcal{I}_2, n_1 \rangle (\exists \alpha \chi),$$

assuming * holds for all sentences of L_{T^+} whose height is less than the height of $\exists \alpha \chi$. (That is our Induction Hypothesis.)

Suppose $\langle \mathcal{I}_1, n \rangle (\exists \alpha \chi) = T$. Then $\langle \mathcal{I}_1, n(\beta \to e) \rangle (\chi \alpha/\beta) = T$ for some $e \in S_1$. Since the height of $\chi \alpha/\beta$ is less than the height of $\exists \alpha \chi$, * holds for $\chi \alpha/\beta$. So, $\langle \mathcal{I}_2, n(\beta \to e) \rangle (\chi \alpha/\beta) = T$, as well. Hence, $\langle \mathcal{I}_2, n \rangle (\exists \alpha \chi) = T$.

Now, suppose $\langle \mathcal{I}_2, n \rangle (\exists \alpha \chi) = T$. Then, by the condition for this direction of the proof, we have that there is some element $e \in S_1$ such that $\langle \mathcal{I}_2, n(\beta \to e) \rangle (\chi[e]) = T$. By * again for sentences of lesser height, $\langle \mathcal{I}_1, n(\beta \to e) \rangle (\chi[e]) = T$. Hence, $\langle \mathcal{I}_1, n \rangle (\exists \alpha \chi) = T$. ∎

Exercises:

1. - 5. Prove * holds for $\neg \chi_1$, $(\chi_1 \to \chi_2)$, $(\chi_1 \vee \chi_2)$, $(\chi_1 \wedge \chi_2)$, and $(\chi_1 \leftrightarrow \chi_2)$.

Note that the above proof shows that the heart of the condition for $\mathcal{S}_1 \leq \mathcal{S}_2$ is:

TARSKI'S CONDITION: If $\langle \mathcal{I}_2, n(\beta \to e) \rangle (\chi[e]) = T$ for some $e \in S_2$, then there is some $e' \in S_1$ such that $\langle \mathcal{I}_2, n(\beta \to e) \rangle (\chi[e']) = T$.

COROLLARY: If TARSKI'S CONDITION holds, then $\mathcal{S}_1 \leq \mathcal{S}_2$ (given the appropriate conditions on all variables).

Exercise:

Provide the correct statement of the conditions for the COROLLARY of TARSKI'S THEOREM and prove that it holds.

Now, let us look back once again at our counter-example, where $S_1 \simeq S_2$ and $S_1 \subseteq S_2$, yet $S_1 \not\preceq S_2$. Remember:

$$S_1 = \langle \mathbb{N}^+, <_1 \rangle, \text{ and}$$
$$S_2 = \langle \mathbb{N}, <_2 \rangle.$$

For $0 \in \mathbb{N}$, this holds:

$$\langle \mathcal{I}_2, n(\beta \to 0) \rangle (\forall y (F[0]y \lor [0] = y)) = \mathbf{T}.$$

But, <u>for</u> <u>any</u> e' such that $e' \in \mathbb{N}^+$:

$$\langle \mathcal{I}_2, n(\beta \to e') \rangle (\forall y (F[e']y \lor [e'] = y)) = \mathbf{F}.$$

Thus, $S_1 \not\preceq S_2$ in this example because the two structures fail to meet TARSKI'S CONDITION.

We now use TARSKI'S THEOREM to show that the rationals are an elementary substructure of the reals. In the theorem below, $\mathbb{Q}$ is the set of rationals, and $<_1$ is "less than" on the rationals; and $\mathbb{R}$ is the set of reals and $<_2$ is "less than" on the reals:

THEOREM: $\langle \mathbb{Q}, <_1 \rangle \preceq \langle \mathbb{R}, <_2 \rangle$.

> PF: Suppose $\langle \mathcal{I}_R, n(\beta \to r) \rangle (\varphi[e_1, \ldots, e_k, r]) = \mathbf{T}$ for some r, where $e_1, \ldots, e_k$ are all rationals, and r is a real that is <u>not</u> rational. All we need to do is find a rational number, say p, such that p bears the same order relation to all of $e_1, \ldots, e_k$ as r does. Since the rationals are everywhere dense, there always is a $p \in \mathbb{Q}$ such that:
>
> $$\langle \mathcal{I}_R, n(\beta \to p) \rangle (\varphi[e_1, \ldots, e_k, p]) = \mathbf{T}$$
>
> This is the needed condition to prove $\langle \mathbb{Q}, < \rangle \preceq \langle \mathbb{R}, < \rangle$ by TARSKI'S THEOREM. ∎

We have already defined what it means for <u>structure</u> S_2 to be an extension of structure S_1 (or S_1 to be a substructure of S_2). Roughly, it means that the universe of S_1 is a subset of the universe of S_2 and all operations and relations on the smaller universe are the restrictions of those operations and relations on the larger one.

<u>Language</u> L_2 is an extension of L_1 if there are at least as many, and possibly more, symbols in L_2 than in L_1. Suppose L_2 is an extension of L_1 and S_2 is a structure appropriate for L_2. When by omitting some relations and/or operations in S_2, we arrive at a structure S_1 that is appropriate for L_1, then S_1 is the **reduction** of S_2 to L_1, written: $S_1 = S_2|L_1$. In the other direction, S_2 is the **expansion** of S_1 to L_2.

If S_2 is an <u>expansion</u> of S_1, then S_2 is identical to S_1 except that S_2 has possibly <u>more</u> operations and relations than S_1. Thus, it is an <u>expanded</u> <u>structure</u> that accommodates an <u>extended</u> <u>language</u>. Note, though, that an expansion and its reduction have exactly the same universe. (An <u>extension</u> of a structure, though, has a (possibly) <u>larger</u> <u>universe</u> for the <u>same</u> language.)

If language L_2 is identical to L_1 except for having more constants, then L_2 is an **extension by constants** of L_1. Thus, we can take an appropriate structure for L_1 and expand it to an appropriate structure for L_2, by providing distinguished elements in the expanded structure for the additional constants.

We can now unify our proof-theoretic and our model-theoretic terminology by re-proving our earlier NEW CONSTANTS THEOREM, this time model-theoretically (see Chapter 8 for the proof in terms of derivations).

NEW CONSTANTS THEOREM: If $L_{T'}$ is an extension by constants of L_T and $\Delta_{T'} = \Delta_T$, then $\mathcal{T}'$ is a conservative extension of $\mathcal{T}$.

PF (model-theoretic): Suppose there are denumerably many additional constants in $L_{T'}$, say $k_1, k_2, \ldots$. Then a model of $\mathcal{T}'$ looks like this:

$$\mathcal{M}' = \langle S', \ldots ; k_1, k_2, \ldots \rangle$$

(where the additional constants have been segregated from the rest by the placement of ';'). Consider any sentence φ of the poorer language L_T. Now, let $\mathcal{M} = \mathcal{M}'|L_T$ (i.e., $\mathcal{M}$ is the reduction of $\mathcal{M}'$ to L_T).

$$\mathcal{M} = \langle S', \ldots \rangle$$

It is clear that for any sentence φ of L_T:

$$M(\varphi) = T \text{ iff } M'(\varphi) = T,$$

which concludes the proof. ∎

Exercise:

Fill in the details of the above proof to make it explicitly satisfy the conditions of the NEW CONSTANTS THEOREM.

10.5 CARDINALITY OF FIRST-ORDER LANGUAGES

The cardinal number of sentences that can be formed in any first-order language discussed up to this point is $\aleph_0$. We can see that there are <u>at least</u> $\aleph_0$ sentences in any first-order theory by noticing that the following set of sentences has cardinality $\aleph_0$: $\{\exists x(x = x), \exists y(y = y), \exists z(z = z), \exists w(w = w), \exists x_1(x_1 = x_1), \ldots\}$. The sentences in that set are some of the sentences of $L_=$, where there are <u>no</u> operation symbols <u>or</u> relation symbols. Clearly, as the number of operation and relation symbols increases, the number of sentences increases as well. $L_=+$ adds $\aleph_0$ names to $L_=$, but the resultant number of sentences in $L_=+$ is still $\aleph_0$. The reason for this is straightforward: sentences are finite strings of symbols. If we have $\aleph_0$ symbols in a language, the number of finite strings of those symbols is $\aleph_0$.

There are occasions when we want a first-order language to contain a non-denumerable number of symbols **k**. Since the cardinality of logical symbols of a first-order language is always $\aleph_0$, we know that when the cardinality of the totality of symbols is **k** and **k** $> \aleph_0$, the cardinality of non-logical symbols is **k** itself. When the cardinality of non-logical symbols of a first-order language is **k** and **k** $> \aleph_0$, the cardinality of the set of all sentences of that language is **k** also. Again, the reason for this is that sentences are finite strings of symbols and thus the cardinality of the set of them cannot exceed **k**. It seems clear there are <u>at least</u> **k** sentences in such a language. Thus, we get two results: (1) that if the cardinality of the set of non-logical symbols of L is $\leq \aleph_0$, then the cardinality of sentences of L = $\aleph_0$ (we write: CARD(L) = $\aleph_0$); and (2) if the cardinality of the set of non-logical symbols for language L is **k**, where **k** $> \aleph_0$, then CARD(L) = **k** (i.e., the cardinality of sentences of language L is **k**).

Thus, when the cardinality of the non-logical symbols of a language exceeds $\aleph_0$, we can identify the cardinality of the set of sentences of the language with the cardinality of its non-logical symbols. We identify the cardinality of a <u>structure</u> S_1 with the cardinality of its universe $\in S_1$.

10.6 LÖWENHEIM-SKOLEM(-TARSKI & VAUGHT) THEOREMS

We now establish some theorems due to Löwenheim (1915), Skolem (1920), and Tarski and Vaught (1957), which, taken together, imply that if a theory has a model of some infinite cardinality, it has a model of *any* infinite cardinality (after one restriction pertaining to the language of the theory has been added). This result separates into two theorems, one called the "Downward Löwenheim-Skolem(-Tarski & Vaught) Theorem", and the other called the "Upward Löwenheim-Skolem (-Tarski & Vaught) Theorem."

The idea of the Downward LSTV Theorem is to begin with a structure S_2 of infinite cardinality **k**. Then, to take some subset **S** of S_2 of any smaller infinite cardinality **c** desired, where **c** is no smaller than the cardinality of the first-order language L that is appropriate for S_2 (this is the added condition referred to above). The subset **S** of S_2 then has at most $\aleph_0$ elements added to it, resulting in S_1. S_1 is then the universe of a new structure S_1 such that $S_1 \leq S_2$.

Before formally stating and proving the theorem, we remark that the cardinality of L, the language appropriate for S_2, serves as a lower bound to the cardinality of any structure we construct. That is, in the following proof, CARD(L) $\leq$ **c**. The reason for this lower bound will become clear <u>after</u> the proof of the <u>Upward</u> Löwenheim-Skolem(-Tarski & Vaught) Theorem.

DOWNWARD LSTV THEOREM: Let S_2 be a structure of infinite cardinality **k**, then for any infinite cardinality **c**, such that **c** < **k**, there is a structure S_1 of cardinality **c**, such that $S_1 \leq S_2$.

PF: Take any $B_0 \subseteq S_2$ such that CARD(B_0) = **c**. Consider any sentence $\exists \alpha \varphi \in L^+$ such that $e_1, e_2, \ldots, e_j \in B_0$ are the denotations for all j distinct names occurring in $\exists \alpha \varphi$ (given by some naming function n on B_0). Whenever it is the case that $\mathcal{I}_2(\exists \alpha \varphi [e_1, e_2, \ldots, e_j]) = \mathbf{T}$, take the *least*

$e \in S_2$ (according to some fixed well-ordering of S_2 [by the axiom of choice]) such that $\langle \mathcal{I}_2, n(\beta \to e) \rangle (\varphi[e_1, e_2, \ldots, e_j, e]) = T$, and put e in set B_1. Hence B_1 is defined as the set of all such elements e. $B_1 \supseteq B_0$, since for any $e_1 \in B_0$, $\mathcal{I}_2(\exists x(x = [e_1])) = T$, and e_1 itself <u>is</u> the <u>least</u> (indeed the only)[2] $e \in S_2$ such that $\langle \mathcal{I}_2, n(\beta \to e) \rangle ([e] = [e_1]) = T$.

More generally, we define B_{k+1} recursively for $k \in \mathbb{N}$, in terms of B_k, exactly as B_1 was defined in terms of B_0 above: for any B_{k+1}, B_{k+1} = the set of all e such that e is the least element belonging to S_2 such that for $e_1, e_2, \ldots, e_j \in B_k$, $\langle \mathcal{I}_2, n(\beta \to e) \rangle (\varphi[e_1, e_2, \ldots, e_j, e]) = T$, for some sentence $\varphi\alpha/\beta \in L_+$.

The same argument as above shows $B_k \subseteq B_{k+1}$ for all $k \in \mathbb{N}$. So, we know that for each k, CARD(B_k) $\geq c$. We show that CARD(B_k) cannot exceed c. Consider all sentences $\exists\alpha\varphi \in L_+$ such that j names occur in φ, there are c^j possible denotations for the j names occurring in φ (there are c possible denotations for <u>each</u> e_j, and since there are j names, there are $c \times c \times c \times \ldots \times c$ total possible denotations, where c is used as a factor j times). But, $c^j = c$, since c is infinite. Thus, for <u>each</u> <u>sentence</u> $\exists\alpha\varphi$ and <u>each</u> <u>naming</u> <u>function</u> n on B_k, there are only c possible e's for B_{k+1}. And, there are only CARD(L) sentences of the form $\exists\alpha\varphi$ in L+ (L+ creates only $\aleph_0 \times c$ new sentences, so CARD(L+) = CARD(L)). Hence, there are only CARD(L) $\times c = c$ possible e's for the totality of sentences $\exists\alpha\varphi$ and each naming function n on B_k. Thus, for each B_k, $k \in \mathbb{N}$, CARD(B_k) $\leq c$. Since we already showed CARD(B_k) $\geq c$, CARD(B_k) = c.

We now let $S_1 = \bigcup_{k \in \mathbb{N}} B_k$. Since $\aleph_0 \times c = c$, CARD(S_1) = c. We show that S_1 is closed with respect to all operations O^k. Assume $e_1, e_2, \ldots, e_k \in B_m$ for some B_m. Then $\mathcal{I}_2(\exists x O^k[e_1, e_2, \ldots, e_k] = x) = T$. By the method of constructing the B's, the <u>least</u> (and only) $e \in S_2$ such that $e = O_2^k(e_1, e_2, \ldots, e_k)$ is a member of B_{m+1}. In particular, for 0-ary constants O^0, we have the sentence $\exists x(x = O^0)$. So, $O_2^0 \in B_1$. So, $\mathcal{S}_2 | S_1$ is a substructure of $\mathcal{S}_2$ (with cardinality c).

Let $\mathcal{S}_1 = \mathcal{S}_2 | S_1$. We show $\mathcal{S}_1 \leq \mathcal{S}_2$. Take any sentence $\exists\alpha\varphi \in L_+$ such that the j distinct names in φ denote $e_1, e_2, \ldots, e_j \in S_1$ and $\mathcal{I}_2(\exists\alpha\varphi[e_1, e_2, \ldots, e_j]) = T$. By the manner of constructing the B's (ensuring that for any B_k, $B_k \subseteq B_{k+1}$), there must be a B_m such that all

---

[2] Note that this could *not* be claimed if '=' were not "True Identity".

$e_1, e_2, \ldots, e_j \in B_m$. Hence by our construction of the B's, for the <u>least</u> e such that $\langle \mathcal{I}_2, n(\beta \to e) \rangle (\varphi[e_1, e_2, \ldots, e_j, e]) = \mathbf{T}$, $e \in B_{m+1}$. Thus, $e \in S_1$. This satisfies TARSKI'S THEOREM and concludes the proof. ∎

COROLLARY (TO DOWNWARD LSTV): If $\mathcal{S}_2$ is any infinite structure, appropriate for L, where CARD(L) = $\aleph_0$, there is a structure $\mathcal{S}_1$ of card $\aleph_0$ such that $\mathcal{S}_1 \leq \mathcal{S}_2$.

What DOWNWARD LSTV shows is that we can trim down any infinite structure to one of any lesser infinite cardinality (down to the cardinality of the language itself). This raises the question whether it is possible to prove a similar theorem for the other direction. Specifically: if we take an infinite structure $\mathcal{S}_1$, can we find another one, $\mathcal{S}_2$, of any larger cardinality we choose, such that $\mathcal{S}_1 \leq \mathcal{S}_2$? Before answering this question, we need to take another look at Compactness.

You will recall from Chapter 7, *Connections Between Syntax & Semantics for First-Order Logic*, that Compactness follows from Strong Completeness. In that chapter Strong Completeness was proved for first-order logic *without* a) equality, and b) operation symbols. The proof of Strong Completeness, and hence, Compactness for the richer language was left as an exercise.

There is another, quite beautiful, purely model-theoretic proof of Compactness that we outline later in this chapter (though some of the concepts involved in the proof go beyond the scope of this book). For the present we wish to <u>assume</u> the truth of Compactness as an important model-theoretic concept and show what follows from it.

The version of Compactness that is most helpful in the present context (two renditions of Compactness were proved equivalent earlier) is:

COMPACTNESS THEOREM: For any first-order language L, if every finite subset of a set Γ of sentences of L has a model, then Γ itself has a model.

PF: Either as a corollary of Strong Completeness (for first-order languages including '=' and operation symbols), or by separate proof sketched toward the end of this chapter. ∎

The following theorem is a finite analog of the Upward Löwenheim-Skolem(-Tarski & Vaught) Theorem, so we dub it the "Upward Finite Satisfaction Theorem":

UPWARD FINITE SATISFACTION THEOREM: If a set of first-order sentences Γ has arbitrarily large finite models, then Γ has an infinite model.

PF: Let '$\exists_n$' abbreviate a sentence saying that there are at least n elements. So:

$\exists_2 = \exists x_1 \exists x_2 (x_1 \neq x_2)$

$\exists_3 = \exists x_1 \exists x_2 \exists x_3 ((x_1 \neq x_2) \wedge (x_1 \neq x_3)) \wedge (x_2 \neq x_3))$

.
.
.

$\exists_n = \exists x_1 \exists x_2 \ldots \exists x_n ((x_1 \neq x_2) \wedge (x_1 \neq x_3) \wedge (x_2 \neq x_3) \wedge \ldots \wedge (x_{n-1} \neq x_n))$.

We then consider the set:

$\Gamma \cup \{\exists_2, \exists_3, \exists_4, \ldots\}$.

By hypothesis, Γ alone has arbitrarily large finite models. Thus, each finite subset of $\Gamma \cup \{\exists_2, \exists_3, \exists_4, \ldots\}$ has a model. By COMPACTNESS, then, $\Gamma \cup \{\exists_2, \exists_3, \exists_4, \ldots\}$ itself has a model. This model has to be infinite because any finite model with n elements will fail to model sentences of the following forms: $\exists_{n+1}, \exists_{n+2}, \exists_{n+3}, \ldots$. ∎

In the proof of the UPWARD FINITE SATISFACTION THEOREM we used COMPACTNESS in the following way. First, we "enlarged" the set Γ so that the enlarged set had the property we desired. In this case the property was: having an infinite model if it has one at all. We then took all finite subsets of the enlarged set, which by hypothesis all had models. Then COMPACTNESS ensured that the enlarged set itself must have a model. The property built into the enlarged set guaranteed that any such model must be infinite. Such an infinite model of the enlarged set is then a model of our original set Γ.

The proof of the full-blown Upward Löwenheim-Skolem (-Tarski & Vaught) Theorem (UPWARD LSTV) demonstrates the use of a more complex variant of the "enlargement plus compactness" technique described in the preceding paragraph. For UPWARD LSTV, the enlarged set contains non-denumerably many sentences saying of non-denumerably many distinct new constants that their denotations all differ. As before, all finite subsets of this enlargement have models. And then COMPACTNESS ensures that the enlarged set itself has a model. Since the enlarged set can be satisfied only by a model <u>at least</u> as large as we want, we take such a model and then find one of <u>exactly</u> the correct cardinality by using the DOWNWARD LSTV. This model restricted to sentences without the new constants then fills the bill.

Let C be a set of constant symbols (0-ary operation symbols) whose cardinality is k, for $k > \aleph_0$ and for any two distinct constants C_a, C_b, take Γ_k to be the set of all sentences of the form: $C_a \neq C_b$. We start out with a set of sentences Γ that belong to L, and suppose Γ has an infinite model. Then, we show that Γ has a model of any infinite cardinality. First we state the theorem more formally:

UPWARD LSTV THEOREM: Assume CARD(L) = $\aleph_0$. If a set of sentences Γ of L has an infinite model, then Γ has a model of card k, $k > \aleph_0$.

PF: If Γ has a model, then by corollary to DOWNWARD LSTV, Γ has a model of card $\aleph_0$. Let $\mathfrak{M}_1$ be such a model of Γ. We first extend the language L to L + C, where C is a set of new constants of card k, as described above (L + C has k sentences). Now, we add the set of sentences Γ_k, as described above, where each sentence says that the denotation of any one new constant is different from the denotation of any other new constant.

Consider a given finite subset of $\Gamma \cup \Gamma_k$, say $\Gamma' \cup \Gamma'_k$. Suppose the number of new constants introduced by Γ'_k is m, say $\delta_1, \delta_2, \ldots, \delta_m$. And, let's suppose the number of constants in Γ' is j, say $\lambda_1, \lambda_2, \ldots, \lambda_j$. From the domain of model $\mathfrak{M}_1$ we assign distinct elements $e_1, e_2, \ldots, e_m$ to the new constants (introduced by Γ'_k) $\delta_1, \delta_2, \ldots, \delta_m$, respectively, where all the e_i are distinct from the j elements of $\mathfrak{M}_1$ that have been assigned to $\lambda_1, \lambda_2, \ldots, \lambda_j$. Since $\mathfrak{M}_1$ is infinite, we are assured that this can be done for any finite $\Gamma' \cup \Gamma'_k$. In each case, $\mathfrak{M}_1$ is simultaneously restricted to the j constants in Γ' and expanded to include the m new constants in Γ'_k. Moreover, $\mathfrak{M}_1$ models $\Gamma' \cup \Gamma'_k$.

$\mathfrak{M}_1$ so restricted and expanded serves as a model of $\Gamma' \cup \Gamma'_k$ because the restricted part models Γ' and when it is expanded, the assignments of the m new constants to distinct elements that differ from those already assigned to constants in Γ' assures that the new sentences of Γ'_k are modeled as well. By compactness, since every finite subset of $\Gamma \cup \Gamma_k$ has a model, $\Gamma \cup \Gamma_k$ has a model itself. Since there are **k** distinct constants and Γ_k says they are all distinct, any model $\mathfrak{M}_2$ of $\Gamma \cup \Gamma_k$ must have CARD $\geq$ **k**. But then by DOWNWARD LSTV, there is a model $\mathfrak{M}_3 \leq \mathfrak{M}_2$ such that CARD($\mathfrak{M}_3$) = **k**.

$\mathfrak{M}_3$ has the requisite cardinality and $\mathfrak{M}_3$ models $\Gamma \cup \Gamma_k$. Thus, $\mathfrak{M}_3$ models Γ alone. But $\mathfrak{M}_3$ is a model for the expanded language L + **c** and contains **k** distinguished elements (as denotations of the **k** constants that were added). So, we take $\mathfrak{M}_4$ as the restriction of $\mathfrak{M}_3$ to the original language L. $\mathfrak{M}_4$ satisfies the theorem. ∎

10.7 FIRST-ORDER CAPTURE & ESCAPE THEOREMS[3]

Let $\mathcal{MOD}(\Gamma)$ be the class of all models of some set of sentences Γ. If there is some property $\mathcal{P}$ possessed by all models in $\mathcal{MOD}(\Gamma)$ and possessed by no other first-order structures, we say that Γ **captures** $\mathcal{P}$. If there is a set of sentences Γ such that Γ captures $\mathcal{P}$, then $\mathcal{P}$ is **first-order capturable**. If no set of first-order sentences captures $\mathcal{P}$, then $\mathcal{P}$ **escapes first-order**.

"FINITUDE" ESCAPES CAPTURE: The property of being finite escapes first-order.

"INFINITUDE" CAPTURED: The property of being infinite is first-order capturable.

---

[3]The terminology of this section was invented because it seemed more intuitive than that used in the literature. According to one alternative definition, an "elementary class" is a class of structures such that they (and no others) model a <u>set</u> of first-order sentences. Then, a "basic elementary class" is a class of structures such that they (and no others) model a <u>single</u> sentence. But this terminology is not uniform. "Elementary class" is also used widely in the sense of modeling a <u>single</u> sentence.

"AT LEAST N" CAPTURED: The property of having at least
n elements for any finite n = 1, 2, 3, ... is
first-order capturable.

"EXACTLY N" CAPTURED: The property of having exactly
n elements for any finite n = 1, 2, 3, ... is
first-order capturable.

Exercises:

1. Prove "FINITUDE" ESCAPES CAPTURE. [Hint: suppose Γ has
   models of any finite cardinality. Look at the UPWARD
   FINITE SATISFACTION THEOREM.
2. Prove "INFINITUDE" CAPTURED. [Consider the set used
   in the proof of the UPWARD FINITE SATISFACTION THEOREM]
3. Prove both "AT LEAST N" CAPTURED and "EXACTLY N" CAPTURED.

If a property $\mathcal{P}$ is first-order capturable by a single sentence χ of L_T, then χ **asserts** $\mathcal{P}$. If there is a sentence of L_T asserting $\mathcal{P}$ then $\mathcal{P}$ is **finitely capturable** (in first-order). (Thus, an assertion of a theory asserts a property.)

NEGATION THEOREM FOR FINITE CAPTURABILITY: $\mathcal{P}$ is finitely
capturable iff $\neg \mathcal{P}$ is finitely capturable.

"INFINITUDE" ESCAPES FINITE CAPTURE: Being infinite is not
finitely capturable.

Exercise:

Prove the NEGATION THEOREM FOR FINITE CAPTURABILITY, and
then prove "INFINITUDE" ESCAPES FINITE CAPTURE as a
corollary.

Obviously, when a concept is finitely capturable we can finitely axiomatize it. To do that we just write the single sentence all of whose models (and no others) possess the property in question. Let's look at some examples.

Consider the "less-than" relation <. Take any set **S** on which < is defined. Call the resulting structure $\mathcal{S}$, where $\mathcal{S} = \langle S < \rangle$. We want to find a single sentence that says **S** is totally ordered by <. Here's our sentence (we call it χ_{ORD}):

$\chi_{ORD} = \forall x(\neg x < x) \land \forall x \forall y((x<y \lor y<x) \lor x=y) \land$
$\forall x \forall y \forall z((x<y \land y<z) \rightarrow x<z)$.

Exercises:

Write a sentence that says what χ_{ORD} says and, in addition, says:
1. There's a minimum element (one less than every other).
2. There's a minimal element (one that no other element is less than).
3. There's an element that is maximal of all elements less than a given non-minimal element.
4. There's an element between any two distinct ones.

As an example of a finite axiomatization, let's look at the axioms for the property of being a group, provided in Chapter 8.

$\mathcal{G}$-A1: $(x \circ y) \circ z = x \circ (y \circ z)$
$\mathcal{G}$-A2: $x \circ I = x$
$\mathcal{G}$-A3: $x \circ x^* = x$

A model for the above three axioms is of the following type: $\langle S, \cdot, *, I \rangle$. Any structure of that type is a group if and only if it satisfies the sentence that is a conjunction of the three axioms above.

Being A Natural Number Is <u>Not</u> First-Order Capturable

We conclude this section with a somewhat bizarre result pertaining to the question of capturing concepts in first-order logic.

Suppose we wish to capture the concept of being a natural number. To be more specific, consider the structure $\mathfrak{N}$, where $\mathfrak{N} = \langle \mathbb{N}, S, +, \cdot, 0 \rangle$. S is the operation of taking the next natural number, and the other elements are the familiar addition operation, multiplication operation and zero. We would like to know whether there is a set of sentences of an appropriate language such that all (and only those) models of the set contain $\mathbb{N}$ as their universe, together with the usual successor addition, and multiplication operations, and 0.

First of all, we admit that the universe of our structure does not have to <u>be</u> $\mathbb{N}$ to satisfy us. It just has to <u>look like</u> $\mathbb{N}$. For example, this set would do: {2, 4, 6, 8, ...}, and so would {-1, -3, -5, ...}, as long as the operations (and 0) correspond. The point is that any structure that is isomorphic to $\mathfrak{N}$ *is* $\mathfrak{N}$ as far as we are concerned. But, then we know immediately *being a natural number is not first-order capturable*, since by the UPWARD LSTV THEOREM if there is a denumerable model, there is one of every infinite cardinality. Since models of any two distinct cardinalities cannot be isomorphic, we have no hope of fully capturing the concept of being a natural number in first-order logic.

We may, however, try for a more limited result. Since the concept "being a natural number" seems to imply denumerability, we could content ourselves with establishing that all *denumerable* models of the truths of $\mathfrak{N}$ (i.e., models of the set of sentences $\mathcal{T}h(\mathfrak{N})$) are isomorphic. If we could establish that all *denumerable* models of $\mathcal{T}h(\mathfrak{N})$ are isomorphic, we could then say that we have "captured the concept with respect to its appropriate cardinality". Compactness, however, dooms this hope forever.

THEOREM: There are denumerable models of $\mathcal{T}h(\mathfrak{N})$ that are not isomorphic to $\mathfrak{N}$.

PF: Consider $\mathcal{T}h(\mathfrak{N})$, which is the set of all sentences true in model $\mathfrak{N}$. Obviously, $\mathfrak{N}$ itself is a model of all the sentences true in $\mathfrak{N}$. We will construct *another* model of all the sentences true in $\mathfrak{N}$ that is denumerable, but is *not* isomorphic to $\mathfrak{N}$. In short, we will construct a model $\mathfrak{N}_1$ that is denumerable and is elementarily equivalent to $\mathfrak{N}$ (i.e., $\mathfrak{N}_1$ will model all the same sentences as $\mathfrak{N}$), and yet $\mathfrak{N}_1$ will *not* be isomorphic to the standard model, $\mathfrak{N}$.

In constructing $\mathfrak{N}_1$, we will use our "enlargement plus compactness" technique in the following way: First, we add a single constant 'k' to L_N. Then we consider the following set of sentences of the new language:

$$\mathcal{T}h(\mathfrak{N}) \cup \{k \neq \underline{0}, k \neq S(\underline{0}), k \neq S(S(\underline{0})), \ldots\}.$$

The set of sentences that we tacked onto $\mathcal{T}h(\mathfrak{N})$ says that k is different from every natural number — different from $\underline{0}$, different from the successor of $\underline{0}$ (i.e., different from $\underline{1}$), and so on.

Now, we claim that every *finite* subset of $\mathcal{T}h(\mathfrak{N}) \cup \{k \neq \underline{0}, k \neq S(\underline{0}), k \neq S(S(\underline{0})), \ldots\}$ has a model. Take any finite subset of $\mathcal{T}h(\mathfrak{N}) \cup \{k \neq \underline{0}, k \neq S(\underline{0}), k \neq S(S(\underline{0})), \ldots\}$. That finite subset can have only a finite number of sentences from the set $\{k \neq \underline{0}, k \neq S(\underline{0}), k \neq S(S(\underline{0})), \ldots\}$. Among those sentences, there must be one sentence of the form $k \neq S(S(\ldots S(\underline{0}))\ldots)$ that has the largest number of occurrences of 'S'. Suppose that number is n. Then, the finite subset of {k ≠ 0, k ≠ S(0), k ≠ S(S(0)), ...} in question "says" that k is different from various numbers between 0 and n. So, we take the model $\mathfrak{N}$ and expand it to $\mathfrak{N}'$ so that $\mathfrak{N}'$ interprets 'k' as standing for n+1. Since $\mathfrak{N}'$ interprets 'k' as standing for n+1, $\mathfrak{N}'$ interprets that finite subset of {k ≠ 0, k ≠ S(0), k ≠ S(S(0)), ...} as true.

This new model, $\mathfrak{N}'$, looks like this:

$$\mathfrak{N}' = \langle \mathbb{N}, S, +, \cdot, 0, n+1 \rangle$$

$\mathfrak{N}'$ models the finite subset of $\mathcal{T}h(\mathfrak{N}) \cup$ {k ≠ 0, k ≠ S(0), k ≠ S(S(0)), ...} that we were considering. (Since $\mathfrak{N}'$ is an expansion of $\mathfrak{N}$, $\mathfrak{N}'$ models $\mathcal{T}h(\mathfrak{N})$, and $\mathfrak{N}'$ models the finite set of sentences "saying" that k is different from various numbers from 0 to n.)

Notice that $\mathfrak{N}'$ is perfectly general. That is, $\mathfrak{N}'$ was constructed as a model of <u>any</u> finite subset of $\mathcal{T}h(\mathfrak{N}) \cup$ {k ≠ 0, k ≠ S(0), k ≠ S(S(0)), ...}. By compactness, there is a model of the <u>entire set</u> $\mathcal{T}h(\mathfrak{N}) \cup$ {k ≠ 0, k ≠ S(0), k ≠ S(S(0)), ...}. Let's call the model of $\mathcal{T}h(\mathfrak{N}) \cup$ {k ≠ 0, k ≠ S(0), k ≠ S(S(0)), ...} $\mathfrak{N}''$. $\mathfrak{N}''$ models $\mathcal{T}h(\mathfrak{N})$ and has a distinguished element **k** assigned to the new constant 'k'. This new number **k** must be <u>greater</u> than <u>all</u> the natural numbers, since it differs from all of them and since **0** is the least natural number (**0** is least because one of the axioms "says" that **0** is the least number). Hence, the universe of the model we have "constructed" looks like this so far:

$$0, 1, 2, \ldots; k$$

That is to say, if we arrange all of our "numbers" in a list from smallest to largest, then **k** appears after <u>all</u> the "natural" ones. Thus **k** is called an "unnatural number". Once **k** appears in the model, <u>other</u> unnatural numbers must appear as well. For example, one axiom ensures that every number has a successor. That means that once **k** is in the model, so are **S(k)**, **S(S(k))**, **S(S(S(k)))**, etc. We won't concern ourselves with this multitutude of unnatural numbers spawned by **k**.

What we do need to do, however, to finish the proof, is to ensure that we have found a denumerable model of $\mathcal{T}h(\mathfrak{N})$ that is the same type as $\mathfrak{N}$, yet is not isomorphic to $\mathfrak{N}$. $\mathfrak{N}''$ is an infinite model of $\mathcal{T}h(\mathfrak{N})$, but it is not

necessarily denumerable. We fix this by using the
DOWNWARD LSTV THEOREM to give us a <u>denumerable</u> model of
$Th(\mathfrak{N})$ that also contains an unnatural number <u>greater</u> than
all the natural ones. Let us call this denumerable model
of $Th(\mathfrak{N})$, $\mathfrak{N}'''$. ($\mathfrak{N}'''$ is an elementary substructure of
$\mathfrak{N}''$.)

$\mathfrak{N}'''$ is still not *exactly* the model we are looking
for, because $\mathfrak{N}'''$ is not exactly the same type as $\mathfrak{N}$.
Recall that $\mathfrak{N}'''$ is appropriate for the language
containing the extra constant 'k', whereas $\mathfrak{N}$ is
appropriate for the language without that constant. So,
we just restrict $\mathfrak{N}'''$ to the original language, L_N
(giving a reduction of $\mathfrak{N}'''$). This, finally, gives us
the "non-standard" model $\mathfrak{N}_1$, which is a model of $Th(\mathfrak{N})$
that is both denumerable and of the same type as $\mathfrak{N}$. ∎

Exercises:
1. State explicitly how the following are used in the above
   proof:
     a) The extension of L_N to $L_N \cup \{\text{'c'}\}$
     b) Compactness
     c) The DOWNWARD LSTV THEOREM
     d) The reduction of $\mathfrak{N}'''$
2. Provide the details showing that $\mathfrak{N}_1$ is not isomorphic
   to $\mathfrak{N}$.

What we have proved is that we cannot capture the concept
of being a natural number in first-order logic even if we
restrict ourselves to denumerable models. We cannot rule out
the possibility that a denumerable model of the truths of $\mathfrak{N}$
contains in its universe all our familiar natural numbers
<u>plus</u> an abundance of "unnatural numbers" as well, which are
all greater than every "natural" number.

10.8 CATEGORICITY & COMPLETENESS

A theory is **categorical** if any two of its models are
isomorphic. For example, consider the first-order theory T
consisting of all consequences of the sentence '$\forall x \forall y (x = y)$'.
Theory T implies that only one object exists. Consider any
two models of T, $\mathfrak{M}_1$ and $\mathfrak{M}_2$. Each must have only one element
in its universe (or else it wouldn't be a model of

'∀x∀y(x = y)'). We now show $\mathfrak{M}_1 \simeq \mathfrak{M}_2$. Just map the single element of the universe of $\mathfrak{M}_1$ to the universe of $\mathfrak{M}_2$. There are no other structural properties to consider. Thus, $\mathfrak{M}_1 \simeq \mathfrak{M}_2$. Since $\mathfrak{M}_1$ and $\mathfrak{M}_2$ were taken to be any two arbitrary models of '∀x∀y(x = y)', theory $\mathcal{T}$ (consisting of all consequences of this single sentence) is categorical.

Recall that a theory $\mathcal{T}$ is mathematically complete if for any sentence φ in L_T, either $\varphi \in \mathcal{T}$ or $\neg\varphi \in T$. Once we know that any two arbitrary models of a given theory are isomorphic, we know that theory is mathematically complete, based on the following theorem:

CATEGORICITY THEOREM: If $\mathcal{T}$ is categorical, then $\mathcal{T}$ is (mathematically) complete.

PF: Suppose $\mathcal{T}$ is not mathematically complete. Then, there is a sentence $\varphi \in L_T$ such that $\varphi \notin \mathcal{T}$ and $\neg\varphi \notin T$. Thus, $\mathcal{T} \not\models \varphi$ and $\mathcal{T} \not\models \neg\varphi$ for sentence φ. Hence for some model $\mathfrak{M}_1$ of T, $\mathfrak{M}_1(\varphi) = T$, and for some other model $\mathfrak{M}_2$ of T, $\mathfrak{M}_2(\neg\varphi) = T$. But then $\mathfrak{M}_1 \neq \mathfrak{M}_2$. Therefore, by the ISOMORPHISM THEOREM, $\mathfrak{M}_1 \not\simeq \mathfrak{M}_2$. But this contradicts the hypothesis that all models of $\mathcal{T}$ are isomorphic. ∎

Returning to our original theory whose language has <u>no</u> non-logical symbols and which consists of all consequences of '∀x∀y(x = y)', we can now say, based on the theorem just proved, that that theory is mathematically complete (since any two of its models are isomorphic). That theory says there is exactly one element. Is a theory (with no non-logical symbols) that says there are exactly two elements complete? That there are exactly three elements? That there are exactly 647 elements?

The answer is 'yes, for any finite number of elements'. The way to prove this in the case of two elements is to (1) exhibit the sentence that says there are exactly two elements, (2) prove that any two models of that sentence are isomorphic, and (3) use the above theorem to establish that the theory of that sentence (in the language containing no non-logical symbols) is mathematically complete.

Exercises:

1. Write a sentence that says there are exactly two elements and prove that the theory of that sentence is mathematically complete.

2. Write a sentence with dots in it (i.e., '. . .') to indicate the general form of a sentence that says there are exactly k elements, where k ≥ 2. Then prove the theory of that sentence is complete.

If '$\exists_k!$' abbreviates a sentence saying there are exactly k elements then, the theories of $\exists_1!, \exists_2!, \ldots, \exists_k!$, (in the language containing no non-logical symbols) are all categorical. And by the CATEGORICITY THEOREM, all these theories are complete.

If we try to extend our categoricity results beyond finite universes to infinite ones, we run into difficulties. We would like to establish the completeness of theories with infinite models by first showing those theories to be categorical. But the Upward & Downward versions of LSTV prevent that. You will recall that the upshot of those two theorems is that if a theory has a model of some infinite cardinality, it has a model of <u>any</u> infinite cardinality (down to the cardinality of the language of the theory). Hence,

THEOREM: No theory with an infinite model is categorical.

Exercise:

Prove the above theorem. ■

So, once we enter the realm of infinite models, the fact that categorical theories are complete does not help us to prove completeness, given LSTV. But, there is a more restricted categoricity result that is helpful.

First, let us define a theory $\mathcal{T}$ to be **k-categorical** for some infinite cardinal **k**, if (1) $\mathcal{T}$ has an infinite model of cardinality **k**, and (2) all models of $\mathcal{T}$ with cardinality **k** are isomorphic. By a result of Łoś (pronounced: wash) and one of Vaught, we have the following theorem:

ŁOŚ-VAUGHT THEOREM: Any **k**-categorical theory $\mathcal{T}$ with no finite models is complete.

Exercise:

Prove the ŁOŚ-VAUGHT THEOREM. [Hint: Suppose $\mathcal{T}$ is not complete and get models of $\mathcal{T} \cup \{\varphi\}$ and $\mathcal{T} \cup \{\neg\varphi\}$ that have CARD(**k**) by LSTV.] ■

We know, however, from both Upward and Downward versions of LSTV, that if a sentence or group of sentences has an infinite model it has a model of every cardinality larger than or equal to the cardinality of the language used to express that sentence or group of sentences. Thus, any axioms of a theory that are true in some infinite model are true in infinitely many non-isomorphic, infinite models (since any two models of different cardinality are non-isomorphic).

Notice that in discussing complete theories nothing has been said about those theories being consistent. So, the theory of any sentence of the form $(\psi \wedge \neg\psi)$ is, by itself, a trivially complete theory, since $\{\psi \wedge \neg\psi\} \models \varphi$ and $\{\psi \wedge \neg\psi\} \models \neg\varphi$ for any φ of the language. Of course such a theory whose sole axiom is $(\psi \wedge \neg\psi)$ has no model, and we want to single out here those theories capable of having a model (i.e., *consistent* theories).

THEOREM: If $\mathcal{T}$ is a consistent theory, the following clauses are all equivalent:
(1) $\mathcal{T}$ is complete.
(2) Every two models of $\mathcal{T}$ are elementarily equivalent.
(3) $\mathcal{T} \equiv \mathcal{T}h(\mathfrak{M})$ for every model $\mathfrak{M}$ of $\mathcal{T}$.

PF: We show (1) $\Rightarrow$ (3) $\Rightarrow$ (2).

First, we show (1) $\Rightarrow$ (3). Assume (1). Suppose $\varphi \in T$. Then for every $\mathfrak{M}$ such that $\mathfrak{M}(T) = \mathbf{T}$, $\mathfrak{M}(\varphi) = \mathbf{T}$. Hence, for every $\mathfrak{M}$ such that $\mathfrak{M}(T) = \mathbf{T}$, $\varphi \in \mathcal{T}h(\mathfrak{M})$. Conversely, suppose $\varphi \in \mathcal{T}h(\mathfrak{M})$ for every model $\mathfrak{M}$ of T. Then, $\mathcal{T} \models \varphi$. Hence, $\varphi \in T$.

We now show (3) $\Rightarrow$ (2). By (3) if $\mathfrak{M}_1$ and $\mathfrak{M}_2$ are any two models of T, $\mathcal{T} \equiv \mathcal{T}h(\mathfrak{M}_1)$ and $\mathcal{T} \equiv \mathcal{T}h(\mathfrak{M}_2)$. Hence, $\mathcal{T}h(\mathfrak{M}_1) = \mathcal{T}h(\mathfrak{M}_2)$. Thus, for any sentence φ of L_T, $\mathfrak{M}_1(\varphi) = \mathfrak{M}_2(\varphi)$. ∎

Exercises:

Finish the proof. That is, show (2) $\Rightarrow$ (1) $\Rightarrow$ (3).

10.9 *** THE ULTRAPRODUCT CONSTRUCTION ***[4]

There is a very beautiful and to some extent intuitive construction in Model Theory that we will now discuss in a somewhat sketchy manner. Unfortunately, a good deal of the scope of this subject is quite advanced for the present book and requires excursions into other areas of mathematics as well. Nonetheless, we will try to capture the intuitive nature of these concepts and recommend other texts for more detailed and precise coverage.

We begin with the desire to prove Compactness model-theoretically. The version of Compactness we will be concerned with here is:

COMPACTNESS: If every finite subset Γ' of an infinite set of sentences Γ has a model, then Γ itself has a model.

Recall that this result is a consequence of the (Henkin) proof of Strong Completeness. But here we wish to prove it without recourse to theorems of proof theory.

When you already know that every finite subset Γ' has a model and you wish to construct a model for all of Γ, it is natural to try to find a model that is somehow composed of the models of all the finite subsets. This is a good intuition, and it turns out that there is a nice way to do this set-theoretically.

The most straightforward idea is to create a Cartesian product of models, $\mathfrak{M}_1 \times \mathfrak{M}_2 \times \ldots$, such that each model $\mathfrak{M}_i$ models some finite set correlated with the index i. Then, we would like to claim that the Cartesian product itself constitutes a model of the entire infinite set of sentences Γ. Unfortunately that won't work. If some structure $\mathfrak{S}_1$ models $\{\varphi\}$ and another structure $\mathfrak{S}_2$ models $\{\psi\}$, both structures may **not** model $\{\varphi,\psi\}$. Thus, the Cartesian product model would be too restrictive.

The next idea is to define a model $\mathfrak{M}$ consisting of the Cartesian product of models of finite subsets of Γ to be true for $\mathfrak{M}$ that does **not** require φ to be true for **every** $\mathfrak{M}_i$. But how can φ be true for $\mathfrak{M}$, while not true for every $\mathfrak{M}_i$ if $\mathfrak{M}$ is defined as $\mathfrak{M}_1 \times \mathfrak{M}_2 \times \ldots$? The answer is to define $\mathfrak{M}$ in such a way as to make φ true for $\mathfrak{M}$ just in case φ is true for a large number of $\mathfrak{M}_i$, but not necessarily true for all $\mathfrak{M}_i$.

---
[4]Material in this section is more advanced and sketchy. For a more thorough treatment, consult Bell & Slomson [1971].

This is where the ultraproduct construction comes to our rescue. An ultraproduct can be said to capture the intuition behind wanting to make a sentence true in a product model whenever it is true in "enough" factors of that model. So, we define $\mathfrak{M}$ as a certain ultraproduct consisting of factors made up of models of finite subsets of Γ. Then, we will show that for any sentence $\varphi \in \Gamma$, φ is true for enough factors of $\mathfrak{M}$ to make φ true for $\mathfrak{M}$ itself. Thus, $\mathfrak{M}$ is a model of Γ.

At this point we diverge from our loose, intuitive explanation of how compactness can be proved model-theoretically, and look at some of the formal features of the proof. Since these details are explained here only sketchily, the reader must consult other, more advanced texts for more particulars.

A family $\mathcal{U}$ of subsets of a given set I is called an **ultrafilter over I** if (where $X, Y \subseteq I$):

(1) $\emptyset \notin \mathcal{U}$
(2) If $X \in \mathcal{U}$ and $X \subseteq Y$, then $Y \in \mathcal{U}$
(3) If $X, Y \in \mathcal{U}$, then $X \cap Y \in \mathcal{U}$
(4) Either $X \in \mathcal{U}$ or $(I-X) \in \mathcal{U}$

FIP THEOREM: If F is a family of subsets of I such that no finite intersection of members of F is empty ("Finite Intersection Property"), then there is some ultrafilter $\mathcal{U}$ over I such that $F \subseteq \mathcal{U}$.

> PF: By Zorn's Lemma (i.e., If A is a non-empty partially ordered set such that each chain in A has an upper bound, then A contains a maximal element).

Exercise:

> Prove that for any non-empty index set I, each $i \in I$, $\{X \subseteq I \mid i \in X\}$ is an ultrafilter.

LEMMA: If I is infinite, the set $\{X \subseteq I \mid I - X \text{ is finite}\}$ has FIP and thus can be extended to an ultrafilter.

Exercise:

> Prove this LEMMA.

The underlying intuition behind ultrafilters is that they enable us to give meaning to the concept of a property holding "almost everywhere". So, for example, if we take some $\mathcal{U}$ that extends $\{X \subseteq \mathbb{N}^+ \mid \mathbb{N}^+ - X \text{ is finite}\}$ (the exercise above proves this set can be so extended), then, we know $\mathbb{N}^+ - \{1,2,3\} \in \mathcal{U}$. Hence, we can say that a property holding of all positive integers except 1, 2, and 3, holds "almost everywhere" with respect to $\mathcal{U}$. Or, we can say more simply, that a property holding in all but a finite number of places holds everywhere MOD $\mathcal{U}$. Similarly, we can say of two denumerable sequences that have the same values in every position except possibly at the 1st, 2nd, and 3rd places that they agree "almost everywhere" and that they are **equal** MOD $\mathcal{U}$.

We can make the above intuition more precise by associating a structure $\mathcal{S}_i$ to each $i \in I$ and then defining $\mathbf{P} = \Pi_{i \in I} \mathcal{S}_i$. That is, $\mathbf{P}$ is the Cartesian product of the universes of each structure $\mathcal{S}_i$. Now, take any two sequences, $\sigma_1, \sigma_2 \in \mathbf{P}$. σ_1 equals σ_2 MOD $\mathcal{U}$ just in case the set of places in the two sequences where they have the same elements belongs to $\mathcal{U}$.
In symbols:

$$\sigma_1 =_{\mathcal{U}} \sigma_2 \text{ iff } \{i \in I \mid \sigma_1(i) = \sigma_2(i)\} \in \mathcal{U}.$$

Exercise:

Prove "$=_{\mathcal{U}}$" is an equivalence relation.

Now, we let $[\sigma]$ be the equivalence class to which σ belongs. Then we define the universe $\mathbf{S}$ of our **ultraproduct** $\mathcal{S}$ to be the set of all equivalence classes of sequences of $\mathbf{P}$. That is,

$$\mathbf{S} = \{[\sigma] \mid \sigma \in \mathbf{P}\}.$$

Now, we must define the relations and operations for our ultraproduct $\mathcal{S}$. To simplify the presentation we will assume that there is only one binary relation R_i^3 in each structure $\mathcal{S}_i$. Then, we define R^3 holding of $[\sigma_1][\sigma_2][\sigma_3]$ just in case the set of $i \in I$ such that R_i^3 holds of $\langle \sigma_1(i), \sigma_2(i), \sigma_3(i) \rangle$ belongs to $\mathcal{U}$. In symbols:

$R^3([\sigma_1],[\sigma_2],[\sigma_3])$ iff $\{i \in I \mid R_i^3(\sigma_1(i),\sigma_2(i),\sigma_3(i))\} \in \mathcal{U}.$

Since $=_\mathcal{U}$ is a congruence relation with respect to R^3, it does not matter which representative we take. That is, $R^3([\sigma_1],[\sigma_2],[\sigma_3])$ holds, and if $\sigma_{1'} \in [\sigma_1]$, $\sigma_{2'} \in [\sigma_2]$, and $\sigma_{3'} \in [\sigma_3]$, then R^3 holds of $([\sigma_{1'}],[\sigma_{2'}],[\sigma_{3'}])$ as well.

Exercise:

Prove that $=_\mathcal{U}$ is a congruence relation with respect to R^3.

So, our **ultraproduct** structure $\mathcal{S}$ is as follows

$\mathcal{S} = \langle S, R^3 \rangle$, where each
$\mathcal{S}_i = \langle S_i, R_i^3 \rangle$, and
$S = \{[\sigma] \mid \sigma \in P\}$, where $P = \Pi_{i \in I} S_i$, and
R^3 is defined as above. $\mathcal{S}$ is an ultraproduct.

Łoš's Theorem

As soon as we explain Łoš's Theorem somewhat, we can prove Compactness using a "quilt model", which is to say, a carefully constructed patchwork of other models — an ultraproduct. Łoš's Theorem gives us just what we want: a sentence holds in an ultraproduct just in case it holds in "enough" of the factors making up the ultraproduct.

First, we state Łoš's Theorem in terms of sentences of L+, which includes names.

Suppose $\sigma = (e_1, e_2, \ldots) \in P$, then $[\sigma] = [(e_1, e_2, \ldots)] \in S$. Then a naming function n for S assigns some $[(e_1, e_2, \ldots)]$ to each name β. If $n(\beta) = [(e_1, e_2, \ldots)]$, then $n_i(\beta) = e_i$, and n_i is a naming function for S_i. We can now state the more general form of Łoš's Theorem:

ŁOŠ'S THEOREM: $\langle \mathcal{S}, n \rangle(\varphi) = T$ iff $\{i \in I \mid \langle \mathcal{S}_i, n_i \rangle(\varphi) = T\} \in \mathcal{U}$.

The proof is by Complete Induction on the height of φ.

COROLLARY to ŁOŠ: If φ is a sentence of L (i.e., φ has no names),

$$\mathcal{S}(\varphi) = T \text{ iff } \{i \in I \mid \mathcal{S}_i(\varphi) = T\} \in \mathcal{U}.$$

Finally, we can prove Compactness using ŁOŚ'S THEOREM:

COMPACTNESS THEOREM: A set Γ of sentences of L has a model iff every finite subset of Γ has model.

PF ($\Leftarrow$.) we construct an ultraproduct of the models of the (infinitely many) finite subsets of Γ such that the ultraproduct itself models the entire set Γ. Define an index set I = set of all finite subsets of Γ. Thus, if Γ' is a finite subset of Γ, $\Gamma' \in I$, where I is our index set. Let $\Gamma_\varphi = \{\Gamma' \in I \mid \varphi \in \Gamma'\}$. Since $\{\varphi\} \in \Gamma_\varphi$, we know each $\Gamma_\varphi \neq \emptyset$. Also, $\{\varphi_1, \varphi_2, \ldots, \varphi_k\} \in \Gamma_{\varphi_1} \cap \ldots \cap \Gamma_{\varphi_k}$.

Thus, $\{\Gamma_\varphi \mid \varphi \in \Gamma\}$ has the Finite Intersection Property. So, $\{\Gamma_\varphi \mid \varphi \in \Gamma\}$ can be extended to an ultrafilter $\mathcal{U}$. We then construct the ultraproduct $\mathfrak{M} = \Pi_{\Gamma' \in I} \mathfrak{M}_{\Gamma'}/\mathcal{U}$. We now wish to show that $\mathfrak{M}(\Gamma) = \mathbf{T}$. Take any $\psi \in \Gamma$. Let $A = \{\Gamma' \mid \mathfrak{M}_{\Gamma'}(\psi) = \mathbf{T}\}$. Then $\Gamma_\varphi \subseteq A$, since for $\psi \in \Gamma'$, $\mathfrak{M}_{\Gamma'}(\psi) = \mathbf{T}$. By construction, $\Gamma_\varphi \in \Gamma$. And since $\Gamma_\varphi \subseteq A$, $A \in \mathcal{U}$. Thus, $\mathfrak{M}(\psi) = \mathcal{T}$ by ŁOŚ'S THEOREM. ■

10.10 CHAPTER POSTSCRIPT

Many mathematicians have created logics with far more expressive power than first-order logic and have examined the relationships between these logics and their models. For example, one logic, $L_{\omega_1\omega}$, permits the formation of sentences with denumerably many conjunctions and disjunctions. Another logic, $L(Q_1)$, contains the quantifier that says "there are non-denumerably many". But, a theorem of P. Lindström [1969] shows that for logics richer than first-order, at least one of these two theorems must fail: the Compactness Theorem and the Downward Löwenheim-Skolem Theorem. For example, for $L_{\omega_1\omega}$, the Compactness Theorem fails. And, for $L(Q_1)$, the Downward Löwenheim-Skolem theorem fails (though a form of Compactness holds). These are two of the many intriguing results in a relatively recent branch of model theory called Abstract Model Theory.

Compactness and Löwenheim-Skolem can, thus, be seen as a kind of upper limit on logic itself. In the following chapter, we examine the theorems of Kurt Gödel, which establish limits on the truths that can be proved within first-order theories.

Chapter Exercises:

1. Suppose S_1 is isomorphic to a subset of S_2 (i.e., there is an *epimorphism* between S_1 and S_2). Prove that there is a structure S_3 such that $S_1 \subseteq S_2$ and $S_3 \cong S_2$.
2. (a) Suppose L_T contains <u>no</u> non-logical symbols. Prove that for any two finite models $\mathfrak{M}_1$, $\mathfrak{M}_2$ of any theory T, $\mathfrak{M}_1 \cong \mathfrak{M}_2$.
   (b) For <u>any</u> language L_T prove that any two finite models of T are isomorphic.
3. Suppose L_T has only finitely many non-logical symbols.
   (a) For any finite structure S_1, prove that $Th(S_1)$ is decidable.
   (b) Prove that the sentences of L_T that have finite models are mechanically listable.
4. For any theory T, if $\Gamma \vdash \varphi$, and for any sentence ψ in Γ, $\mathfrak{M}(\psi) = T$, then $\mathfrak{M}(\varphi) = T$.
5. Suppose T_1 and T_2 are theories with the same language. Suppose any structure S models T_1 iff S does <u>not</u> model T_2. Prove that both T_1 and T_2 are finitely axiomatizable.
6. A set of sentences is independent iff no sentence in the set is a consequence of the set of the rest of the sentences. Prove that for any theory T there is an independent set of sentences Γ_T such that T is the theory for Γ_T.
7. Prove that the following sentence is true for all finite structures of the appropriate type, but false for some infinite structures.

$$[\forall x \forall y \forall z (Fxx \wedge (Fxy \wedge Fyz \rightarrow Fxz) \wedge (Fxy \vee Fyx)] \rightarrow \exists y \forall x Fyx.$$

CHAPTER 11

GÖDEL'S THEOREMS

Chapter Abstract:

This chapter is somewhat idiosyncratic. First, there is an attempt to understand Gödel's philosophical outlook and how that outlook facilitated his arriving at the conclusion that (a first-order theory of) arithmetic must be incomplete. Then there is an explanation of why he proved the theorem in the way he did once he realized intuitively that the theorem must hold. This historical account of what Gödel thought at the time is purely speculative, possibly not fully accurate, and is included to make the reasoning behind the theorem easier to follow. It is also thought that trying to follow the steps Gödel himself may have taken makes the reasoning livelier, more accessible, and possibly more interesting.

The presentation of Gödel's theorem in this chapter, besides possessing the quasi-historical overlay explained above, is kept as intuitive as possible as long as possible, postponing the introduction of technical details until they seem absolutely necessary. Even when technical definitions are required, they are first presented informally and then later more formally. In general, the chapter becomes more formal as it continues, and the most formal matters are pushed towards the end. The earlier intuitive emphasis is intended to motivate Gödel's Theorem and make it more comprehensible.

Besides presenting Gödel's Theorem several times, based on different levels of technicality, this chapter also contains different sorts of proofs of Gödel's results from different perspectives. It is hoped that these proofs help to illustrate the multidimensionality of Gödel's remarkable results.

11.1 PLATONISM AND SEMANTIC PROOFS

In 1931, Kurt Gödel, a twenty-five year old mathematician living in Vienna, wrote a paper that was to stun the mathematical world. In very loose terms, the paper showed that no first-order formal system strong enough to produce derivations of ordinary statements of arithmetic could be complete and consistent. Further, the paper established that if a system of arithmetic were consistent, a statement of its consistency could not be derived within the system itself. These results dealt a powerful blow to the prevailing efforts at the time to place all of mathematics on a secure foundation, and since that time have had a tremendous impact on the development of logic. We spend this entire chapter trying to understand the theorems in that remarkable paper of 1931.[1]

An article by Feferman [1988] suggests that at an early age Gödel was a mathematical Realist, believing that things like numbers and sets had an <u>independent</u> existence. Sometimes this view is called Platonic Realism after the Greek philosopher Plato who first said that the "real world" contained, among other things, mathematical objects. Plato contrasted that real world of immutable objects with our "world of appearance" where objects seem to change from day to day. To know mathematical truth for Plato was to know the properties possessed by mathematical objects in the real world.

Gödel may have become interested in Platonism as early as age 15, and by 19 accepted the independent reality of mathematical objects and the "objective truth" of mathematics. In fact, he credits his early mathematical discoveries to his outlook at that time, which sharply contrasts with that of many of his contemporaries, according to whose work truth could be identified with derivability. For Gödel the truth of statements of arithmetic was an independent objective fact, which he distinguished clearly from the question of the derivability of such statements.

Gödel was also interested in the concept of truth in ordinary language. Of particular interest to him were the "semantic paradoxes", all of which seemed to show some sort of inconsistency, based on reasonable assumptions about the relationship between language and the world. The oldest and most familiar of the semantic paradoxes Gödel considered was the Paradox of the Liar, or just the "Liar Paradox". The paradox is arrived at by asking whether the Liar Sentence is true or whether it is false. Here is the Liar:

*   this sentence is false.

---

[1] The title of Gödel's paper in German was, "Uber formal unentscheidbare Sätze der Principia Mathematica und verwandter Systeme I", which, translated into English is, "On Formally Undecidable Propositions of Principia Mathematica and Related Systems I".

We put the symbol '*' to the far left of the page as a name for the sentence 'this sentence is false'. This allows us to substitute '*' for 'this sentence' and rewrite the Liar like this:

*   * is false.

Looked at in one way, * expresses the inconsistency of the English language. How?

English Is Inconsistent

PF: Well, suppose * is True, then since it <u>says</u> it's False, it is False. But, if it's False, then since it <u>says</u> it's False, <u>that</u> must be True. Hence, it is True. Thus, we have a proof that * is True if and only if * is False. ∎

Let us assign a unique number to each sentence of Number Theory, N, in some way. The exact way won't matter (though certain things to be explained later will matter). This enables us to speak of "sentence n", meaning: the sentence assigned the number n. For example, take the (false) sentence '$\underline{7} = \underline{8}$' and suppose it happens to be assigned the number p. Sentence p is: $\underline{7} = \underline{8}$. Now, we assume there is a formula of N with one free variable that holds of numeral term $\underline{n}$ if and only if the sentence assigned the number n is false (when symbols of N are interpreted in their normal way). We'll call this formula the Falsity Formula and write '$\mathcal{F}$' to abbreviate it. For example, $\mathcal{F}\underline{p}$ is true, since sentence p ('$\underline{7} = \underline{8}$') is false. For any sentence (whose number is) n, then, this holds of the Falsity Formula $\mathcal{F}$:

        sentence n is false   iff $\mathcal{F}\underline{n}$ is true.

Suppose we could find some instance of the Falsity Formula $\mathcal{F}\underline{k}$ having k itself assigned to it. Then $\mathcal{F}\underline{k}$ would "say" 'this sentence is false' within Number Theory. And, we could write it like this:

k $\mathcal{F}\underline{k}$

This is starting to look a lot like the Liar Paradox. But, instead of arguing that English is inconsistent, we can now argue that N is inconsistent:

N Is Inconsistent
assuming Falsity can be defined in N

PF: Suppose sentence k is true. Sentence k is: $F\underline{k}$. So, $F\underline{k}$ is true. But F holds of $\underline{k}$ only when k is false. So k is false. Hence, if sentence k is true, then it is false. Now, let's suppose sentence k is false. Then $F\underline{k}$ is false (since $F\underline{k}$ is sentence k). But, if F is false of $\underline{k}$, then sentence k is true. So, k is true. Thus, if k is false, then it is true. Given two assumptions about Number Theory N, we have just arrived at the following result:

Sentence k is true   iff   sentence k is false.   ∎

Truth Is Not Definable In N

What were the two assumptions that led to the above result? One of them was that there was a Falsity Formula F. The second assumption was that we could find some $F\underline{k}$ where k was the assigned number to that sentence. The second assumption may not seem plausible at all, but it is actually the first assumption that leads to the Liar Paradox in N. Once there is a Falsity Formula, the above contradiction can be arrived at in Number Theory. And, of course, there is a Falsity Formula in N if and only if there is a Truth Formula in N, since that would just be the negation of the Falsity Formula. Hence, there cannot be any Truth Formula in N.

There is evidence that Gödel's reasoning proceeded along these lines (See Feferman [1988]). He realized he could say 'this sentence' within Number Theory and create an analog of the Liar Paradox, on the supposition that there was a Truth Formula in N. Since this leads to the inconsistency of N, and since N is clearly consistent, there must be no Truth Formula in N. Another way to put this is to say: **Truth cannot be defined in** N.

Derivability Is Definable In N

But, Gödel realized somehow that Derivability can be defined in N. That is, some formula of N can be used to correctly express the idea that a given sentence can be derived in N. To do this, he first assigned a unique number to each derivation in N (The formal details of "Gödel Numbering" are provided at the end of this chapter). Assigning numbers to derivations may seem puzzling at first. But, if you think of a derivation as a series of sentences of N and bear in mind that each sentence of a derivation has a unique number assigned to it, then it shouldn't seem too bizarre to somehow put together the numbers of the sentences to get one big number for the derivation itself. Supposing that i is the derivation number of sentence n, then Gödel's "Derivability Formula", $Der(\underline{i},\underline{n})$, says that i is a derivation of sentence n. To say, then, that sentence n has a derivation, we write: $\exists y Der(y,\underline{n})$. Let's highlight the assumption that Derivability is definable in N:

Sentence n is derivable in N iff $\exists y Der(y,\underline{n})$.

Once Gödel saw that Derivability could be defined in N, he realized N must be incomplete. Let's look at the reasoning for this conclusion:

N Must Be Incomplete
since Truth is not definable, but Derivability is

PF: If Derivability can be defined in N, but Truth in N cannot, then Truth in N cannot be identical to Derivability in N. Since the axioms of N are all true of the natural numbers, and since all of our inference rules are truth-preserving (by soundness), any derivable sentence is true of the natural numbers. That is: Derivability $\subseteq$ Truth. But, since Truth and Derivability are distinct, there must be at least one true sentence that is not derivable. N must be incomplete! ∎

A Sentence Of N "Says": 'I am not derivable'

Once Gödel realized that there had to be at least one true sentence that is not derivable, he took pains to construct such a sentence. We supposed above that if Truth could be defined in N (which it can't), then there would be a sentence $\mathcal{F}\underline{k}$ expressing its own falsity, on a par with the Liar Paradox. Since Derivability <u>can</u> be defined in N, can we find a sentence expressing its own non-derivability, leading to a "Non-Derivability Paradox"?

Take the "Self" operation (or function) on numbers that works like this: if n is the number of any formula $\varphi(x)$ with one free variable, then Self(n) is the number of $\varphi(\underline{n})$. (That is, Self(n) is the number of the sentence resulting from the substitution of '$\underline{n}$' for all free occurrences of 'x' in $\varphi(x)$.) Suppose that N contains the term $Self(\underline{n})$[2] that defines the "Self" function Self(n). Now, let's consider the formula '$\neg \exists y Der(y, Self(x))$' and assume its number is k. The sentence '$\neg \exists y Der(y, Self(\underline{k}))$' is obtained from '$\neg \exists y Der(y, Self(x))$' by substituting '$\underline{k}$' for 'x'. Therefore, the number of the resultant sentence is Self(k). Hence '$\neg \exists y Der(y, Self(\underline{k}))$' "says" of itself that it is not derivable. More succinctly:

Self(k) $\neg \exists y Der(y, Self(\underline{k}))$.

Exercise:

   Expand on the reasoning in the above paragraph to establish clearly that the number of the sentence '$\neg \exists y Der(y, Self(\underline{k}))$' is Self(k), assuming that the number of the formula '$\neg \exists y Der(y, Self(x))$' is k.

Using the fact that '$\neg \exists y Der(y, Self(\underline{k}))$' says it is not derivable, together with the definability of Derivability from above:

    Sentence n is derivable in N iff $\exists y Der(y, \underline{n})$.

---

[2]That there is a <u>term</u> Self(<u>k</u>) is a simplifying assumption we use to make the reasoning in the proofs of (A) and (B) easier to follow and less technical. Gödel showed that there is a <u>sentence</u> representing the numeric function Self(n) = m that suffices for the proof of (A) and (B). (Constructing that sentence is an exercise in the section "Representability For Functions", later in this chapter.)

N Is Incomplete
because a true sentence says it is not derivable

PF: Suppose Self(k) is true. That is, $\neg\exists y Der(y, Self(\underline{k}))$. Then, Self(k) is not derivable in *N*. Hence, we have a sentence, namely Self(k), that is true but not derivable. Suppose sentence Self(k) is false. That is, suppose '$\neg\exists y Der(y, Self(\underline{k}))$' is not true. That means that '$\exists y Der(y, Self(\underline{k}))$' is true. Then sentence Self(k) is derivable in *N*. Hence, self(k) is false, yet derivable in *N*. We know by the soundness of *N* that this latter possibility cannot hold — we cannot derive anything false of the natural numbers in *N*. So it must be the case that the earlier possibility holds: Self(k) is true, yet Self(k) cannot be derived in *N*. Thus *N* is incomplete. ∎

11.2 SYNTACTIC PROOF OF INCOMPLETENESS

Banishing Truth From The Argument

The above argument shows that Number Theory must be incomplete, but it is based on reasoning about the Truth of sentences of *N*. Assuming the truth of sentence Self(k), we have a true but underivable sentence. Assuming the falsity of Self(k), we have a false but derivable sentence. Gödel wished to banish truth assumptions from his proof and to reason solely about the finite strings of symbols that can appear in a derivation.[3] To do that, he first proved that Derivability is representable in *N* (the representability of numeric relations in *N* is explained in Chapter 9):

(1) If i is a derivation of a sentence n,
    then $N \vdash Der(\underline{i}, \underline{n})$,

(2) If i is not a derivation of sentence n,
    then $N \vdash \neg Der(\underline{i}, \underline{n})$.

---

[3] Banishing truth assumptions makes the resulting argument completely syntactic and eradicates the appearance of any Platonistic insights. This would make the proof acceptable to the "Formalists" at the time, the most prominent of which was the mathematician David Hilbert.

Once Gödel proved (1) and (2), he could turn semantic reasoning (about Truth) into syntactic reasoning (about derivations, which are just sequences of strings of symbols). But, in order to arrive at a purely syntactic proof modeled on the Liar, he needed to prove that "Self-Reference" is representable as well.

Assuming that n is the number of formula $\varphi(x)$, here are representability clauses for the Self(n) function (together with a simplification that we explain later):

(1)′ If Self(n) = m, then $N \vdash$ *Self*($\underline{n}$) = $\underline{m}$

(2)′ If Self(n) ≠ m, then $N \vdash$ *Self*($\underline{n}$) ≠ $\underline{m}$

Given the representability of Derivability and Self(n), one-half of the incompleteness of Number Theory follows from assuming only the consistency of N. Consistency, you'll recall, is purely syntactic, pertaining to the strings that can appear in derivations.

**(A) If N Is Consistent, Then $N \nvdash \neg \exists y Der(y, Self(\underline{k}))$
(assuming the representability of Derivability & Self)**

PF: Assume N is consistent. Now, assume $N \vdash \neg \exists y Der(y, Self(\underline{k}))$. Suppose i is the number of the derivation of '$\neg \exists y Der(y, Self(\underline{k}))$', and suppose the number of the sentence '$\neg \exists y Der(y, Self(\underline{k}))$' is m. By (1), $N \vdash Der(\underline{i}, \underline{m})$. We also know that the number of '$\neg \exists y Der(y, Self(\underline{k}))$' is Self(k) (by explanation of the "Self" function above). Since Self(k) = m, we have by (1)′, $N \vdash Self(\underline{k}) = \underline{m}$. By Leibniz's Law, then, $N \vdash Der(\underline{i}, Self(\underline{k}))$. But, since $N \vdash \neg \exists y Der(y, Self(\underline{k}))$, we get $N \vdash \forall y \neg Der(y, Self(\underline{k}))$ by QE. Then, by UI, we have $N \vdash \neg Der(\underline{i}, Self(\underline{k}))$, which contradicts $N \vdash Der(\underline{i}, Self(\underline{k}))$. Thus: $N \nvdash \neg \exists y Der(y, Self(\underline{k}))$. ∎

We can extend the above reasoning to arrive at the Incompleteness of N this way: since '$\neg \exists y Der(y, Self(\underline{k}))$' "says" 'I am not derivable', it is in fact true. So, we have a true sentence that is not derivable, assuming the consistency of N. But, this line of argument makes an assumption about the <u>truth</u> of the sentence '$\neg \exists y Der(y, Self(\underline{k}))$'. To <u>completely</u> banish Truth from the proof and make it purely syntactic (based on the strings of symbols that constitute derivations in N) requires us to establish the existence of a sentence such that neither it nor its negation is derivable in N. We have just proved (informally) by purely syntactic means that '$\neg \exists y Der(y, Self(\underline{k}))$' is not derivable in N. What about the sentence '$\exists y Der(y, Self(\underline{k}))$' (or, equivalently, the sentence '$\neg\neg \exists y Der(y, Self(\underline{k}))$')? Since it "says" sentence Self(k) is derivable and we have just shown that it is not, '$\exists y Der(y, Self(\underline{k}))$' is false. Thus, we should be able to prove $N \not\vdash \exists y Der(y, Self(\underline{k}))$, or else a false sentence would be derivable in N. But, again, we are basing our conclusion that we should be able to prove $N \not\vdash \exists y Der(y, Self(\underline{k}))$ on the fact that '$\exists y Der(y, Self(\underline{k}))$' is false. To actually prove it, however, is a more difficult matter.

In his proof of (A), Gödel was able to banish all semantic assumptions and to obtain $N \not\vdash Der(y, Self(\underline{k}))$ by purely syntactic means. But, he was not able to obtain $N \not\vdash \exists y Der(y, Self(\underline{k}))$ from the assumption of simple consistency alone, as he was able to do in the proof of (A). He needed to assume that N was "ω-consistent" for the second part of the proof. (A few years later, J. Barkley Rosser showed a way to eliminate "ω-consistency" in favor of simple consistency, but Rosser's improvement renders the argument more complex.) Following Gödel's original argument, we first need to define "ω-consistency" for N:

ω-Consistency

A first-order theory $\mathcal{T}$ with numeral terms (i.e., having '$\underline{0}$' and 'S') is ω-**consistent** iff for any formula φx with one free variable it is not the case that both $\mathcal{T} \vdash \exists x \varphi x$ and for each numeral term $\underline{i}$, $\mathcal{T} \vdash \neg\varphi(\underline{i})$. Applying this to N and stating it less technically, N is ω-consistent if it is not possible to derive in N an existential sentence $\exists x \varphi x$ and to also derive in N the negation of each instance, $\neg\varphi(\underline{i})$, where $\underline{i}$ is a numeral term. The idea here is that if $N \vdash \exists x \varphi x$, then some number n has property φ, though we may not be able to derive: $N \vdash \varphi(\underline{n})$. Even so, if at least one number has φ, we certainly do not wish to be able to derive $\neg\varphi(\underline{0})$, $\neg\varphi(\underline{1})$, $\neg\varphi(\underline{2})$, ... for every numeral term. If that were the case, then which number could it be that does have property φ? (Answer: none of them, since for each one it is derivable that that one does not have φ.)

THEOREM: N is (simply) consistent if it is ω-consistent.

Exercise:

Prove this theorem.

Now, using ω-consistency, Gödel was able to furnish a syntactic proof for the other half of the incompleteness of Number Theory:

(B) If N Is ω-Consistent, Then $N \not\vdash \exists y Der(y, Self(\underline{k}))$ (assuming the representability of Derivability & Self)

PF: Assume N is ω-consistent. Next, assume $N \vdash \exists y Der(y, Self(\underline{k}))$. Since N is ω-consistent, the following is <u>false</u>: For every $\underline{i}$, $N \vdash \neg Der(\underline{i}, Self(\underline{k}))$. Since N is ω-consistent, N is (simply) consistent. Hence, by (A) $N \not\vdash \neg \exists y Der(y, Self(\underline{k}))$. That is, there is <u>no</u> derivation in N of the sentence '$\neg \exists y Der(y, Self(\underline{k}))$'. So, for <u>all</u> derivation numbers i, i is <u>not</u> a derivation of sentence Self(k) (since sentence Self(k) <u>is</u> '$\neg \exists y Der(y, Self(\underline{k}))$'). But, then, by clause (2) of the representability of Derivability, the following is <u>true</u>: For every $\underline{i}$, $N \vdash \neg Der(\underline{i}, Self(\underline{k}))$. This contradicts the ω-consistency of N. ∎

Putting (A) and (B) together, we get:

GÖDEL'S INCOMPLETENESS THEOREM: If N is ω-consistent, then there is a sentence φ of L_N such that $N \not\vdash \varphi$ and $N \not\vdash \neg\varphi$.

Gödel-Rosser Theorem

In 1936 J. Barkley Rosser sharpened Gödel's result by taking a somewhat more complicated sentence than our '$\neg \exists y \neg Der(y, Self(\underline{k}))$' that "says" 'There is no derivation of the sentence whose number is Self(k)', and of course Self(k) <u>is</u> the number of '$\neg \exists y \neg Der(y, Self(\underline{k}))$'. Rosser's sentence "says" 'If there is a derivation of the sentence whose number is Self(k), then there is a shorter derivation of its negation'.

Following through with the same sort of reasoning used in showing Gödel's theorem, it is possible to obtain:

GÖDEL-ROSSER THEOREM: If N is consistent, N is incomplete,

though we will not prove it here.

11.3 SUMMARY OF PROOF OF INCOMPLETENESS

Let us summarize the steps we took to arrive at an informal proof of the INCOMPLETENESS THEOREM. We began with the inconsistency of English based on the Liar Sentence. Then, we translated the Liar into N, assuming the existence of a "Falsity Formula" in N. This showed N to be inconsistent, which we "know" not to be the case. Hence, we reasoned, there is no "Falsity Formula" in N. Gödel realized, though, that there is an analogous "Non-Derivability Formula" in N. The presence of a Non-Derivability Formula together with the absence of a Falsity Formula in N, led us to a proof that there is at least one true but underivable sentence in N. The sentence we chose, modeled on the Liar, translates as 'this sentence is not derivable (in N)'. The presence of this sentence in N establishes the incompleteness of N, just as the presence of the Liar establishes the inconsistency of English. But, all of our arguments up to that point — from the inconsistency of English to the incompleteness of Number Theory — relied on assumptions about the Truth or the Falsity of various sentences in the language under consideration. (In the proof of the inconsistency of English, we needed first to assume the Liar was true, and then to assume it was false. To prove the incompleteness of Number Theory, we first assumed sentence Self(k) was true, and then we assumed it was false.) To eliminate all reference to Truth and Falsity in the argument, and to make the reasoning pertain to finite strings of symbols in derivations, we assumed that Derivability and Self-Reference were both representable in N. We then found a sentence $\mathcal{G}$ (short for '$\neg \exists y Der(y, Self(\underline{k}))$') whose number is Self(k). Then we proved informally that $\mathcal{G}$ is not derivable in N, based on the assumption of the consistency of N. We then showed that $\neg \mathcal{G}$ ('$\neg\neg \exists y Der(y, Self(\underline{k}))$' or equivalently '$\exists y Der(y, Self(\underline{k}))$') is also not derivable in N, based on the assumption of the ω-consistency of N.

11.4 "MECHANICAL", REPRESENTABLE AND UNDECIDABLE

We equate "mechanical procedure", "mechanical process", mechanical method", and so on. When we say that a relation is "mechanical" we mean that there's a mechanical procedure that determines when the relation holds. When we say that a function (or operation) is "mechanical", we mean that the result of applying it is determined by a mechanical procedure. "Mechanical functions" are also referred to as "mechanically computable functions", or as just "computable functions", or sometimes "effectively computable functions". (See Chapter 8, *First-Order Theories*, for a discussion of the meaning of "mechanical procedure" and how it relates to the concepts of Decidability and Axiomatizability.)

In order to simplify the evaluation of functions and relations, we use a gimmick that associates a "characteristic function" with any relation. This allows us to dispense with relations entirely and to consider only functions. (Later in this chapter we explain characteristic functions in detail. You can turn to that explanation now, or tentatively accept the reduction of relations to functions for the moment, and wait for the details later.)

"All mechanical functions are representable" is a helpful slogan to keep in mind. It is not an exact statement, since being mechanical is an intuitive, rather than a mathematically precise, concept. If we were to define mechanical functions as a certain class of mathematically exact functions, we would have the following exact test:

REPRESENTABILITY TEST: If a function on natural numbers is "mechanical", then it is representable in N.

Let us apply the REPRESENTABILITY TEST to features of theory N. Take the example of checking whether a string of symbols is a formula of L_N. Since numbers have been assigned to formulas in some unique way (we have not yet said precisely how this can be done), there is some set of numbers, Form-Nums, containing all and only those numbers that have been assigned to formulas. Now, here's the question: is there a mechanical test for whether a given number is in Form-Nums?

The answer is 'yes'. We have assumed that there is a mechanical procedure for finding the string of symbols (or the sequence of strings of symbols) that a number is assigned to. Given a string of symbols, we know we can determine whether it is a formula of L_N without using any ingenuity. Therefore, it passes the REPRESENTABILITY TEST and is representable in N. So, there must be some formula, say $Form(x)$, such that:

(1) if n ∈ Form-Nums, then $N \vdash Form(\underline{n})$, and

(2) if n ∉ Form-Nums, then $N \vdash \neg Form(\underline{n})$.

We have claimed above that the numeric relation "i is a derivation number of the sentence whose number is n" is representable. Let's abbreviate that relation by '$Der(i,n)$'. Suppose we wanted to find out whether $Der(i,n)$ is representable by using the REPRESENTABILITY TEST. How would we go about it?

To test whether $Der(i,n)$ is mechanical, we first obtain the sequence of strings of symbols whose number is i. Let's suppose i is a sequence of strings of symbols, say <$String_1$, $String_2$,..., $String_k$>. (If i is not a sequence of strings of symbols, it fails the test immediately.) Then we check whether n is the number of $String_k$. If that's true, we then look at $String_1$ to determine whether it is a sentence of N. If $String_1$ passes the test of sentence-hood, we check it against the axioms of N. If it is one of them, then it is certified as OK, so we then begin testing $String_2$. We check whether $String_2$ is a sentence of the wider language that includes logical names (recall that <u>within</u> a derivation names can occur). If it is a sentence in this wider sense, we check to see whether it is itself an axiom or whether it has been obtained from earlier sentences (really only one sentence at this point) by an application of any of our inference rules (T, UI, UG, QE, or SL).

Checking whether a sentence has been obtained from earlier ones on the basis of an inference rule is a more complicated task than merely checking that it is a sentence, but the task is still mechanical. If $String_2$ passes the test, we move on to $String_3$ and continue until we get to $String_k$. (The last string of a derivation must be a sentence of the language without names.) If any string fails the test, the sequence fails to be a derivation in N. If all strings pass the test, the sequence <u>is</u> a derivation in N of $String_k$, the sentence of L_N whose number is n, and $Der(i,n)$ is true. Then, by the REPRESENTABILITY TEST, $Der(i,n)$ is representable in N.

Exercises:

1. The test of Der(i,n) in the above paragraph is not particularly efficient. Write an improved test of Der(i,n).
2. Let Self(n) be m if and only if m is the number of the sentence of L_N resulting from the substitution of '$\underline{n}$' for the free variable 'x' in the formula $\varphi(x)$ (as above). Argue that there is a mechanical test for determining whether Self(n) = m. [If any of the conditions fail to be met, then Self(n) ≠ m.]

Church's Thesis

According to our REPRESENTABILITY TEST, as explained above, every "mechanical function" on natural numbers is representable in N. We said that our test was vague because the notion of "mechanical" is itself vague. We also said that if mechanical functions were <u>defined</u> to be an exact class of mathematical functions, the test would thereby become an exact one. In 1936, Alonzo Church proposed such a definition. He proposed that we define the class of "mechanical functions"[4] to be the "recursive functions", a well-defined class of functions that we formally characterize later in this chapter.

We characterize Church's Thesis only in one direction, and prove the other direction after recursive functions have been rigorously defined:

CHURCH'S THESIS: For any "mechanical" procedure[5] that operates on natural numbers, $n_1, n_2, \ldots, n_k$, and arrives at the number m, there is a recursive function f such that $f(n_1, n_2, \ldots, n_k) = m$.

Once the mechanical functions have been defined as the recursive functions, our REPRESENTABILITY TEST is susceptible of proof. Given Church's identification of the mechanical functions with the recursive functions, our test becomes the following theorem:

---

[4]Church called them 'effectively computable functions'.
[5]Church actually refers to 'effectively computable functions' on the natural numbers rather than to 'mechanical procedures operating on natural numbers'.

REPRESENTABILITY THEOREM: All recursive functions are representable in N.

Gödel proved the REPRESENTABILITY THEOREM and indirectly justified our REPRESENTABILITY TEST, assuming, of course, the acceptance of Church's proposed identification. Though Church proposed it as a definition, it has come to be called "Church's Thesis", since it appears to make the claim that a certain well-defined mathematical class contains the same elements as would be arrived at by applying an intuitive concept. Since all recursive functions are clearly "mechanical", the real issue is whether someone could discover a "mechanical function" that is not recursive. (We discuss this in greater detail later.)

By now Church's Thesis is almost universally accepted by mathematicians. So much so, that it has sometimes been referred to as Church's Law.[6]

Later on in this chapter we define the recursive functions and exhibit part of the proof of the REPRESENTABILITY THEOREM.

The Undecidability of First-Order Logic

We chose N as our first-order theory of arithmetic because it allowed us to obtain derivations relatively easily in the chapter, *First-Order Number Theory*. One disadvantage to N, pointed out at the end of *First-Order Theories*, is that it contains infinitely many axioms. You'll recall that our equality "axiom" that we used to express Leibniz's Law is not one but infinitely many axioms, and our induction "axiom" is really infinitely many axioms as well. Theory Q, presented briefly toward the end of *First-Order Theories*, supplemented by only finitely many equality axioms, is strong enough to give us all of Gödel's results that apply to N, though not as conveniently. We now know that the theorems of N are not representable and thus, not recursive. (Assuming Church's Thesis, they are not "mechanical".) This means that N is not decidable. Now, assuming we use some finite theory to obtain

---

[6]The phrase 'Church's Law' comes from Howard DeLong [1970] (p. 195) who cites Emil Post as referring to Church's Thesis as a "natural law". For an extensive discussion of Church's Thesis, see Epstein & Carnielli [1989], especially Chapter 25.

Gödel's results (such as Q supplemented by finitely many equality axioms), we can arrive at Church's Theorem (1936):

CHURCH'S THEOREM: First-Order Logic (for L_N) is undecidable.

PF: Assume the existence of some finite theory $\mathcal{T}$ whose language is L_N with k axioms, say $\varphi_1, \varphi_2, \ldots, \varphi_k$, such that $\mathcal{T}$ is undecidable. Let *Conjunct* be the conjunction of those k axioms. Now, assume (by way of contradiction) that the set of theorems of L_N (without any axioms) is recursive. Then, for any sentence ψ, there would be a mechanical test to determine whether (*Conjunct* $\rightarrow \psi$) is a theorem of L_N. Since there is a mechanical test for checking *Conjunct* as well, that would yield a "mechanical" test for ψ itself. Thus, ψ would be recursive as well. (Note: this uses Church's Thesis.) Hence, $\mathcal{T}$ itself would be decidable, which we know is not the case. ∎

Exercises:

1. Spell out the details in the above proof using Church's Thesis.
2. Use the Gödel Numbering scheme presented at the end of this chapter to assign a number to (*Conjunct* $\rightarrow \psi$) in order to show <u>without</u> the use of Church's Thesis that if (*Conjunct* $\rightarrow \psi$) is recursive, then so is ψ itself.

Representability for Functions

A 1-place function $f(x) = y$ is representable in $\mathcal{T}$ whose language is L_N if there is a formula $\varphi_f(x,y)$ of L_N with two free variables such that for any two natural numbers n, and m:

(1) If $f(n) = m$, then $\mathcal{T} \vdash \forall y(\varphi_f(\underline{n},y) \leftrightarrow y = \underline{m})$.

Exercises:

1. Extend clause (1) above to apply to any k-ary function Fn^k.
    Call any function that is representable by clause (1) above a Rep_1 function. The following two clauses define a Rep_2 function:
    (2) If $Fn(n) = m$, then $\mathcal{T} \vdash \varphi_f(\underline{n},\underline{m})$.
    (3) $\mathcal{T} \vdash \forall y(\varphi_f(\underline{n},y) \rightarrow y = \underline{m})$.

2. Prove $\text{Rep}_1 \iff \text{Rep}_2$.
   Here is a third definition of representability, Rep_3:
   (4) If $\text{Fn}(n) = m$, then $\mathcal{T} \vdash \varphi_f(\underline{n},\underline{m})$.
   (5) If $\text{Fn}(n) \neq m$, then $\mathcal{T} \vdash \neg\varphi_f(\underline{n},\underline{m})$.
3. Prove $\text{Rep}_3 \Rightarrow \text{Rep}_1$ (or: Prove $\text{Rep}_3 \Rightarrow \text{Rep}_2$).
4. Prove $\text{Rep}_1 \Rightarrow \text{Rep}_3$ (or: Prove $\text{Rep}_2 \Rightarrow \text{Rep}_3$), assuming $\mathcal{T}$ is a theory such that: if $m \neq n$, then $\mathcal{T} \vdash \underline{m} \neq \underline{n}$. ($\mathcal{N}$ is such a theory.)
5. Assuming that the "Self" function (used in the first proof above of (A) and (B) of Gödel's First Theorem) is a Rep_3 function, there is a <u>formula</u> $\varphi_{\text{Self}}(\underline{n},\underline{m})$ that represents $\text{Self}(n) = m$, for any numbers n and m.
   (a) Use the formula φ_{Self} to construct a sentence that expresses the same thing as '$\neg\exists y \text{Der}(y, \text{Self}(\underline{k}))$', but in which '$\text{Self}$' does not occur.
   (b) Prove that (A) and (B) hold for the constructed sentence. (That there was a <u>term</u> $\text{Self}(\underline{k})$ was a simplifying assumption in order to make the two proofs less technical.)

We now provide the details for associating a relation with its "characteristic function". This association makes it possible for us to consider a relation as a special kind of function.

Characteristic Functions & Representability

Consider an n-ary relation R^n whose arguments are natural numbers. Suppose some n-tuple of values, $\langle a_1, a_2, \ldots, a_n \rangle$ is in the relation R^n. That is, suppose: $\langle a_1, a_2, \ldots, a_n \rangle \in R^n$. We define the **Characteristic Function of R**, C_R to be equal to 1 in this case, the case when the relation holds of the ordered n-tuple. In the case when the relation does <u>not</u> hold (i.e., when $\langle a_1, a_2, a_3, \ldots, a_n \rangle \notin R^n$), then $C_R = 0$. For example, let R be the '<' relation. Then $\langle 1,2 \rangle \in <$, because $1 < 2$. On the other hand, $\langle 2,1 \rangle \notin <$, since $2 \not< 1$. We can express these two facts using the Characteristic Function for <, $C_<$, this way:

$C_<(1,2) = 1$ since $\langle 1,2 \rangle \in <$ (i.e., $1 < 2$)
$C_<(2,1) = 0$ since $\langle 2,1 \rangle \notin <$ (i.e., $2 \not< 1$)

Defining the characteristic function of a relation allows us to restrict our attention to representable

<u>functions</u>, since a relation is representable if and only if its characteristic function is. Restricting our attention to functions (and not considering relations) is formally justified by the following theorem:

CHARACTERISTIC FUNCTION THEOREM: A relation R is representable in N iff C_R is representable in N.

PF: $\Rightarrow$. Assume $\varphi(x,y)$ represents R (R is binary only for notational convenience; this doesn't affect the proof). Then, take the formula $(\varphi(x,y) \land z = \underline{1}) \lor (\varphi(x,y) \land z = \underline{0})$ as C_R.
$\Rightarrow$. Assume $\psi(x,y,z)$ represents C_R. Then, take the formula $\psi(x,y,\underline{1})$ as R. (Note: this requires $N \vdash \underline{0} \neq \underline{1}$.)

<u>Exercises</u>:

1. Fill in the details of the above proof. ∎
2. Write the characteristic function for Der(x,y).

On the basis of the above theorem, we can now work with functions, which are more convenient. Now, all our results pertain to relations as well.

11.5 THE FIXED POINT THEOREM, FORMAL DEFINABILITY, AND TRUTH

Another way to prove Gödel's Theorem is to prove that the truths of N (in the standard model $\mathfrak{N}$) are <u>not</u> definable. We reasoned informally that this was so early in this chapter when we supposed that there was a "Falsity Formula" in N. That supposition led to the inconsistency of N. Assuming N is consistent, we reasoned that there could not be such a Falsity Formula, and hence there was no "Truth Formula" either. The formal theorem that Truth is undefinable in N was first proved by Alfred Tarski in 1936. We provide some of the details for the Truth Theorem by first proving the Fixed Point Theorem.

The Fixed Point Theorem says that for any formula $\varphi(x)$ with just one free variable, there exists a sentence ψ such that φ holds of the number of ψ just in case ψ itself holds. Thus, in a sense, sentence ψ "says" 'φ is true of me'.

All we really need for the proof of this theorem is that the language of the theory for which it applies is L_N and the theory is strong enough to represent Self(n). But, we'll prove it for N.

First, if m is the number of sentence ψ, then we will write ⌜ψ⌝ for the numeral term $\underline{m}$. This notation makes it easier to think of a sentence as having a certain property. For example, we will want to indicate below that sentence ψ has the property associated with the formula φ. We could write $\varphi(\underline{m})$ and then recall that m is the number of sentence ψ. But, φ⌜ψ⌝ portrays sentence ψ as having property φ more graphically.

FIXED POINT THEOREM: For any formula $\varphi(x)$ containing only 'x' free, there is a sentence ψ such that:

$$N \vdash \varphi\ulcorner\psi\urcorner \longleftrightarrow \psi.\ {}^{7}$$

PF: Begin with the following formula and suppose its number is m:

m $\qquad\qquad \forall y(Self(x) = y \rightarrow \varphi(y))$.

Then, obtain the sentence Self(m) from sentence m: (Recall that to get sentence Self(m) you substitute '$\underline{m}$' for the free 'x' in the formula whose number is m.)

Self(m) $\qquad\qquad \forall y(Self(\underline{m}) = y \rightarrow \varphi(y))$.

Now, let ψ be the sentence Self(m) and

PROVE: $\quad N \vdash \forall y(Self(\underline{m}) = y \rightarrow \varphi(y)) \longleftrightarrow \varphi\ulcorner\psi\urcorner$.

Exercise:

Finish the proof by using simple properties of equality possessed by any first-order theory. ∎

The Fixed Point Theorem (FPT) says that for any "property in N" there's a sentence "saying" 'that property is true of me'.

---

[7] This is often referred to as the 'Diagonalization Lemma'.

Undecidability of Arithmetic

To prove Tarski's Truth Theorem, which formally establishes that Truth in N is not definable, we first prove that the theorems of N are not representable. Informally, that means there is no mechanical test for theoremhood in N. Another way to state this is that N is undecidable (See Chapter 8 for Decidability).

UNDECIDABILITY OF ARITHMETIC: The Theorems of N Are Not Representable.

PF: Assume there is some formula *Theorem*(x) in N that represents being a theorem of N. That is, *Theorem*($\underline{n}$) represents the numerical property Theorem(n) that holds of a number n just in case n is the number of a sentence of L_N that is a theorem of N.

Take the formula ¬*Theorem*(x) and get a sentence ψ by The Fixed Point Theorem, which "says" 'I am not a theorem of N'. This yields a contradiction.

Exercise:

SHOW A CONTRADICTION by applying the Fixed Point Theorem to the formula ¬*Theorem*(x). Then, assuming that all recursive relations are representable, conclude that being a theorem is not a recursive relation. ■

The intuitive basis for saying that being a theorem is not "decidable" is that it is not recursive. Recall that the very notion of a recursive function or relation is intended to capture the intuitive idea of "mechanical". The claim, then, that there is no "mechanical procedure" for determining whether a sentence is or is not a theorem of N is based on our assuming that for any mechanical procedure there is a recursive function. So, to claim there is no "mechanical method" for distinguishing a theorem from a non-theorem requires the acceptance of Church's Thesis.

Incompleteness of N (Following from FPT)

Take $\varphi(x) = \neg \exists y Der(y,x)$. By FPT (the Fixed Point Theorem) there is a sentence $\mathcal{G}$ (for 'Gödel') such that:

$$N \vdash \neg \exists y Der(y, \ulcorner \mathcal{G} \urcorner) \longleftrightarrow \mathcal{G}$$

$\mathcal{G}$ is any sentence found by FPT that satisfies the above equivalence. Thus, $\mathcal{G}$ "says" the same thing as the sentence we identified as $\neg \exists y Der(y, Self(\underline{k}))$ (and abbreviated by '$\mathcal{G}$') earlier. In the present case $\mathcal{G}$ (gotten by FPT) "says" 'there is no derivation of me'.

GÖDEL'S INCOMPLETENESS THEOREM:
  (A) If N is consistent, then $N \not\vdash \mathcal{G}$.
  (B) If N is ω-consistent, then $N \not\vdash \neg \mathcal{G}$.

Exercises:

1. Use $N \vdash \neg \exists y Der(y, \ulcorner \mathcal{G} \urcorner) \longleftrightarrow \mathcal{G}$ from FPT to prove Gödel's Incompleteness Theorem.
2. Suppose we add the Gödel sentence $\mathcal{G}$ to the axioms of N to obtain a new theory N', which is an extension of N. Argue that N' is incomplete as well. (This shows that N is <u>incompletable</u>.) (This argument can be used later to show "Gödel's Second Theorem" holds for N' as well.)

Definability In N

Before proving Tarski's Theorem from FPT, we wish to explain the concept of Definability more formally. Previously, we wrote that a formula $\varphi(x)$ defining a property P 'says P' or 'expresses P'. Or, more generally, a formula $\varphi(x,y)$ defining a relation R 'says R' or 'expresses R'. (Sometimes we put double quotes around 'says' or 'expresses' to call special attention to this usage.) We clarified this somewhat by indicating that 'says' and 'expresses' used in that way always imply the existence of an intended interpretation. For theory N the intended interpretation is the **standard interpretation**, $\mathfrak{N}$, which interprets the symbols of the language in their normal way. (That is, when interpreted by $\mathfrak{N}$, '<u>0</u>' <u>really</u> means 0, 'S' is <u>really</u> the operation that gets the next natural number, etc.) So, the intuition was that a formula, say '∃z(S(z) + x = y)' <u>defines</u> the relation "x < y" because when the symbols of '∃z(S(z) + x = y)' are interpreted normally, '∃z(S(z) + <u>n</u> = <u>m</u>)' comes out true just in case n < m. But, we now wish to define the notion of "definability in interpretation $\mathfrak{N}$" more precisely.

A formula $\varphi(x)$ of L_N with one free variable **defines the property P in interpretation** $\mathfrak{N}$ iff

(1) If n ∈ P, then $\mathfrak{N}(\varphi(\underline{n})) = \mathbf{T}$, and
(2) If n ∉ P, then $\mathfrak{N}(\varphi(\underline{n})) = \mathbf{F}$.

A property is **definable in** $\mathfrak{N}$ iff there is a formula that defines it.

Exercises:

1. Based on the above definition, complete the following so that it applies to any k-ary relation and to any interpretation $\mathcal{I}$ (appropriate for L_N):
 A formula $\varphi(x_1, x_2, \ldots, x_k)$ of L_N with k distinct free variables **defines the k-ary relation R in interpretation** $\mathcal{I}$ iff
 _____.

2. Prove that if relation R is representable in $\mathcal{N}$, then R is definable in $\mathfrak{N}$ (or in any model of $\mathcal{N}$).
3. Prove that if a theory $\mathcal{T}$ is (mathematically) complete, then all definable relations (in any model of $\mathcal{T}$) are representable.

Tarski's Truth Theorem

Now that the definability of a relation (in an interpretation) means having a formula that defines it, we can prove Tarski's Truth Theorem from FPT:

TARSKI'S TRUTH THEOREM: Truth Is Not Definable in $\mathfrak{N}$.

PF: Assume there is a formula *Truth*(x) that defines the set of all true sentences of $\mathfrak{N}$. Let Truth be the set of all numbers whose sentences are evaluated as true by the standard interpretation $\mathfrak{N}$. Hence:

(1)' If n ∈ Truth, then $\mathfrak{R}(Truth(\underline{n})) = \mathbf{T}$
(2)' If n ∉ Truth, then $\mathfrak{R}(Truth(\underline{n})) = \mathbf{F}$.

Consider the formula ¬Truth(x). By FPT there is a sentence ψ such that:

$$N \vdash \neg Truth\ulcorner\psi\urcorner \longleftrightarrow \psi$$

Since $\mathfrak{R}$ is a model of N, the following holds by soundness:

$$\mathfrak{R}(\neg Truth\ulcorner\psi\urcorner \longleftrightarrow \psi) = \mathbf{T}$$

Hence, $\mathfrak{R}(\neg Truth\ulcorner\psi\urcorner) = \mathfrak{R}(\psi)$. Let the number of ψ be k. Then:

$$\mathfrak{R}(\neg Truth(\underline{k})) = \mathfrak{R}(\psi).$$

Assume k ∈ Truth, and then assume k ∉ Truth. In each case a contradiction is reached. Thus, Truth is not definable in $\mathfrak{R}$.

Exercise:

Fill in the details of the rest of the above proof. ∎

Notice that sentence ψ in the above proof "says" 'I am not true'.

11.6 MOORE'S PRESENTATION OF GÖDEL'S THEOREM

Another informal approach to understanding Gödel's Theorem is provided by A. W. Moore [1990][8]. Recall from the previous chapter that a complete, axiomatized theory is decidable. This means that if N <u>were</u> complete, there would be some <u>mechanical</u> way to test whether a given sentence of L_N is a theorem of N, and hence whether it is true in <u>all</u> models of N. We are going to <u>assume</u> that the truth of any sentence (in all models) can be mechanically determined (i.e., that truth for Number Theory is decidable) and we will infer a contradiction from this assumption. This shows N is <u>not</u>

---

[8]Notice that this presentation is similar to the argument of the Fixed Point Theorem or Diagonalization Lemma (by which it is also known), leading to Tarski's Truth Theorem, and then to Incompleteness.

decidable, and, since we have an axiomatization of N, that establishes that N must be <u>in</u>complete.

Exercise:

Prove that any axiomatic theory whose truths are undecidable is incomplete.

Now, we use some mechanical method to order all formulas of the form φx, where 'x' is the only free variable. We list all the formulas φx like this:

$$\varphi_0 x, \varphi_1 x, \varphi_2 x, \ldots$$

Our mechanical method for ordering these formulas must be such that given a number n, we can mechanically determine the one associated with n: $\varphi_n x$. And, it must be such that if we are given a formula φx, we can mechanically determine the number n, such that $\varphi x = \varphi_n x$. Now we set up a table having 'Y' stand for 'Yes' and 'N' stand for 'No' as follows:

	0	1	2	3	4 ...
$\varphi_0 x$	Y	N	Y	N	Y ...
$\varphi_1 x$	N	N	N	N	N ...
$\varphi_2 x$	N	Y	N	Y	N ...
$\varphi_3 x$	Y	N	N	N	Y ...
$\varphi_4 x$	N	N	Y	Y	Y ...

For example, $\varphi_0 x$ could be the formula '$\exists y(x = \underline{2} \circ y)$', which says that x is even. Hence, our chart shows that 0 has the property of being even, 1 is <u>not</u> even, 2 is even, and so forth. The set of even numbers is then <u>defined</u> by $\varphi_0 x$ since all (and only) even numbers have the property that $\varphi_0 x$ singles out. Similarly, if 'N' and 'Y' continue to alternate in the $\varphi_2 x$ row, then $\varphi_2 x$ defines the set $\{1, 3, 5, \ldots\}$ — the set of odd numbers.

Now, we use Cantor's method of Diagonalization to arrive at a formula $\varphi_D x$ (subscript 'D' is for 'Diagonalization') that cannot be on our list. The way we do this is to take the diagonal values, which we highlight below as:

	0	1	2	3	4	
$\varphi_0 x$	**Y**	N	Y	N	Y	...
$\varphi_1 x$	N	**N**	N	N	N	...
$\varphi_2 x$	N	Y	**N**	Y	N	...
$\varphi_3 x$	Y	N	N	**N**	Y	...
$\varphi_4 x$	N	N	Y	Y	**Y**	...

Reading down the diagonal in the above chart gives us: **Y, N, N, N, Y**. The set of values defined by the bold diagonal letters above is: {0, 4, . . .}, since there is a '**Y**' under 0 and 4, but there is an '**N**' under 1, 2, and 3. You would think that this set should be called the Diagonal Set, since the diagonal **Y**'s and **N**'s determine its membership. But the term 'Diagonal Set' is reserved for the set of numbers determined by the result of <u>switching</u> the diagonal values (i.e., switching **N** for **Y** and **Y** for **N**).

Here's what we do. We start out with the bold diagonal values '**Y, N, N, N, Y**, . . .', and we <u>switch</u> them. Our <u>switched</u> diagonal values are:

N   Y   Y   Y   N . . .,

which determine the **Diagonal Set**, **D** = {1, 2, 3, . . .}. 1, 2, and 3 all belong to **D**, since they all have a '**Y**' below them, whereas 0 and 4 have an '**N**' below them and so 0 and 4 do <u>not</u> belong to **D**. (Notice that **D** is the <u>complement</u> of the earlier set that was defined by the <u>old</u> diagonal values.) Now, we have a <u>mechanical</u> <u>method</u> for deciding whether a number n is in **D** or not. First, we use our mechanical method for finding $\varphi_n x$ if we are given n. Then, we mechanically check whether n has the property singled out by $\varphi_n x$. If n <u>does</u> have P_n (the property singled out by $\varphi_n x$), then n does <u>not</u> belong to **D**. If n does <u>not</u> have P_n, then n <u>does</u> belong to **D**. We can sum this up as:

n ∈ **D**   iff   n does <u>not</u> <u>have</u> P_n.

The Diagonal Set is Not Definable in N
(assuming N is Decidable)

PF: Assume **D** is definable in N. Then there is some formula $\varphi_D x$ that defines **D**. This means that $\varphi_D x$ is somewhere on our list of formulas of the form φx, and thus $\varphi_D x$ must equal $\varphi_k x$ for some k. If we check whether k itself has the property P_k determined by $\varphi_k x$ we get contradictory results, since there are two incompatible ways of determining whether k has P_k. If k is in **D**, then by the displayed equivalence above: (1) k does not have P_k. But, **D**'s defining formula $\varphi_k x$ determines the property P_k that all (and only the) numbers in **D** have. So, by this method, if k is in **D**, then: (2) k does have P_k. This contradictory result establishes that **D** is not definable in N. ∎

Therefore: N is Incomplete

Since our chart enables us to mechanically check whether a given k belongs to **D**, **D** is representable (because it is mechanical) and therefore it should be definable (because all representable sets are definable by an earlier exercise). But, by the above reasoning, **D** is not definable. Hence, N is not decidable (our initial assumption), and thus N is not complete. ∎

11.7 GÖDEL'S SECOND THEOREM

Usually Gödel's Incompleteness Theorem is referred to as "Gödel's Theorem". But, when his second theorem is about to be mentioned, it is then usually referred to as "Gödel's First Theorem".

Gödel's first theorem is: Number Theory (N) is incomplete (provided N is ω-consistent). Since the theorem applies not only to N, but to any first-order theory rich enough to represent all "computable" functions and relations, Number Theory is incompletable. Since N is intended to capture all the truths of arithmetic, it is sometimes said that Gödel's Theorem shows that arithmetic is incomplete. But, that doesn't mean "arithmetic" in the abstract —

whatever that is — it means arithmetic taken as a first-order theory.

Gödel's second theorem is that the consistency of arithmetic (as a first-order theory) cannot be derived within the resources of first-order arithmetic. That is, Gödel <u>proved</u> that a sentence in L_N expressing the consistency of N itself cannot be derived in N — assuming N is consistent. This is sometimes taken to mean that the derivation of the consistency of N would have to be done within a more powerful theory whose consistency would be in more doubt than the consistency of N itself. It is often said that Gödel's Second Theorem shows that it is impossible to prove the consistency of arithmetic without presupposing a more powerful — and hence more dubious — theory than arithmetic itself. (Of course, phrases like 'more powerful' in the above statement can lead to philosophical disputation.)

More formally, here is what the proof amounts to: We find a way to "express" the consistency of N within N itself. That should not be too hard to imagine. Remember, consistency is just the inability to derive a sentence and its negation. And we know that we can express facts about which sentences can be derived in N <u>within</u> N <u>itself</u>. So, we construct a sentence, *Cons*, that "says" there is no derivation of the conjunction of a sentence and its negation.

Exercises:

If n is the number of formula φ, let Neg(n) be the number of $\neg\varphi$ (i.e., the negation of φ).
1. Argue (informally) that there must be a formula $Neg(\underline{n},\underline{m})$ in N representing Neg(n) = m.
2. Using **1**, write a sentence that "says" 'there are not two derivations in N such that one of them is the derivation of a given sentence and the other is the derivation of its negation'.

Let *Cons* be the sentence in *exercise* **2.** above. We then have:

GÖDEL'S SECOND THEOREM: If N is consistent, then $N \not\vdash Cons$.

PF: Gödel first proved $N \vdash Cons \rightarrow \mathcal{G}$, assuming the consistency of N, from which the result follows.

Exercise:

Write the details of the above proof. ■

You will notice that the second theorem does not answer the question of the consistency of *N*. It just states that *Cons* is not derivable in *N* — assuming *N* is consistent in the first place. Of course, if *N* is not consistent, then *Cons* is derivable in *N*, since every sentence is derivable in an inconsistent theory.

It can be argued that the sentence *Cons* does not really "express" the consistency of *N*. Recall that a sentence of a theory does not "say" or "express" anything until the symbols of the language of the theory have been given assignments by an interpretation. So, when we say that *Cons* "expresses" the consistency of *N*, we are presupposing the existence of an interpretation $\mathcal{I}$ for which $\mathcal{I}(Cons) = \mathbf{T}$ iff *N* is consistent. The interpretation $\mathcal{I}$ we use here is the so-called "standard interpretation" $\mathfrak{R}$, according to which all the symbols stand for exactly what we think they stand for. So, for interpretation $\mathfrak{R}$, '0' stands for 0, 'S' stands for the successor operation, and so forth. Hence, when we say that *Cons* expresses the consistency of *N*, we really mean that for the standard interpretation $\mathfrak{R}$, if *Cons* is true then *N* is consistent. That is, if $\mathfrak{R}(Cons) = \mathbf{T}$, then *N* is consistent. We know in particular that the intended model $\mathfrak{R}$ — arithmetic in the Platonic "real world" — is in fact a model of *N*. Gödel's Incompleteness Theorem tells us there are truths in this realm that cannot be captured in *N*, and one of these truths is *Cons* itself. Gödel's Second Theorem tells us that if we take a sentence of Number Theory that expresses its consistency in the standard interpretation, that sentence is true but not derivable. Hence, it can be said that the Incompleteness Theorem applies directly to the "real model" of Number Theory, not to just any model, since it is the "real model" that is used to interpret *Cons*.

But, the "real model" of *N* is not the only possible interpretation of L_N for which all the axioms of *N* are true. (See the chapter on Model Theory for other models of *N*.) This raises the question of whether *Cons* expresses the consistency of *N* in all models of *N*.

Cons Does <u>Not</u> Express The Consistency of N in <u>All</u> Models of N

PF: Let $Cons^*$ be any sentence of N that <u>does</u> express the consistency of N in <u>all</u> models of N. We argue that $Cons \neq Cons^*$. If $Cons^*$ is true for <u>all</u> models of N, then, by Strong (Logical) Completeness, $N \vdash Cons^*$. But, since $N \nvdash Cons$, it must be the case that $Cons \neq Cons^*$. Hence, $Cons$ cannot express the consistency of N in <u>all</u> models of N. That is, on the assumption of the consistency of N, there's at least one model of N in which $Cons$ is <u>false</u> (and the consistency of N is <u>true</u>). ∎

Hence, if N is consistent, then $Cons^*$ is true in <u>all</u> models of N and $N \vdash Cons^*$. And, if N is <u>not</u> consistent, then $N \vdash Cons^*$ as well. Thus, $N \vdash Cons^*$ whether N is consistent or not. It could be argued that the expression of consistency in "unnatural interpretations" of N is not really meaningful, since those interpretations are aberrations in the first place. The claim, then, is that we do not care whether for some grotesque interpretation of the symbols of L_N all the axioms of N are true, <u>and</u> $Cons$ is false. But, that argument surely conflicts with the formalistic spirit behind the attempt to establish the consistency of N by finite means alone.

This raises the question of whether Gödel's Second Theorem generalizes readily in the following way to <u>any</u> model of N:

¿GENERALIZATION OF GÖDEL'S SECOND THEOREM: For any model $\mathfrak{M}$ of N, if sentence φ "expresses the consistency of N in $\mathfrak{M}$", then $N \nvdash \varphi$?

I do not know the answer to this question.

Gödel's two theorems sounded the death knell for the effort to place all of mathematics on the secure foundation of a finite theory of signs. Number Theory itself, having just "plus", "times", "0", and "the successor operation" seemed to be a simple place to start in proving all of mathematics to be both consistent and complete. But Gödel showed that even for ordinary Number Theory this could not be done.

11.8 PRIMITIVE RECURSIVE AND (GENERAL) RECURSIVE FUNCTIONS

In this section we shift our focus of interest to properties pertaining to the natural numbers. We momentarily detach ourselves from any concern with first-order theories (such as N) in order to look at functions on the natural numbers. We do not care for now how much of what we say in this section can be carried out in N — or in any other such theory — we care only about the "objective facts" about the "real world" of functions on natural numbers.

The functions we define as "recursive" will all be "obviously mechanical" in some intuitive sense, but in order to see that this is true, we must look at the <u>ideas</u> behind the formal definitions, rather than looking just at the definitions themselves. The reason for saying this is to warn you that to make our definitions completely rigorous required us to introduce a bit of machinery which somewhat obscures their intuitive meaning. After presenting the formal definition of recursive functions and proving that several familiar functions are recursive, we will pause to argue informally that every recursive function is intuitively mechanical, which is, after all, the *raison d'être* of the notion of "recursive function" to begin with.

We now formally define the class of "primitive recursive" functions on the natural numbers. The class we define will include all the "basic primitive recursive" functions and will also include any of those that can be generated from ones we already have by using two rules for generating new from old ones. The "recursive functions" will be all primitive recursive functions plus those obtainable by the use of a third rule for generating new functions from old ones.

Definitions of Primitive & (General) Recursive Functions

We now define the **basic primitive recursive functions**:
(1) **The zero function**: $Z(n) = 0$, for all n.

(2) **The successor function**: $S(n)$ = the natural number that follows n.

(3) **The projection functions**: $\Pi_i^k(x_1, x_2, \ldots, x_k) = x_i$ for $1 \le i \le k$.

(1) just returns 0 for any argument. (2) gets the next number following n. (3) are a class of functions. The superscript k indicates the number of arguments, and the subscript i indicates which of those k arguments is "projected". For example, $\Pi_1^1(x) = x$; Π_1^1 is the identity function.

Rules For Obtaining New Recursive Functions From Old Ones

(4) **Composition:**

If g is a function of m arguments and each of $h_1, h_2, \ldots, h_m$ is a function of n arguments, then

$$f(x_1, x_2, \ldots, x_n) = g(h_1(x_1, \ldots, x_n), \ldots, h_m(x_1, \ldots, x_n))$$

is the function f obtained by composition from $g, h_1, h_2, \ldots, h_m$.

(5) **Primitive Recursion:**

$$f(x_1, x_2, \ldots, x_n, 0) = g(x_1, x_2, \ldots, x_n)$$

$$f(x_1, x_2, \ldots, x_n, m+1) = h(x_1, x_2, \ldots, x_n, m, f(x_1, x_2, \ldots, x_n, m)).$$

If f is a function of one variable (i.e., n = 0), we have:

$$f(0) = k \quad \text{(for some number k)}$$
$$f(m+1) = h(m, f(m))$$

(6) **The μ-Operator:**

$$\mu y(g(x,y)) = \text{the least y such that } g(x,y) = 0.$$

Assuming that for any x there is at least one y such that $g(x,y) = 0$, f is obtained from g by the μ-operator if f is defined as follows:

$$f(x) = \mu y(g(x,y)).$$

Note that f is not defined if for some x there is no y such that $g(x,y) = 0$.

A sequence of lines each consisting of a function being defined on the left of an equals sign followed by its definition on the right will be called a **sequence of defined functions**. Then a sequence of defined functions is a **primitive recursive derivation** if each line has a basic primitive recursive function on the right or has been arrived at by using Rules (4) or (5) above (Composition or Primitive Recursion) on the definitions of earlier lines. Then we can say that a function (from natural numbers to natural numbers) is **primitive recursive** if it can appear in a primitive recursive derivation.

A **(general) recursive derivation** can contain anything permitted in a primitive recursive derivation plus the use of Rule (6), the μ-operator, to define new functions. A **(general) recursive function** is any function that can be obtained by a (general) recursive derivation. Thus, all primitive recursive functions are (general) recursive functions. But, those recursive functions requiring the use of the μ-operator are recursive but not primitive recursive functions.

Proofs of Recursive Functions

Let us now formally show that some ordinary functions are primitive recursive (and thus recursive, as well):

1: "x + y" is primitive recursive.

PF: Instead of writing 'x + y' we will write 'Sum(x,y)'. We want to PROVE:
$$\text{Sum}(x,0) = x \quad \text{and}$$
$$\text{Sum}(x, m+1) = S(\text{Sum}(x,m)),$$

for the two clauses above provide the ordinary recursive definition of '+'. In order to show that '+' is primitive recursive rigorously, however, we must always have a primitive recursive function on the right. The following is a primitive recursive derivation of '+':

1. $\Pi^1_1(x) = x$ by Projection.

2. $f(x_1, x_2, x_3) = S(\Pi^3_3(x_1, x_2, x_3))$ by Composition of Successor and Projection.

3. $\text{Sum}(x,0) = \Pi^1_1(x)$ and

Sum $(x, m+1) = f(x, m, \text{Sum}(x,m))$ by (1),(2) using Primitive Recursion. ∎

Since the sequence of lines 1, 2, and 3 all constitute a primitive recursive derivation, the Sum function "$x + y$" is primitive recursive.

2: "$x \cdot y$" is Primitive Recursive.

PF:
1. $Z(x) = 0$ by Zero Function

2. $f(x_1, x_2, x_3) = \text{Sum}(\Pi_2^3(x_1, x_2, x_3), \Pi_3^3(x_1, x_2, x_3))$
    by Composition of Sum with two Projection Functions

3. Times $(x, 0) = Z(x)$ and
    Times $(x, y+1) = f(x, y, \text{Times}(x,y))$ by (1),(2), using Primitive Recursion.

This shows "$x \cdot y$" is Primitive Recursive. ∎

We will use the more familiar way of writing the definition of "$x \cdot y$", which is:

Times$(x, 0) = 0$ and
Times$(x, y+1) = \text{Times}(x, y) + y$

Derivations 1 and 2 are somewhat unintuitive and unwieldy, due to our sticking strictly to the formal definition of a primitive recursive derivation. From now on we derive new primitive recursive functions informally, sometimes pausing to indicate the formal steps that were left out and sometimes leaving those steps as exercises.

In order to establish that the relation $x < y$ is primitive recursive, we show a few more functions to be primitive recursive. First, we show that a limited Predecessor Function Pre(x) is Primitive Recursive, where:

$$\text{Pre}(x) = \begin{cases} x - 1 & \text{if } x > 0 \\ 0 & \text{if } x = 0 \end{cases}$$

1. $Pre(0) = 0$ and $Pre(x + 1) = x$ by Primitive Recursion.

Exercise:

Write a formal derivation of $Pre(x)$.

The idea of Pre is that it "almost" gets the Predecessor (we take 'Pre' to be the almost-Predecessor, assuming the real predecessor to be 'Pred'). That is, it gets the real Predecessor if it can, but it goes no lower than 0.

Now, we show that a limited subtraction function $Sbt(m,n)$ is Primitive Recursive, where

$$Sbt(x,y) = \begin{cases} x - y & \text{if } x \geq y \\ 0 & \text{if } x < y \end{cases}$$

1. $Sbt(x,0) = x$ and $Sbt(x,y + 1) = Pre(Sbt(x,y))$

Exercise:

Provide a formal derivation of $Sbt(x,y)$.

Sbt is the reverse of Sum. Sum breaks down the sum of two numbers into the successor of the sum of the first number together with one less than the second number, and continues to take successors while taking one away from the second number until it is zero. Sbt reverses Sum by taking the almost-predecessor of the Sbt of the first number together with one less than the second number, until the second number is zero. Once it is zero, the first number's almost-predecessor is taken the number of times it took to reduce the second number to zero. That is, almost-predecessor is applied the second number of times to the first number.

The function $Nz(x)$, which returns 0 if $x = 0$ and returns 1 otherwise is primitive recursive:

1. $Nz(0) = 0$ and $Nz(x + 1) = S(0)$

Exercise:

Show that $Nz(m)$ is formally derivable.

Nz is the "not-zero" function. The function Nz yields 1, the value for "true", when the argument is <u>not</u> 0. $Nz(0) = 0$ because it is false that 0 is not 0. But $Nz(m + 1) = 1$ because any number gotten by adding 1 to a number is not zero.

Armed with the Primitive Recursive functions we have obtained so far, we wish to show "greater than" is a primitive recursive relation. Remember, to show a relation is primitive recursive it is necessary to show that its Characteristic Function is primitive recursive. The Characteristic Function of ">" is $C_>$, where this is its definition:

$$C_>(x,y) = 1 \quad \text{if } x > y, \text{ and}$$
$$C_>(x,y) = 0 \quad \text{if } x \le y.$$

Consider $Sbt(x,y)$. When $x > y$, $Sbt(x,y) > 0$. When $x \le y$, $Sbt(x,y) = 0$. That is,

$$Sbt(x,y) \ne 0 \text{ iff } x > y.$$

We now use our not-zero function, Nz, to say that when the difference between x and y is not zero, then x is greater than y:

$$Nz(Sbt(m,n)) = 1 \text{ iff } m > n.$$

Thus $Nz(Sbt(x,y))$ is the characteristic function of the relation ">". Symbolically:

$$C_>(x,y) = Nz(Sbt(x,y))$$

Hence, the "greater than" relation is primitive recursive.

All Recursive Functions Are "Mechanical"
(the converse of Church's Thesis)

When the formal apparatus for primitive recursive functions has been stripped away, we can see that every primitive recursive function is indeed mechanical. That is, it is a purely mechanical task to calculate values for the basic primitive recursive functions. For example, the zero function, returning 0 for any numeric input, is clearly mechanical. And, the successor function is also clearly mechanical. The fact that the projection functions are mechanical may be more difficult to see since they are a class of functions, not just one. So, let's look at an example, say, $\Pi_3^5(2,7,8,5,1)$. In this case, Π_3^5 tells us that to pluck out — or to "project" — the 3rd element in the 5-element sequence that follows. By inspection, we see that the 3rd element is 8. So: $\Pi_3^5(2,7,8,5,1) = 8$. After looking at a couple of examples, it becomes clear that all projection functions are mechanical, as well as the zero function and the successor function.

It is also a mechanical task to calculate the values of functions obtained from earlier functions by using Composition and Primitive Recursion. Primitive Recursion may be less easy to see because there are two clauses to look at. But, remember, we keep applying the second clause until the value of m winds down to 0, which we are guaranteed will happen since the second clause always pushes us closer to 0. When 0 is reached, we then use the first clause. So, Primitive Recursion is intuitively mechanical too.

Obtaining a recursive function by the μ-operator is a mechanical task as well. To see this, suppose $g(x,y)$ is such that for every x there is at least one y such that $g(x,y) = 0$. Then, to calculate $f(n)$, where $f(x) = \mu y(g(x,y))$, we first test $g(n,y)$ for the case when $g(n,y) = 0$. We try $g(n,0)$, $g(n,1)$, $g(n,2)$, and so forth, until we arrive at $g(n,k) = 0$. We know there must be at least one k because that was assumed in order for f to be well-defined. Since our mechanical method found the first such k, $f(n) = k$. Therefore, obtaining a recursive functions by using the μ-operator is also intuitively mechanical too.

Recall that according to Church's Thesis every mechanical function on natural numbers is recursive. We have argued in the paragraph above that all recursive functions are mechanical. Thus, we have the following result:

~~THEOREM~~: A function on natural numbers is "mechanical" iff it is recursive.

PF: **Impossible!** See next section.

Impossibility of Proving "Mechanical" = "Recursive"

When we say that something is mechanical, like checking whether a string is a formula of N, we have an intuitive concept in mind, not a mathematically rigorous one. The mathematically rigorous notion that almost all mathematicians think <u>explicates</u> the concept of a "mechanical procedure", is the notion of a "recursive function". But the two are different <u>kinds</u> of things. One is a common-sense concept and the other is a mathematically defined one. One could never <u>prove</u> that the two notions are equivalent, since to do that would require us to compare a vague notion with a precise one. In order for the comparison to be an exact one, we would first need to make the vague notion precise. But, <u>how</u> to make the vague notion precise is the very point at issue. So, we'd have to beg the question if we tried to prove that the notion of "recursive function" singles out the same class of things as "mechanical procedure".

Another way to see this is to suppose that a single mathematician were to believe that a certain procedure was "mechanical" and proved that the results of this procedure could <u>not</u> be reached by any recursive function. Now, let us suppose the mathematician were to show the procedure to several other mathematicians as a proposed counter-example to the thesis that all mechanical procedures can be expressed as recursive functions. But, suppose none of them agreed with him. Suppose all of them thought that the procedure was not <u>intuitively</u> mechanical at all. Who would be right, and how could we tell?

It seems that the only thing we would be left with in such a case would be the fact that almost all mathematicians believe that "recursive function" and "mechanical procedure" single out the same class of things. Then, since "recursive function" is mathematically precise, as opposed to the notion of "mechanical procedure", they would prefer to use "recursive function" instead of "mechanical procedure" where the precision of the concept matters. This is essentially the situation today. Most mathematicians think no counter-examples have ever been presented to the thesis that the two concepts single out the same things. Thus, they take them to be equivalent (though they agree that this equivalence can never be <u>proved</u>).

The notion of "recursive function" could be "proved" to be equivalent to "mechanical procedure" if "mechanical procedure" just comes to <u>mean</u> the same thing as "recursive function". That is, if the intuitive concept of "mechanical procedure" evolves in peoples' minds over time so that at a certain point it simply <u>means</u> the same thing as "recursive function", then pointing out this meaning-identification could be thought of as a "proof" of their equivalence.

All Recursive Functions Are Representable

To show that all recursive functions are representable requires an inductive proof on the conditions of representability. There are three basic kinds of functions that are primitive recursive, and thus recursive. These three basic functions are (1) the zero function, (2) the successor function, and (3) the projection functions. We begin an inductive proof by first showing that all three kinds of functions are representable. Then, to complete the proof requires that the three additional clauses for obtaining new recursive functions be shown to preserve representability. We will <u>assume</u> this part of the proof has been provided and conclude that all recursive functions are representable. A relation is recursive if and only if its characteristic function is. Thus, all recursive relations are representable as well.

We now exhibit the first part of the proof that all recursive functions are representable in $\mathcal{N}$.

PF:
IDEA OF PROOF: Show Z, S, Π_i^k are all representable. Then we need to show that composition, primitive recursion, and the μ-operator all preserve representability. This shows that all recursive functions are representable.

(1) Show Z is representable, where $Z(x) = 0$. Consider the formula '$x = x \wedge y = \underline{0}$'. The claim is that this formula represents the zero function, Z.

PF: We have to PROVE:

If $Z(n) = m$, then $\mathcal{N} \vdash \forall y(\varphi_f(\underline{n}, y) \longleftrightarrow y = \underline{m})$,

where $\varphi_f(x,y)$ is '$x = x \wedge y = \underline{0}$'. Hence, we have to PROVE:

If $Z(n) = m$, then $\mathcal{N} \vdash \forall y((\underline{n} = \underline{n} \wedge y = \underline{0}) \longleftrightarrow y = \underline{m})$.

1. $Z(n) = m$. Therefore,
2. $m = 0$.   (by definition of Z)   So, we need to PROVE:

$$N \vdash \forall y((\underline{n} = \underline{n} \land y = \underline{0}) \leftrightarrow y = \underline{0}).$$

Instead, we will prove (1) and (2) of Rep_2 (which, by the above exercises, is equivalent to Rep_1). That is, we PROVE:

$$N \vdash \underline{n} = \underline{n} \land \underline{0} = \underline{0}, \text{ and}$$
$$N \vdash \forall y((\underline{n} = \underline{n} \land y = \underline{0}) \rightarrow y = \underline{0}).$$

Since $N \vdash \underline{0} = \underline{0}$ and for every n, $N \vdash \underline{n} = \underline{n}$. So:

3. $\underline{n} = \underline{n} \land \underline{0} = \underline{0}$ by SL. This is the first clause.

To prove the second clause,
$\sqrt{}$PROVE: $(\underline{n} = \underline{n} \land a = \underline{0}) \rightarrow a = \underline{0}$.
4. $\underline{n} = \underline{n} \land a = \underline{0}$
5. $a = \underline{0}$.   This shows that Z is representable. ∎

Now, we show S is representable, where for any n, $S(n)$ is the natural number following n. ('S' is the symbol both for the function on numbers and for the one-place operation letter <u>inside</u> N.) Consider '$y = S(x)$' we show that this formula represents the successor function. For all n,m, we need to PROVE:

If $m = S(n)$ then $N \vdash \underline{m} = S(\underline{n})$.

PF: Suppose $m = S(n)$. Then $\underline{m}$ is $S(\underline{n})$. And, when $\underline{m}$ is $S(\underline{n})$, $N \vdash \underline{m} = S(\underline{n})$. We also need to PROVE:

$$N \vdash \forall y(y = S(\underline{n}) \rightarrow y = \underline{m}).$$

$\sqrt{}$PROVE $a = S(\underline{n}) \rightarrow a = \underline{m}$
1. $a = S(\underline{n})$
2. $\underline{m} = S(\underline{n})$ by first part.
3. $a = \underline{m}$ Subst of =. So, S is representable. ∎

We now show that the projection functions Π_i^k are all representable. $\Pi_i^k(x_1, x_2, \ldots x_k) = x_i$ for $1 \leq i \leq k$, are representable in N by '$(x_1 = x_1 \land x_2 = x_2 \land \ldots \land x_k = x_k) \land y = x_i$'. For, if $\Pi_i^k(n_1, n_2, \ldots n_k) = m$, then $m = n_i$. So, for the first clause, we look at '$(\underline{n}_1 = \underline{n}_1 \wedge \underline{n}_2 = \underline{n}_2 \wedge ... \wedge \underline{n}_k = \underline{n}_k) \wedge \underline{m} = \underline{n}_i$', which is straightforward, since $\mathcal{N} \vdash \underline{m} = \underline{n}_i$.

Exercise:

Finish the proof that the projection functions are representable. ∎

To continue the proof that <u>all</u> recursive functions are representable requires that we now show that the method for obtaining new recursive functions from old ones preserves Representability. We omit the details of this part of the proof.

11.9 GÖDEL NUMBERING

We now present one formal way to assign numbers to the symbols of L_N so as to satisfy our earlier constraints. The assigned numbers are called Gödel Numbers, abbreviated GN. Our main concern is that given a GN we can tell whether it is the GN of a symbol, a term, a formula, a sentence, or a derivation of $\mathcal{N}$, and we can recapture that symbol, term, formula, or derivation. Similarly, given a specific symbol, term, formula, or derivation, we can write the GN of that item.

Here are the GN's of symbols of L_N:

$\forall, \exists, \wedge, \vee, \neg, \rightarrow, \leftrightarrow, (,), E^2, A, B^1, C^2, D^2, x, a, y, b,$
↓ ↓ ↓ ↓ ↓ ↓ ↓ ↓ ↓ ↓ ↓ ↓ ↓ ↓ ↓ ↓ ↓ ↓
3, 5, 7, 9, 11, 13, 15, 17, 19, 21, 23, 25, 27, 29, 31, 33, 35, 37,

$z, c, w, d, x_1, ...$
↓ ↓ ↓ ↓ ↓
39, 41, 43, 45, 47, ...

Note that every Gödel number (GN) of a symbol is odd and that each symbol of L_N has a unique GN. Each formula defines a unique sequence of GNs.

For example, consider the formula '∀x(x + 0 = x)', which is A5. "Officially", this formula really is:

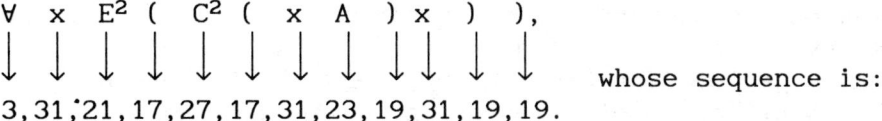

whose sequence is:

3, 31, 21, 17, 27, 17, 31, 23, 19, 31, 19, 19.

We can turn the above sequence into a unique single GN by doing this:

$$2^3 \cdot 3^{31} \cdot 5^{21} \cdot 7^{17} \cdot 11^{27} \cdot 13^{17} \cdot 17^{31} \cdot 19^{23} \cdot 23^{19} \cdot 29^{31} \cdot 31^{19} \cdot 37^{19}.$$

This product is a GN that can be factored uniquely to regain the sequence, which in turn gives us back '∀xE²(C²(xA)x)'. That is, no other sequence of symbols can have the same GN. Also, no GN of a formula is the GN of a single symbol because GNs of single symbols are all odd, whereas the GN of an expression is even because it always has some power of 2 as a factor. (It is even true that the GN of a single symbol, say 'A', which is 23, differs from its GN when taken as an expression, which is 2^{23}.) We also will assign a GN to any sequence of expressions. The idea is to assign a unique GN to a derivation, which is a finite sequence of sentences. For example, suppose we have a derivation consisting of the sentences $\varphi_1, \varphi_2, \varphi_3 \ldots \varphi_n$. That is, it is a derivation of φ_n in N. Where 'GN(φ_i)' stands for the GN of sentence φ_i the GN of the sequence of sentences in the derivation is:
$2^{GN(\varphi_1)} \cdot 3^{GN(\varphi_2)} \cdots P_n^{GN(\varphi_n)}$. ($P_n$ is the nth prime).
Note that the GN of a sequence of expressions is even, since it has a power of 2 in it as a factor. The power of 2 to which it is raised is also even, since it is the GN of an expression, whose GN is even. Thus, the GN of a sequence of expressions always contains 2 to an even power, while the GN of an expression always contains 2 to an odd power. So the GN of a sequence of expressions can never be the same as the GN of a single expression. (Also, the GN of a one-line derivation of '∀xE²(C²(xA)x)' — since it is an axiom, its derivation consists solely of the sentence itself — differs from the GN of the sentence. If the GN of the sentence is k, then the GN of the derivation is 2^k.)

Mathematically, GN is a function from the set of symbols of L_N, expressions of L_N, and finite sequences of expressions of L_N to the set of natural numbers. The function is 1-1 but not onto, since not every natural number is a Gödel number (1, 2, 4, 6, ... are not GNs). Also, there is a mechanical procedure, an algorithm, for obtaining the GN of any symbol, expression, or sequence of expressions. Similarly, one can go in the reverse direction: Given any GN, there is a mechanical method for finding out (1) what <u>kind</u> of thing it is the GN of (a symbol, expression, or sequence of expressions), and (2) which <u>particular</u> thing of that kind it is (i.e., <u>which symbol,</u> <u>which expression,</u> or <u>which sequence</u> of expressions it is).

Now that we have assigned Gödel numbers to symbols of L_N as given above, we can now say things about the syntax of L_N in terms of Gödel numbers. For example, we can say what it means to be a symbol of the language — it means to have a GN. So, we can say that 3 is a symbol of L_N, meaning that the symbol with GN 3 is a symbol of L_N. We can say what a numeral term in L_N is. A numeral term is either 2^{23}, the expression consisting of 'A' alone, or $2^{25} \cdot 3^{17} \cdot 5^{23} \cdot 7^{19}$, the expression 'S(A)', etc. With a little thought we could probably write an algorithm for this. Let us try it. The idea is to get the GN for: A, B(A), B(B(A)), B(B(B(A))), etc. How about this as a specification of an algorithm: take the first prime to the 25th power followed by the second prime to the 17th power, followed by the third prime to the 25th power, followed by the fourth prime to the 17th power, and continue this way with $P_{2n-1}^{25} \cdot P_{2n}^{17}$. Then multiply this by P_{2n+}^{23}, then by $P_{2n+2}^{19} \cdot P_{2n+3}^{19} \cdot \ldots \cdot P_{2n+(n+1)}^{19}$.

Specifically, for any n, GN($\underline{n}$) = $(P_1^{25} \cdot P_2^{17}) \cdot (P_3^{25} \cdot P_4^{17})$ $\cdot \ldots \cdot (P_{2n-1}^{25} \cdot P_{2n}^{17}) \cdot P_{2n+1}^{23} \cdot P_{2n+2}^{19} \cdot P_{2n+3}^{19} \cdot \ldots \cdot P_{2n+(n+1)}^{19}$

Let us look at the above product for n = 0,1,2. For n = 0, GN($\underline{0}$) is : 2^{23}, which is the GN of 'A'. For n = 1, GN($\underline{1}$) = $2^{25} \cdot 3^{17} \cdot 5^{23} \cdot 7^{19}$, which is the GN of 'S(A)'.

For n = 2, GN($\underline{2}$) = $(P_1^{25} \cdot P_2^{17}) \cdot (P_3^{25} \cdot P_4^{17}) \cdot P_5^{23} \cdot P_6^{19} \cdot P_7^{19}$, which = $(2^{25} \cdot 3^{17}) \cdot (5^{25} \cdot 7^{17}) \cdot 11^{23} \cdot 13^{19} \cdot 17^{19}$, which = the GN of 'B(B(A))'. We can see that this works. So, we can say that an expression, e, of length (3n + 1) is the numeral term n if and only if GN(e) = GN($\underline{n}$).

We can continue to explore the properties of Gödel Numbers, but this seems like enough detail to support the assumptions made in the early part of this chapter that symbols of L_N, formulas of L_N, and derivations of sentences of L_N can be coordinated with unique numbers so that the arguments for Gödel's Theorems succeed.

11.10 EPILOG

That the truths of a simple system for ordinary arithmetic should outrun the proofs in the system shocked and dismayed members of the mathematical community who had hoped all of mathematics could be formally secured. Gödel's results seemed to show that very little of mathematics could be placed on a secure foundation, though they opened up the possibility that "securing mathematics" could be interpreted in many different ways. On the positive side, many persons have interpreted Gödel's results as indicating that there is a creative side to our thinking that can never be totally captured by machine thought. Whether that more positive evaluation of Gödel's results is correct or not, it is surely the case that his results have stimulated further interest in the foundations of mathematics and have led directly to the establishment of many new branches of thought.

For us in this book, Gödel's Theorems signify the end of a journey that began with us accepting ordinary set-theoretical notions and then formalizing principles of logic and mathematics within that framework. Along the way, a number of logical and mathematical principles were defined and their "correctness" justified. Just as Gödel's Theorems represent a kind of defeat in going further with that justification, they represent a kind of victory as well. The principles formalized were so rich and interesting that we did not simply <u>fail</u> <u>to</u> <u>prove</u> the completeness and consistency of Arithmetic, but we were able, thanks to Gödel, to actually **prove**: there can be no proofs of the completeness and consistency of Arithmetic, meeting certain conditions. Hence, there is a sense in which the defeat of formalism contains the seeds of its own victory.

POSTSCRIPT: CHAITIN'S THEOREM

This postscript is being added to reflect some exciting new ideas currently being developed by Gregory Chaitin. 'Chaitin's Theorem' refers to a result in an emerging field called "Algorithmic Information Theory", which he co-founded, together with the Russian mathematician, Kolmogorov. We have space here to only allude to some of the fascinating ideas in this new discipline.

Chaitin begins with the notion of the complexity of a string of 0's and 1's. For example, the complexity of the string '01010101010101' is not very great, since we can perceive that the string is just '01' written seven times. The high degree of order to the string can be reflected by the simplicity of the command 'print "01" 7 times'. The string consisting of 10,000 repetitions of '01' is not much more complex than the first string, as it can be generated by the command 'print "01" 10,000 times'. On the other hand, the string '110101001100110101010011100' seems much more complex, as the simplest way to generate it seems to be the command 'print "110101001100110101010011100"'.

To be slightly more precise, the command to generate the first string required only 18 symbols (including spaces), to generate the second took 23, and the third required 36. Supposing that these are in fact the shortest commands in a given language for generating these strings, the complexity of the three strings is 18, 23, and 36, respectively — for that language. The language in which commands like this are written could be a computer language, but for now we will stick with ordinary English.

Even though Chaitin's Theorem is a mastery of technical detail, we can already appreciate some of his ideas in an intuitive way. Suppose in our "command language" we permit expressions like 'print the first string of complexity > 41'. Since the complexity of the command itself is 41 (i.e., there are 41 characters in the command), we are saying to print the first string that <u>cannot</u> be generated by this very command.

Note the similarity to the Liar Paradox in the previous chapter. We will call this paradox, "Chaitin's Paradox". The paradox is to command that something be done that cannot be consistently done. Suppose such a string <u>were</u> generated. Then, its complexity would be 41 or less, since a command of length 41 can generate it. On the other hand, if the command is carried out faithfully, then the string's complexity must be <u>greater</u> than 41 (since the command <u>says</u> that the string must have complexity greater than 41).

Formally, Chaitin's Paradox can be carried out by coding the symbols of N into binary strings, and then by defining a mechanical derivation-checker ('d-c') to print a number if it is a code number of a derivation in N. We know from the previous chapter that the process of coding and decoding sentences of N is mechanical, and so is the process of checking whether a string is a derivation. Further, this d-c (derivation checker) has a unique length that can be measured in terms of the number of the symbols the procedure employs.[1] Suppose that the length (somehow measured) of d-c is ℓ. Then, define the **complexity** of a coded theorem number to be the length of the shortest procedure that prints that number.

Since there are infinitely many theorems, there must be theorem numbers whose complexity is greater than ℓ. In fact there are infinitely many true sentences of the following form 'the complexity of k is greater than ℓ'. Let j be the number of one such sentence, where the complexity of j itself is greater than ℓ. (Since there are infinitely many true sentences of that form, there must be infinitely many of them whose code numbers are more complex than ℓ.) So, suppose that d-c checks j. Since d-c does not make errors in checking theorems (by Strong Soundness), it will correctly arrive at the fact that the complexity of k is greater than ℓ. Or will it?

For d-c to carry out its task, it must print j, which can only be printed by procedures of greater length than d-c itself (since the complexity of j is greater than ℓ). Thus, it can't consistently print j, and thus it cannot determine whether or not the complexity of k is in fact greater than ℓ. Since there are infinitely many such true sentences, there are infinitely many of them whose code numbers are greater than ℓ that d-c cannot analyze.

Thus, Chaitin's Paradox can be produced in N, showing, as Gödel did, that N is incomplete. Chaitin's Paradox, however, gives us different information than Gödel's First Theorem. It tells us that every procedure has a given complexity which it cannot go beyond. This fuels more epistemic speculation about the limitations of our own capacity to understand the world in which we live. Suppose that our degree of complexity is h (for humans). Chaitin's Theorem seems to tell us that we cannot correctly analyze the complexity of anything beyond h. This should <u>not</u> be

---

[1] This can be coded so that the number for each procedure is unique.

confused, however, with our being able to understand that some fact of complexity greater than h is <u>true</u>. We may be able to "understand" something much more complex than ourselves. There are infinitely many theorems of N that can be derived in N, and thus there are infinitely many of them beyond any particular complexity. What we cannot determine in N is that the complexity of a theorem is beyond the complexity of the procedure used to check its complexity. And, thus, Chaitin's Theorem seems to tell us that we are in no position to evaluate the complexity of anything greater than our own.

Recently, Chaitin has extended this train of thought to show that in a sense there is randomness in arithmetic. He has also argued that this complexity-based approach to incompleteness tends to support a "quasi-empirical" view of the foundations of mathematics.

REFERENCES

(intended to be helpful; by no means inclusive)

Bell & Slomson [1971], Models and Ultraproducts, North-Holland, Amsterdam.

Boolos and Jeffrey [1989], (3rd edition), Computability And Logic, Cambridge Univ. Press, Cambridge, Mass.

Chaitin, Gregory [1990], Information, Randomness, & Incompleteness, World Scientific, Singapore; [1992], Information-Theoretic Incompleteness, World Scientific, Singapore.

Chang & Keisler [1977], Model Theory, 2nd edition, North-Holland, Amsterdam.

Chihara & Schaffter, Unpublished manuscript, Logic: An Introduction to Computability and Completeness.

Davis, Martin [1965], The Undecidable, Raven Press, New York.

Delong, Howard [1970], A Profile of Mathematical Logic, Addison-Wesley, Reading, Mass.

Enderton, Herbert [1972], A Mathematical Introduction to Logic, Academic Press, New York.

Epstein & Carnielli [1989], Computability: Computable Functions, Logic, and The Foundations of Mathematics, Wadsworth, Pacific Grove, Cal.

Feferman, Solomon [1988], "Kurt Gödel: Conviction and Caution," in Shanker [1988], Gödel's Theorem, 96-114, Routledge, London.

Gödel, Kurt [1931], Über formal unentscheidbare Sätze der Principia Mathematica und verwandter Systeme I; translated in Davis [1965], by Elliott Mendelson, as: On Formally Undecidable Propositions of Principia Mathematica and Related Systems I, 4-38.

Halmos, Paul [1960], Naive Set Theory, Van Nostrand, Princeton, N.J.

Henkin, Leon [1950], The Completeness of the First-Order Functional Calculus, JSL, **14**, 159-166.

Kalish & Montague [1964], Logic: Techniques of Formal Reasoning, Harcourt, Brace & World, New York.

Mates, Benson [1972], Elementary Logic, 2nd edition, Oxford University Press, New York.

Mendelson, Elliott [1987], Introduction To Mathematical Logic, 3rd edition, Wadsworth, Pacific Grove, Cal.

Moore, A. W. [1990], The Infinite, Routledge, London.

Rogers, Hartley [1967], *Theory of Recursive Functions and Effective Computability*, McGraw-Hill, New York.

Rosser, John Barkley [1936], "Extensions of Some Theorems of Gödel and Church, in Davis [1965], 230-235 (from *JSL*, **1**, 87-91).

Rucker, Rudy [1987], *Mind Tools*, Houghton Mifflin, Boston.

Shoenfield, Joseph [1967], *Mathematical Logic*, Addison-Wesley, Reading, Mass.

Tarski, Alfred [1953], *Undecidable Theories*, in collaboration with Mostowski & Robinson, North-Holland, Amsterdam; [1954], "Contributions to the theory of models I, II," *Indag. Math.* **16**, 572-588; [1956], *Logic, Semantics, Metamathematics*, Oxford Univ. Press, Oxford; and Vaught [1957], "Arithmetical Extensions of relational systems," *Compositio Math.* **13**, 81-102; [1965], *Introduction To Logic*, 3rd edition, Oxford University Press, New York.

Vaught, Robert [1954], "Applications of the Löwenheim-Skolem-Tarski Theorem to problems of completeness and decidability," *Indag. Math.* **16**, 467-472.

INDEX

A

Addition, Theory of, 241
Adjunction (ADJ), 62
Alphabet of a language, 28
Antecedent, 30
Appropriate language (for a structure), 280
Arbitrary instance, 141
Argument of a function, 14
Arithmetic, theories of, 243
Assertion (of a theory), 283
Asserts (a property), 307
Assignment
    to sentence letters, 84
    Tree, 161
Associated interpretation (of a structure), 281
Assumption, 32
    Existential, 149
    number, 32
    Rule (A), 32
Atomic
    components, 181
    formula, 134
    sentence, 30
Axiom(s), 158
    of choice, 302
    non-logical, 216
Axiomatic theories
    Arithmetic, 243
    Group Theory, 236
    Less-Than, 158
    Number Theory, 255
    Orderings, 228
    Sentential Logic, 78
Axiomatic theory, 248
Axiomatizable, 247
    finitely, 307
Axiomatized theory, 248

B

β-change, 161
Basic rules,
    of L_{PL}, 140
    of L_{SL}, 50
Belongs to (a theory), 222
Biconditional, 30
    -Conditional (BC), 61
    Detachment (BD), 69
Binary Relation, 10
Boolean Interpretation, 84
Bound occurrence of a variable, 135
Bracket notation, 166

C

Cantor, 344
Capturable (in first-order), 306
    Finitely, 307
Cardinal, 16
    of languages, 300
Cartesian Product, 10
Categorical (theories), 311
    k-, 313
Chaitin, G.
    Chaitin's Paradox, 364
    Chaitin's Theorem, 364
Change (β-), 161
Characteristic function, 327
    and representability, 337
    Theorem, 338
Church, A., 247
    Church's Law, 335
    Church's Theorem, 336
    Church's Thesis, 334
Class, 1

Compactness, 105, 129, 214
   (proved model-
     theoretically), 319
Complement, 7
Complete
   Decidable Extension
     Theorem, 254
   Extension Theorem, 254
   Induction, 16
   Theory, 120
Completeness
   mathematical, 120, 225
   Strong (for L_{SL}), 119
    (for L_{PL}), 209
   theorem (for L_{SL}), 121
    (for L_{PL}), 211
   Weak, 131
Complexity, 364
Composition
   of functions, 15
   Rule, 351
Computable function, 332
Conclusion of argument, 73
Conditional, 30
Conditional Derivation
 (CD) Rule, 32
Conditionals To
 Biconditionals (CTB), 60
Conjunct, 30
Conjunction, 30
 Rule (CONJ), 61
Connective, 29
   main, 31
Connectivity, 218
Consequence
   of Δ_T, 235
   Logical, 186
   Tautological, 90
Consequent, 30
Conservative extension, 252
Consistent
   Rules for L_{PL}, 208
   Rules for L_{SL}, 117
   set of sentences, 118

Constant
   logical, 134
   non-logical, 134
CONTRA
   Rule, 32
   Strategy, 40
Contradiction, 3
Contradictory, 30
Contrapositive, 40, 101
Correspondence (1-1), 13
Countable set, 16
Counterpart, 190
Cross product, 11

D

Decidability Theorem, 251
Decidable, 247
   set (of sentences), 248
   theory, 248
Decision procedure, 251
Deduction Theorem, 81
Definability (in N), 342
Denote, Denotation, 160
Denumerable
   model, 309
   sequence, 317
   set, 16
Derivability Formula, 325
Derivable, 110
Derivation, 28
   in first-order logic, 138
   in first-order theories,
    231
   form, 60
   full, 53
   from a set, 51
    (of assumptions),
     in L_{SL}: 73, in L_{PL}: 158
   machine, 249
   of a sentence, 35
   in sentential logic, 32
   short, 53
   Technical, 109

Derived rules,
    of L_{PL}, 140, 187
    of L_{SL}, 58, 76
Diagonal(ization)
    Argument, 16, 344
    Lemma, 339
    Set, 345
Difference, 6
Disjoint sets, 2
Disjunct, 30
Disjunction, 30
Divisibility, 270
Domain
    of discourse, 140, 160
    of a function, 12
    of an interpretation, 160
Double negation, 30
    (DN) Rule, 62
Downward Löwenheim-
  Skolem(-Tarski & Vaught)
    Theorem, 301

E

EG Rule, 147
EG Theorem, 188
Element, 1
Elementarily equivalence, 289
Elementary
    extension, 295
    substructure, 295
Empty set, 1
Epimorphism, 288
    φ-, 198
Epstein & Carnielli, 335
Equality
    realm, 226
    relation, 11
    Theory of, 220
    theory with, 223
Equinumerous, 16

Equivalence
    class, 11
    elementary, 289
    relation, 10
ES
    Argument, 147
    Rule, 149
    Theorem, 189
Escapes (first-order), 306
Existential
    Assumption (EA), 149
    formula, 136
    Generalization, 147
    quantifier, 136
    Switch (ES) Rule, 149
Expansion (of a structure), 299
Extension, 252
    conservative, 252
    by constants, 252, 299
    elementary, 295
    of a language, 299
    by names, 209
    of a structure, 294
    Theorem(s), 254

F

False, 14
    for an interpretation, 85
Falsity Formula, 323
Feferman, S., 322, 324
Finite
    intersection property, 316
    number of axioms, 246
Finitely Axiomatizable, 246
FIP Theorem, 316, 339
First new name (to a
  formula), 164
First-order theory
    defined model-
      theoretically, 282
    defined proof-
      theoretically, 219, 233

Fixed Point Theorem, 338, 339
Form
   derivation, 60
   formula, 220
   of a sentence, 59
Formal system, 142
Formal theory, 216, 243
Formalize, 216, 245
Formula (of L_{PL}), 135
   atomic, 134
   existential, 136
   molecular, 136
   monadic, 136
   quantified, 136
   universal, 136
FPT (Fixed Point Theorem), 339
Free
   occurrence of a variable, 135
   term, 232
   for a variable in a formula, 220
Full derivations, 53
Function, 12
   characteristic, 327
   domain, 12
   general recursive, 352
   partial, 13
   primitive recursive, 352
   range, 12
   symbols, 230
   total, 13
   truth, 14

G

Generalization
   Existential (EG), 147
   Universal (UG), 139
General recursive function, 352

Gödel, K., 321
   Incompleteness, 330
   Numbering, 360
   Platonism, 321
   -Rosser Theorem, 330
   Second Theorem, 346
Groups
   Theory of, 236

H

Halmos, P., 1, 278
Height (H)
   of a formula, 23, 124
Henkin, L., 120, 212
Herbrand, J., 81
Hilbert, David, 327
Homomorphism, 286

I

Identity
   relation, 11
Image, 15
   inverse, 15
Inclusion, 2
Incompleteness (of N)
   because a true sentence says so, 327
   because Truth not Definable, 325
   following from FTP, 341
   Gödel-Rosser, 330
   Moore's presentation, 344
   statement, 330
   summary of argument, 331
   syntactic proof, 328, 330
Inconsistent
   set of sentences, 51
Individual
   constants, 134
   symbols, 134
   variables, 134

Induction
    complete, 16
    mathematical, 24
    principle (for $\mathcal{P}$), 245
    rule (MATH-IND), 257
Inductive Hypothesis, 18
Inference
    rules for L_{PL}, 138
    rules for L_{SL}, 32, 50
Infinite
    model, 304
    number of axioms, 245
    number of initial sentences, 243
    number of objects, 226
    number of sentences (for LL), 222
Informal meanings, 140
Initial sentences (of a theory), 222
Inserting Antecedents, 82
Instance
    arbitrary, 141
    of a theorem, 55
    of a universal generalization, 140
Instantiation
    Universal (UI), 139
Interchange Rule (INT), 33
Interpretation
    Boolean, 84
    Lemma, 127
    for L_{PL}, 163
    naming, 164
    of relation letters, 163
Intersection, 6
Invalidity, 178
Inverse
    function, 15
    image, 15
Isomorphism, 288

K

k-ary operation, 14
k-ary relation, 12
k-Categorical (theories), 313
k-tuple, 19
Kalish & Montague, 36
Kleene, 245

L

Language
    of first-order logic, 134
    of sentential logic, 28
Law
    Church's, 335
    of the Excluded Middle, 9
    Leibniz', 222
    of Non-Contradiction, 9
    Peirce's, 42
    of Trichotomy, 230
Least Counterexample Principle (LCP), 22
Leibniz' Law (LL), 222
Less-than, (Theory of), 158
Liar
    Paradox, 322
    Sentence, 322
Lindenbaum-Henkin Theorem, 212
Lindenbaum's
    Lemma, 122
    Theorem, 210
Listable (mechanically), 250
Listability Theorem, 251
LL (Leibniz' Law), 222
Logical
    consequence, 186
    constants, 134
    names, 134
    symbols, 134
Łoš
    Theorem, 318
    -Vaught Theorem, 313

Löwenheim, L.
  Downward Löwenheim-
    Skolem(-Tarski &
    Vaught) Theorem, 301
  Upward Löwenheim-
    Skolem(-Tarski &
    Vaught) Theorem, 305

M

μ-operator, 351
Main connective, 31
Map, mapping, 12
Mathematical
  completeness, 225, 312
  constants, 134
  Induction, 24
MATH-IND Rule, 257
Mechanical, 247, 332
Mechanically
  listable, 250, 251
Member, 1
Mendelson, E., 246
Metatheorem
  Syntactic, 110
Model
  Boolean (for sentences
    of L_{SL}), 88
  denumerable, 309
  finite, 304
  nonstandard, 311
  for sentences of L_{PL}, 176
  standard, 309
Modus Ponens (MP), 32
Modus Tollendo Ponens
  (MTP), 63
Modus Tollens (MT), 62
Molecular
  formula, 136
  sentence, 30
Monadic formula, 136
Monomorphism, 288,
MP (Rule), 32

N

Name (logical), 134
  To name, 160
Naming Interpretation, 164
  Truth for, 165
Naming function, 160
  Thorough, 160
Natural number, 8
Negation, 30
New (logical name)
  First (new), 164
  to a derivation, 138
  to a formula, 138, 164
New Constants Theorem, 253
Non-logical symbols, 134
Non-standard models, 311
Null set, 1
Number
  cardinal, 16
  Gödel (GN), 360
  natural, 8
  Theory, 255
  unnatural, 310
Number Theory, 255
Numeral term, 264

O

0-ary operation, 15
Occurrence, 18
  (free, bound), 135
Official
  formulas (of L_{PL}), 137
  sentences (of L_{SL}), 32
1-1
  correspondence, 13
  function, 13
Onto function, 13
Operation, 12
Operation symbols, 230
Order
  of an expression, 23
Ordered pair, 19
Orderings, theory of, 228

P

φ-epimorphism, 198
Pair (ordered), 19
Paradox
   Chaitin's, 364
   Liar, 322
   Non-Derivability, 326
   Russell's, 155, 183
   of Universal Set, 155
Parentheses, 17, 31
Partial
   function, 13
   interpretation, 171
Partial Interpretation
 Theorem, 172
Particularizations, 180
   finite, infinite, 185
Peano Arithmetic, 244
Platonism, 321
Power set, 6
Predicate calculus
   monadic, 136
Premises of an argument, 73
Prime, 270
Primitive recursive
   function, 352
   relation, 337
Primitive Recursion Rule, 351
Principle
   Least Counterexample, 22
   of Complete
    Induction, 19
Principle (continued)
   of induction (for $\mathcal{P}$), 245
   of Mathematical
    Induction, 24
Product, Cartesian (or Cross
 Product), 10
Projection functions, 350
Proof
   formal, 91
   informal, 91
   semi-formal, 91
Proper subset, 2
Property (capturing), 306

Q

QE (Rule), 139
Quantified formula, 136
Quantifier, 133
   existential, 136
   universal, 136
Quantifier Exchange (QE)
 Rule, 139
Quasi-Equivalence, 217
Quasi-Reflexivity, 217

R

Range
   of a function, 12
Realism (Platonic), 322
Realm
   arithmetic, 227
   equality, 226
Recognizable, 216, 222, 243, 248
Recursive functions
   derivations of, 352
   (General), 350, 352
   Primitive, 350, 352
Recursively enumerable
 (mechanically listable), 250, 251
Reduction (of a structure), 299
Reflexive, 10
Relation
   binary, 10
   equivalence, 10
   identity, 11
   k-ary, 12
   letters, 134
   symmetric, 10
   transitive, 10
Representability
   Test, 332
   Theorem, 335
Representable (in N)
   function, 332, 336
   property, 274
   relation, 274

Representation Theorem
　　for '=', 271
　　for '<', 276
Restriction (of a relation), 13
Robinson's system, 246
Rosser, J. B., 329
Rule(s)
　　basic (for L_{SL}), 32
　　　(for L_{PL}), 138
　　derived (for L_{SL}), 76
　　　(for L_{PL}), 187
　　Induction (MATH-IND), 257
Russell's Paradox, 155

S

Satisfiable, 88, 176
"Says", 142, 163, 341
Schema (see Form), 245
Scope of a variable, 135
Self
　　formula, 337
　　operation, 326
　　-Reference, 328
Semantic, 84, 108
Sentence
　　of L_{PL}, 136
　　of L_{SL}, 29
Sentence letters
　　(of L_{PL}), 135
　　(of L_{SL}), 28
Sentential logic, 27
　　language of, 28
Separation of Cases (SOC), 66
Sequence, 10
Set, 1
Short Derivations, 53
SHOW lines, 36

Simplification Rule (SIMP), 60
SL (Rule), 138
Soundness
　　metatheorem for, 206
　　　sentential logic, 115
　　Strong, 113
　　Weak, 118
Standard
　　Interpretation, 243, 342
　　Model, 309
Stand-ins, 192
Stands for, 160
Structure, 279
　　appropriate language for, 280
　　reduction, expansion, 299
　　specified (by an interpretation), 281
　　type, 280
　　vs. Interpretation, 281
Subset, 2
　　Elementary, 295
Subst (for LL), 238
Substitution
　　instance in first-order logic, 153
　　instance in sentential logic, 56
Substructure, 294
Successor function, 350
Symbols
　　logical, 134
　　non-logical, 134
Symm (axiom), 222
Symmetric, 10
Symm-Trans, 255
Syntactic, 84, 108
　　Metatheorem, 110
Syntax of a language, 28

T

T (Theorem) Rule, 55, 57, 155
Tarski, A.
    Deduction Theorem, 81
    Mostowski & Robinson, 243
    Truth Theorem, 342
Tautological consequence, 90
Tautology, 89
Technical Derivation, 109
Term
    free, 232
    of L_{T^+}, 230
Theorem, 35
    Rule, 55, 57, 155
    of a theory, 221
Theories of Arithmetic, 243
Theory
    of Addition, 241
    axiomatic, 248
    axiomatized, 248
    of class of structures, 284
    complete, 225
    of Equality, 220
    with equality, 223
    first-order, 219, 233
    formal, 216
    of Groups, 236
    of "less-than", 158
    Number Theory, 255
    of Orderings, 228
    of Peano Arithmetic, 241
    of Quasi-Equivalence, 217
    undecidable, 246
Thesis (of a theory), 222
Thorough (naming function), 160
Total
    function, 13
    interpretation, 172
Trans (axiom), 222
Transitive relation, 10
Trichotomy (Law of), 230
True, 14
    for a boolean interpretation, 85, 87
    for '=' (in L_{T^+}), 234
    for an interpretation, 173
    for a naming interpretation, 165
    for a sentence letter, 84
Truth Formula, 324
Tuple, 10
Type (of structure), 280

U

UG
    Rule, 139
    Rule (Modified), 231
    Strategy, 145
UI (Rule), 139
Ultrafilter, 316
Ultraproduct, 318
Unary operation, 14
Undecidability
    of arithmetic, 340
    of first-order logic, 335
Union, 6
Unique readability, 31, 136
Universal
    formula, 136
    Generalization (UG) Rule, 139
    Instantiation (UI) Rule, 139
    quantifier, 136
Universalization, 172, 174
    Metatheorem, 178, 205
Universe
    of a set, 7
    of a structure, 279
Upward Finite Satisfaction Theorem, 304
Upward Löwenheim-Skolem (-Tarski & Vaught) Theorem, 305

V

Vacuous truth, 2
Valid, 177
Value of a function, 14
Variable, 134
    bound, 135
    free, 135
    scope, 135
Vaught, R., 304, 305, 313

W

ω-Consistency, 329
Witness, 209

Z

Zero function, 350
Zorn's Lemma, 316